Springer Proceedings in Physics **316**

Indexed by Scopus

The series Springer Proceedings in Physics, founded in 1984, is devoted to timely reports of state-of-the-art developments in physics and related sciences. Typically based on material presented at conferences, workshops and similar scientific meetings, volumes published in this series will constitute a comprehensive up to date source of reference on a field or subfield of relevance in contemporary physics. Proposals must include the following:

- Name, place and date of the scientific meeting
- A link to the committees (local organization, international advisors etc.)
- Scientific description of the meeting
- List of invited/plenary speakers
- An estimate of the planned proceedings book parameters (number of pages/articles, requested number of bulk copies, submission deadline).

Please contact:

For Americas and Europe: Dr. Zachary Evenson; zachary.evenson@springer.com

For Asia, Australia and New Zealand: Dr. Loyola DSilva; loyola.dsilva@springer.com

Reshmy Rajasekharan · Sindhu Raveendran · Libin P. Oommen
Editors

Proceedings of The International Conference on Material Science

MatZone-2024, 17–19 December, Kerala, India

Springer

Editors
Reshmy Rajasekharan
Department of Science and Humanities
Providence College of Engineering
Alappuzha, Kerala, India

Sindhu Raveendran
Department of Food Technology
TKM Institute of Technology
Kollam, Kerala, India

Libin P. Oommen
Department of Mechanical Engineering
Providence College of Engineering
Alappuzha, Kerala, India

ISSN 0930-8989 ISSN 1867-4941 (electronic)
Springer Proceedings in Physics
ISBN 978-981-96-6540-2 ISBN 978-981-96-6541-9 (eBook)
https://doi.org/10.1007/978-981-96-6541-9

This Springer imprint is published by the registered company Springer Nature Singapore Pte Ltd.
The registered company address is: 152 Beach Road, #21-01/04 Gateway East, Singapore 189721, Singapore

"This book is dedicated to the glory of God and all the enthusiastic readers"

Foreword

New perspectives and a route to a better future are offered by challenges. MatZone-2024, an international conference on material science, was organized by Providence College of Engineering to provide an extremely promising platform for recent multi-disciplinary research sharing focuses on a wide range of areas namely material science, nanoscience, environmental, biomedical, biotechnology, and electronics. This conference targets the fast-growing technologies-related research to impart an extra step towards sustainable developments together with the Government of India launched sustainability promoting goals. Tremendous amounts of research are being carried out in the area of material science in many institutes and research centers across the globe, but a very few number only achieving the extended outcome of developing value added products. In this context, we would like to organize an International Conference on Material Science 'MatZone-2024' at Providence College of Engineering, Chengannur, India. It brings me great pleasure to introduce the compiled Proceedings of the International Conference on Materials MatZone-2024. This book consists of 23 full length articles submitted to the conference after completing the peer review processes.

I hope this conference proceedings provide you all an insight into the advancements in the material science field and the interdisciplinary researches ongoing across the globe. This book will motivate you all to explore and investigate the numerous opportunities in this research area.

Reshmy Rajasekharan
Convenor, International Conference on
Materials (MatZone-2024)

Sindhu Raveendran
Professor & Dean of Research (Advisory
Committee MatZone-2024)

Libin P. Oommen
Co-convenor, International Conference on
Materials (MatZone-2024)

Preface

It brings us great pleasure to introduce the Proceedings of the International Conference on Materials (MatZone-2024). The event was organized jointly by the Department of Mechanical Engineering and Science & Humanities of Providence College of Engineering, Chengannur, Kerala, India, from December 17–19, 2024. This conference MatZone-2024 aims to provide a platform for researchers and students to explore the applications of different materials on recent developments in the field of material science. This conference will be beneficial to post graduate students, research scholars, scientists, and teaching faculty to refine their ideas, enhance their knowledge base, and broaden their horizons. The conference will feature thematically arranged invited lectures of eminent researchers from top institutions in India and abroad.

More than seventy five researchers, hundred post-graduate and graduate students contributed their work to the conference, out of which only 23 submissions were accepted as full papers for publication in the Proceedings of MatZone-2024. These chapters were successfully presented as oral or poster presentations which were carefully chosen based on the quality and relevance to the conference.

The success of this conference was due to the collective efforts of all the organizing committee, reviewers, and advisory members. We would like to express our sincere gratitude to all faculty members in the committee for their timely support for the successful conduct of the conference. We acknowledge all the authors for their valuable contributions and sponsors for their support. Our sincere gratitude to each and every member of the college staff who helped makes the event a success. At last, we hope the contents of this volume will provide useful information for researchers and students for developing and applying new technologies and processes in the field of material science.

Reshmy Rajasekharan
Sindhu Raveendran
Libin P. Oommen

Theme of the Book

This Proceedings aims to bring together novel and valuable scientific contributions to address several relevant issues in the areas of renewable energy, green technologies, sustainable building, water purifications, and other fast-emerging techniques in academia and industries.

The areas to be covered include:

- Sustainability science and technologies
- Climate change and mitigation, carbon footprints, global warming
- Smart building technologies
- Green synthesis
- Smart materials
- Nanocomposites
- Biomaterials
- Energy storage materials
- Magnetic nanomaterials
- Biosensors
- Optical devices
- Nanostructured materials
- Nonlinear materials
- Inorganic nanomaterials
- Sustainable transport, smart vehicles, and smart roads
- Information technology and artificial intelligence applied to sustainability.

Contents

About Editors

Dr. Reshmy Rajasekharan is an eminent researcher with a strong focus on materials science, nanomaterials, and environmental remediation. Her work spans crucial areas such as the development of nanocomposites, biopolymers, and sustainable materials for various industrial and environmental applications. She has an impressive citation count of h- index of 25 (Google Scholar Citations) and publication record, underscoring her impact in the field. She is serving as the convenor of MatZone-2024 and has successfully organized various conferences and workshops. With her extensive research experience and expertise in materials engineering and environmental solutions, she would be an excellent candidate to serve as an editor for the Proceedings of the International Conference on Materials.

Dr. Sindhu Raveendran is a distinguished researcher in the field of biofuels, biopolymers, and microbial enzymes. She has about 357 publications/communications with an h-index of 48 (Google Scholar Citations). She is one among the top 2% scientists in the world consistently for the last five years based on the study conducted by Stanford University. She is a recipient of several awards and fellowships. She is an editorial board member of five journals–BioEnergy Research, Annals of Agricultural and Crop Sciences, Journal of Environmental Sciences and Renewable Resources, EC Microbiology and Journal of Innovation and Applied Research. She is a review editor for Frontiers in Food Biotechnology and Frontiers in Microbiotechnology. She is a reviewer of 108 SCI Journals in the field of bioprocesses and products. She is one of the advisory committee members of MatZone-2024. She has edited three books including the one entitled "Green Bio-processes: Industrial Enzymes for Food Applications" (Springer Nature Singapore Pte Ltd, 2019) ISBN 978-981-13-3262-3. She is a recipient of Outstanding Reviewer Certificate from leading publishers like ACS, RCS, Wiley, Springer, and Elsevier.

Dr. Libin P. Oommen, an accomplished scholar with a strong research background in Mechanical Engineering and materials science, has published extensively on the application of neodelta magnetic materials in altering the combustion behavior of liquid and gaseous phase hydrocarbons. His expertise in engineering materials, particularly in the integration of metal organic frameworks for carbon capture under temperature swing technique, positions him as an ideal editor for the Proceedings of the International Conference on Materials. Dr. Libin is a recipient of an outstanding reviewer certificate from Wiley and Elsevier. His experience and academic contributions reflect a deep understanding of interdisciplinary material research, making him highly suitable for the role.

Upcycling Agricultural and Industrial Waste into High-Value Bioproducts: The Versatility of Carboxymethylcellulose in Food Systems and Beyond

Sathaiah Gunaseelan and Prasanth K. S. Pillai(✉)

Department of Food Science and Nutrition, University of Minnesota, St. Paul, MN 55108, USA
psasidha@umn.edu

Abstract. The global shift towards sustainability has driven the exploration of renewable materials to replace non-biodegradable substances. Carboxymethylcellulose (CMC), a versatile cellulose derivative, has emerged as a key biopolymer due to its water solubility, biodegradability, and functional properties, enabling applications across food systems, packaging, textiles, pharmaceuticals, and environmental remediation. This review highlights the potential of upcycling agricultural and industrial waste into CMC, promoting resource efficiency and advancing the circular economy. Bioconversion of waste materials into CMC leverages biological, chemical, and thermal methods, transforming cellulose-rich byproducts into high-value materials while addressing waste management challenges. Agricultural residues such as corn husks, sugarcane bagasse, and cotton waste serve as renewable feedstocks, offering cost-effective and sustainable pathways for CMC production. Wastepaper also presents an opportunity for upcycling, with advanced recycling processes yielding high-quality cellulose for diverse applications. CMC's versatility enables numerous sustainable applications. In food systems, it enhances texture, stability, and shelf life, supporting the development of eco-friendly packaging and healthier formulations. Its biocompatibility allows advanced biomedical applications, including drug delivery and wound healing. Furthermore, CMC is instrumental in environmental remediation, facilitating pollutant removal and wastewater treatment. This review emphasizes the transformative potential of CMC, derived from upcycled waste, to foster sustainable innovation and reduce environmental impact across industries.

Keywords: Carboxymethylcellulose · Upcycling · Biowaste valorization

1 Introduction

The increasing global emphasis on sustainability has driven the exploration of renewable and biodegradable materials as alternatives to petroleum-based and non-biodegradable substances. Among such materials, polysaccharides, including cellulose derivatives, have emerged as versatile candidates for diverse industrial applications due to their abundance, biodegradability, and biocompatibility [1, 2]. Carboxymethylcellulose (CMC), a

R. Rajasekharan et al. (Eds.): MatZone 2024, SPPHY 316, pp. 1–19, 2025.
https://doi.org/10.1007/978-981-96-6541-9_1

derivative of cellulose, is widely utilized in biomedical, environmental, food, and industrial sectors due to its functional properties such as water solubility, gelling ability, and rheological behavior [3, 4].

CMC is traditionally derived from cellulose sources like cotton linters and wood pulp through reactions with chloroacetic acid in an alkaline medium, making it a renewable and commercially valuable biomaterial [5]. It serves as a thickener, stabilizer, and emulsifier in food, pharmaceuticals, and personal care industries. However, the increasing demand for CMC raises environmental concerns, particularly regarding its reliance on virgin cellulose sources, which contribute to deforestation, resource depletion, and pollution [6, 7]. Addressing these challenges, researchers have explored the conversion of agricultural and industrial waste into high-value CMC, promoting sustainable material development while enhancing the economic viability of agricultural practices [18–21].

The emergence of novel biomaterials from renewable feedstocks has amplified research outcomes across various technological domains. Polysaccharides and proteins, key components of natural polymers, play a pivotal role in applications spanning packaging, food, paper, textiles, and medicine due to their bioactivity and biodegradability [8, 9]. These materials also contribute to carbon capture, offering an eco-friendly approach to reducing atmospheric CO_2 levels. Material scientists are leveraging nanotechnology and polymerization techniques to enhance the molecular architecture and functionality of biopolymers, reinforcing their role in sustainable development. Furthermore, research indicates that agricultural waste, such as grape marc and olive pomace, can be repurposed to improve soil quality and plant nutrition, supporting sustainable agricultural practices [22, 23].

Upcycling presents a promising solution to mitigate the environmental impact of CMC production by transforming waste materials into higher-value products while adhering to circular economy principles. Unlike recycling, which often results in material degradation, upcycling enhances the functional value of cellulose-based derivatives. CMC derived from agricultural residues, industrial byproducts, and post-consumer paper can be repurposed into advanced materials such as hydrogels, superabsorbents, and biodegradable films, contributing to sustainable innovation in structural, environmental, and biomedical applications [11, 12].

In the food industry, CMC enhances texture and stability in formulations and serves as a key ingredient in low-fat and gluten-free products, acting as a fat replacer and thickener while preserving product quality [26, 27]. Additionally, its emulsifying properties improve the stability of food emulsions, essential for maintaining sensory attributes [28, 29]. Studies have shown that CMC can enhance the rheological properties of food systems, facilitating better handling and processing characteristics [26, 30].

Various biological, chemical, and thermal strategies have emerged for upcycling cellulose-rich waste into CMC. Biological methods, such as enzymatic hydrolysis, provide renewable feedstocks by breaking down complex cellulose structures [13]. Chemical techniques, including alkali and acid treatments, enable efficient cellulose extraction while minimizing toxic byproducts [14]. Meanwhile, thermal processes like pyrolysis and hydrothermal carbonization offer innovative pathways for recovering valuable fractions from lignocellulosic materials [15]. These advancements underscore the potential of upcycling strategies to redefine CMC production, integrating sustainability with material innovation.

Furthermore, the tunable properties of CMC—such as molecular weight, carboxyl content, and substituent distribution—enable its adaptation across industries. The incorporation of sodium carboxymethyl groups (CH_2COONa) enhances its water solubility, making it a critical ingredient in soaps, detergents, and personal care products [16]. Its high viscosity and non-toxicity also make it an ideal gelling agent and stabilizer in biomedical and environmental applications [17]. Additionally, CMC is utilized in pharmaceuticals and cosmetics, where it functions as a viscosity enhancer and stabilizer, and in drug delivery systems requiring controlled release properties [19, 31, 32].

To address the challenges and opportunities associated with CMC production, this review explores recent advancements in its production and modification from renewable feedstocks (Fig. 1), evaluates the role of CMC in sustainable upcycling processes and highlights emerging applications that align with global sustainability goals. By synthesizing current research, this review positions CMC as a cornerstone for advancing eco-friendly technologies and fostering a circular economy.

2 Bioconversion of Waste Materials into CMC: Advancing Sustainability and Circular Economy Practices

The bioconversion of waste materials into valuable products, particularly carboxymethyl cellulose (CMC), represents a significant advancement in sustainable practices. CMC is a cellulose derivative with numerous applications across various industries, including food, pharmaceuticals, and textiles. The production of CMC from waste materials, particularly cellulose-rich agricultural and industrial by-products, not only addresses waste management issues but also contributes to the circular economy by transforming waste into high-value products.

The initial step in producing CMC involves the extraction of cellulose from waste materials. Various waste sources, such as agricultural residues (e.g., corn peels, oil palm empty fruit bunches, and cotton waste), have been identified as potential feedstocks for cellulose extraction. For instance, Zainol et al. demonstrated that CMC can be synthesized from oil palm empty fruit bunches using ionic liquids and hydrogen peroxide, achieving a degree of substitution (DS) of 0.6 to 0.67, which indicates a significant transformation of cellulose into CMC [33]. Similarly, Bahlool and Kamel highlighted the potential of utilizing Egyptian cotton waste fibers for CMC production, emphasizing the underutilization of agricultural waste in the textile industry [34].

The extraction process typically involves pretreatment methods that enhance cellulose accessibility. For example, Rojsitthisak et al. utilized a response surface methodology to optimize the preparation of CMC from corn peel agricultural waste, demonstrating that proper pretreatment can significantly improve the yield and quality of CMC produced [35]. The use of alkaline solutions, such as sodium hydroxide, is common in the mercerization stage, which modifies cellulose to facilitate further chemical reactions necessary for CMC synthesis [36]. This process is crucial, as it alters the crystalline structure of cellulose, making it more amenable to subsequent carboxymethylation reactions.

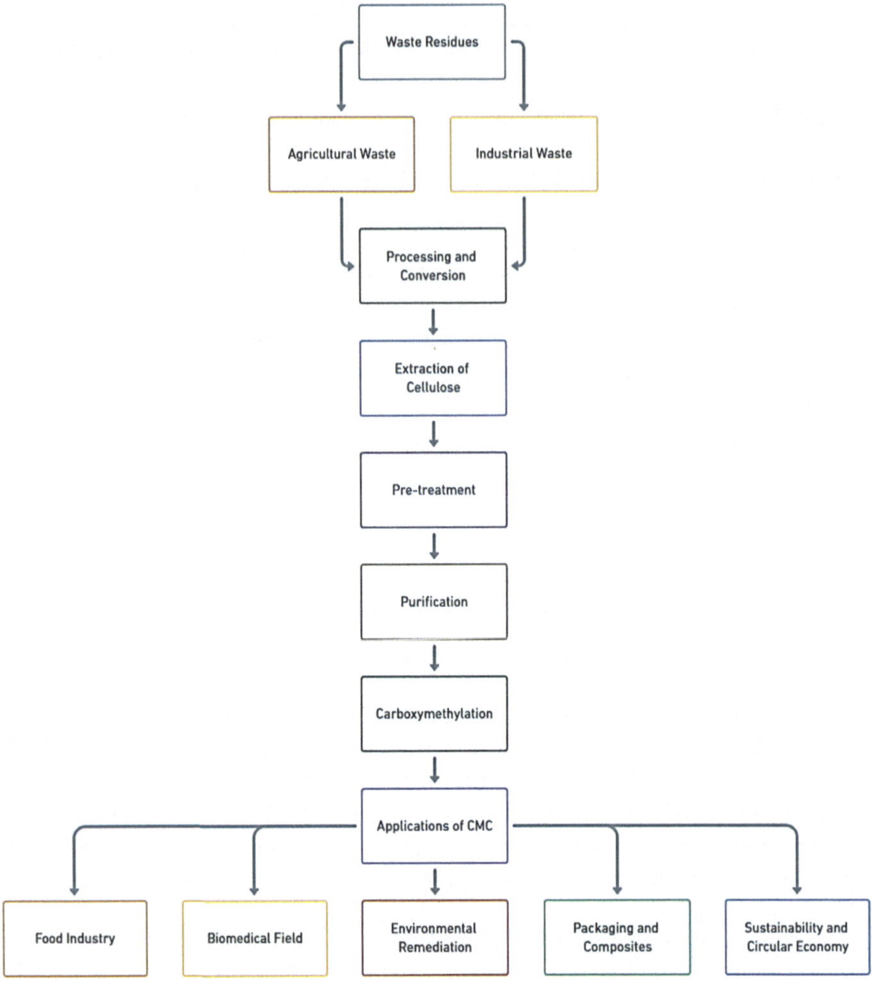

Fig. 1. Flow chart illustrating upcycling of agricultural and industrial waste into CMC

The bioconversion process can also be enhanced by incorporating biological methods. The use of cellulolytic microorganisms, such as Trichoderma viride, has been investigated for their ability to degrade cellulose in waste materials, thereby facilitating the extraction of fermentable sugars that can be further processed into CMC [38]. Mokatse et al. highlighted the effectiveness of cellulase enzymes derived from T. viride in saccharifying various types of waste paper, underscoring the potential of using fungal enzymes to improve cellulose accessibility and conversion efficiency [38]. This biological approach not only aids in cellulose extraction but also contributes to the reduction of waste volume, aligning with environmental sustainability goals.

Furthermore, the integration of bioconversion techniques with traditional chemical methods can lead to more efficient CMC production processes. For instance, the combination of enzymatic hydrolysis followed by chemical carboxymethylation has been proposed as a synergistic approach to enhance CMC yield and quality [39]. The enzymatic step can effectively break down complex lignocellulosic structures, making the cellulose more reactive in the subsequent chemical treatment.

The environmental benefits of utilizing waste materials for CMC production are significant. The conversion of agricultural and industrial waste into valuable products reduces the burden on landfills and minimizes the environmental impact associated with waste disposal [40]. Additionally, the use of renewable biomass as a feedstock for CMC production contributes to a more sustainable and circular economy, reducing reliance on fossil fuels and promoting resource efficiency [41].

3 Agricultural Residues for Sustainable CMC Production

Agricultural residues represent a significant yet underutilized resource for the production of CMC. The conversion of agricultural waste into CMC not only addresses waste management issues but also contributes to sustainable practices by providing an alternative to synthetic polymers. The potential of agricultural residues such as cotton waste, corn husk, and sugarcane bagasse for CMC production has been documented in recent studies, demonstrating the feasibility of utilizing these materials as raw sources of cellulose (Table 1).

Table 1. Agricultural residues as cost-effective & sustainable feedstocks for CMC production

Agro residue	Characteristics	Applications/significance
Corn husk	High cellulose content, widely available	Produces high-purity CMC with excellent film-forming properties
Sugarcane bagasse	Rich lignocellulosic material, abundant	Suitable for biodegradable films and composites
Rice husk	Abundant byproduct of rice cultivation, renewable	Used for food-grade CMC and hydrogel applications
Wheat straw	Low lignin content, easy extraction	Enhances CMC performance in composites
Banana peel	Renewable, cellulose-rich	Produces biodegradable films and food-safe hydrogels
Olive pomace	Residue from olive oil production	Eco-friendly food coatings and packaging
Grape marc	Byproduct of winemaking	Source for bio-based polymers

The extraction of cellulose from agricultural residues is a critical first step in CMC production. Various methodologies have been employed to isolate cellulose from different agricultural by-products. For instance, Sritapunya et al. highlighted the extraction of cellulose from Cavendish banana peel, corn silk, and bagasse, which was subsequently converted into CMC through carboxymethylation using monochloroacetic acid in the presence of sodium hydroxide [26]. Similarly, Rojsitthisak et al. reported the use of corn peel, rice straw, and papaya peel as viable sources for cellulose extraction, emphasizing the versatility of agricultural waste in CMC synthesis [43]. The diversity of agricultural residues available for cellulose extraction underscores the potential for large-scale CMC production, as these materials are often abundant and underutilized.

The optimization of extraction and synthesis processes is essential for enhancing the yield and quality of CMC derived from agricultural residues. Bahlool and Kamel investigated the use of Egyptian cotton waste fibers for CMC production, demonstrating that various agro-cellulosic sources, including paper sludge and wood residues, can be effectively utilized [44]. Their findings indicate that optimizing the conditions for cellulose extraction and subsequent carboxymethylation can significantly improve the properties of the resulting CMC. Moreover, Rachtanapun et al. explored the production of CMC from durian rind, further illustrating the potential of diverse agricultural waste materials in CMC synthesis [45]. The application of response surface methodology has also been employed to optimize the preparation of CMC from corn peel, highlighting the importance of systematic approaches in enhancing production efficiency [43].

The environmental benefits of utilizing agricultural residues for CMC production are substantial. By converting waste materials into valuable products, the process not only reduces the volume of agricultural waste but also mitigates the environmental impact associated with waste disposal. Landolsi et al. demonstrated a novel methodology for producing CMC from olive tree pruning residues, employing a chlorine-free bleaching process to enhance cellulose purity [46]. This approach aligns with sustainable practices and minimizes the environmental footprint of CMC production. Furthermore, Churam's research on sugarcane leaves emphasizes the potential for sustainable practices in CMC synthesis, as the study identified optimal conditions for carboxymethylation that maximize yield while minimizing resource use [47].

The physicochemical properties of CMC derived from agricultural residues can vary significantly based on the source material and the processing conditions employed. Studies have shown that the degree of substitution (DS) and viscosity of CMC can be influenced by factors such as the concentration of sodium hydroxide used during synthesis. For example, Churam reported a DS of 0.86 for CMC produced from sugarcane leaves under optimized conditions, indicating a high level of carboxymethyl substitution [47]. Similarly, the research by Suriyatem et al. on palm bunch and bagasse highlighted the effects of delignification on the physical properties of CMC, demonstrating that treatment with hydrogen peroxide can enhance the quality of the final product [48]. These findings underscore the importance of tailoring processing conditions to achieve desired properties in CMC derived from agricultural waste.

The applications of CMC produced from agricultural residues are diverse and expanding. CMC is widely used as a thickening agent, stabilizer, and emulsifier in food products, pharmaceuticals, and cosmetics. The versatility of CMC makes it an attractive alternative to synthetic polymers, particularly in the context of increasing environmental concerns regarding plastic waste. Yaradoddi et al. explored the use of CMC derived from sugarcane bagasse in biodegradable packaging materials, demonstrating its potential as a sustainable alternative to conventional plastics [49]. This aligns with the growing trend towards eco-friendly materials in various industries, as consumers and manufacturers alike seek to reduce their environmental impact.

The synthesis of CMC from agricultural residues also opens avenues for innovation in material science. Research has indicated that CMC can be combined with other biopolymers to enhance its properties for specific applications. For instance, the study by Tohamy et al. on biodegradable films highlighted the potential for CMC to be blended with gelatin and agar to create materials with improved mechanical properties [50]. This suggests that the integration of CMC with other natural polymers can lead to the development of advanced materials suitable for various applications, including packaging and biomedical uses.

4 Upcycling Wastepaper into Carboxymethylcellulose

Waste paper, a largely underutilized resource, can be transformed through various processes into CMC. The process of upcycling wastepaper involves several stages, including deinking, refining, and chemical modification, which can enhance the properties of the resultant cellulose derivatives.

Wastepaper recycling is a critical first step in the upcycling process. The deinking of waste paper is essential to remove inks and other impurities that can adversely affect the quality of the recycled pulp. Fang et al. highlight that the presence of residual ink is a significant challenge in the recycling process, as it influences both the quality of the deinked pulp and its subsequent applications [51]. This process not only recycles the cellulose fibers but also contributes to the reduction of waste in landfills, thereby promoting environmental sustainability. The effective removal of impurities can lead to higher quality cellulose, which is crucial for the production of CMC.

The technological advancements in recycling processes play a crucial role in enhancing the quality of the recycled material. Marchenko discusses the use of knifeless grinding technology, which minimizes damage to the fibers during recycling, thus preserving their properties for further processing into CMC [53]. This gentle refining process is particularly important for maintaining the integrity of the cellulose fibers, which are essential for high-quality CMC production. The ability to produce high-performance materials from waste paper not only reduces environmental impact but also opens new avenues for innovation in material science.

Furthermore, the extraction of cellulose from waste paper through upcycling processes can lead to materials with enhanced functionalities. Ye et al. note that when cellulose is derived from waste sources, the resultant materials often exhibit improved properties compared to their original forms [54]. This transformation is particularly relevant for CMC, which benefits from the enhanced characteristics of upcycled cellulose. The circular economy model, which emphasizes the reuse and recycling of materials, is thus effectively supported by the upcycling of waste paper into CMC, contributing to both economic and environmental sustainability.

The ecological implications of waste paper recycling cannot be overlooked. Dai highlights the importance of assessing the ecological costs associated with material flows in recycling systems, suggesting that incorporating recycled resources can mitigate environmental impacts [55]. By utilizing waste paper as a raw material for CMC production, the ecological footprint of the manufacturing process can be significantly reduced.

Moreover, the versatility of CMC derived from waste paper extends to various applications, including its use as a thickening agent in food products, a stabilizer in pharmaceuticals, and a binder in personal care items. The potential for high-absorbency products from upcycled materials has been explored by Zambrano et al., who developed tissue paper from upcycled corrugated packaging waste, demonstrating the feasibility of creating high-value products from waste streams [56]. This indicates that the upcycling of waste paper can lead to innovative applications that meet consumer demands while promoting sustainability.

The economic benefits of upcycling waste paper into CMC are also significant. By creating a market for upcycled products, new business opportunities can arise, fostering job creation and economic growth. The systematic review by Caldera et al. identifies various barriers and enablers for closing the loop through waste upcycling, emphasizing the potential for businesses to capitalize on upcycled materials [57].

5 CMC Extraction and Pre-treatment Processes

The extraction and pre-treatment processes of CMC can be categorized into mechanical, chemical, and enzymatic methods, each contributing to the yield and quality of the final product.

5.1 Mechanical Methods

Mechanical extraction methods often involve physical processes such as grinding or milling to break down the cellulose structure, thereby facilitating the release of CMC. These methods can be effective but may require subsequent chemical or enzymatic treatments to enhance extraction efficiency. For instance, the mechanical disruption of plant materials can increase the surface area for subsequent enzymatic action, as noted in studies focusing on the extraction of polysaccharides from various sources [58]. However, mechanical methods alone may not achieve the desired purity or yield without additional treatments.

5.2 Chemical Methods

Chemical pre-treatment processes typically involve the use of alkaline or acidic solutions to modify the cellulose structure, making it more amenable to extraction. For example, alkaline treatments can solubilize hemicellulose and lignin, which can enhance the extraction of CMC from lignocellulosic biomass [59]. The use of hydrogen peroxide in alkaline media has been reported to effectively isolate hemicellulose, which can be further processed to yield CMC [59]. Additionally, the optimization of chemical conditions, such as pH and temperature, is crucial for maximizing the extraction yield while minimizing degradation of the polysaccharide [60].

5.3 Enzymatic Methods

Enzymatic extraction is increasingly favoured due to its specificity and milder conditions, which help preserve the integrity of the extracted compounds. Enzymes such as cellulases and xylanases play a pivotal role in breaking down cellulose and hemicellulose, facilitating the release of CMC [61]. Studies have demonstrated that the application of cellulolytic enzymes can significantly enhance the yield of CMC from various substrates, including agricultural waste [62]. Furthermore, the use of ultrasonic-assisted enzymatic extraction has been shown to improve the efficiency of polysaccharide extraction by enhancing enzyme accessibility to the substrate [58].

6 Carboxymethylation Process—Chemical Modification of Cellulose into CMC

Carboxymethylation transforms cellulose into carboxymethyl cellulose (CMC), a derivative known for its solubility and versatility in industries like food, pharmaceuticals, and textiles. The process introduces carboxymethyl groups into cellulose via a heterogeneous reaction involving monochloroacetic acid (MCA) and sodium hydroxide (NaOH) [63]. Initially, cellulose undergoes mercerization with NaOH, forming alkaline cellulose that reacts with MCA [64]. A degree of substitution (DS) of at least 0.4 is required for water solubility [66].

Recent advancements have improved the process through homogeneous solvent systems like NaOH/urea, which ensure uniform carboxymethylation and better control over DS, enhancing CMC solubility [66, 67]. One-pot reactions have also minimized solvent use and environmental impacts [68, 69]. Optimal conditions, including NaOH concentration, temperature (35 °C–55 °C), and alternative solvents like DMSO or molten salt hydrates, are crucial to achieving efficient synthesis without degradation [67, 71, 72].

7 Sustainable Applications of Carboxymethylcellulose

CMC is a versatile biopolymer that has garnered significant attention in various fields, including environmental remediation, biomedical applications, and food technology. Its unique properties, such as biocompatibility, biodegradability, and ability to form hydrogels, make it an ideal candidate for upcycling strategies aimed at addressing environmental challenges and enhancing material performance.

7.1 Applications of CMC in the Food Industry

CMC, a cellulose derivative, is highly regarded for its versatile properties, including its role as a thickener, stabilizer, and emulsifier. It is extensively used in food formulations to enhance texture, stability, and shelf life, significantly improving product quality. For instance, CMC effectively modifies the flow properties of dairy systems, stabilizing and refining textures in skimmed and whole milk [73].

7.1.1 CMC in Food Packaging

CMC has been successfully incorporated into biodegradable food packaging, particularly in edible coatings, to extend the freshness of perishable goods while reducing spoilage (Table 2). These coatings not only preserve quality but also ensure food safety [74, 75]. Additionally, its use in packaging materials improves mechanical properties and barrier capabilities, making it an eco-friendly alternative to conventional plastics [76, 77].

Table 2. Applications of upcycled CMC in the food industry

Application	Functionality
Edible films and coatings	Agricultural residues, such as corn husks and wheat straw, can be processed into CMC for edible films that extend the shelf life of fresh produce and baked goods. These films reduce oxygen and moisture permeability, enhancing food preservation
Functional additives	Incorporating bioactive compounds, such as antioxidants or antimicrobial agents, into upcycled CMC films boosts their utility in food packaging while reducing agricultural waste
Composite films	Development of CMC-based films infused with pomegranate peel extract improves their mechanical and antimicrobial properties, suitable for sustainable food packaging
Bilayer edible coatings	CMC-based coatings with antimicrobial additives like *lactobacillus acidophilus* enhance shelf life of fruits and vegetables while reducing food waste

7.1.2 CMC in Upcycled Foods

The upcycling of food waste into valuable products aligns with sustainability goals and supports a circular economy. CMC enhances the sensory and physical attributes of upcycled foods, making them more appealing to consumers. For example, it has been used in edible coatings for meat products, extending shelf life while minimizing packaging waste [78, 79].

Consumer awareness of sustainability claims has further bolstered the acceptance of upcycled products. Research shows that informing consumers about the nutritional and environmental benefits of upcycled foods positively influences their willingness to pay, thereby fostering market growth [80, 81]. Innovations in food processing have expanded CMC's role in the encapsulation of bioactive compounds, such as probiotics, to enhance the nutritional value of upcycled foods. This has led to the development of functional foods tailored to specific health benefits, meeting diverse consumer preferences [82].

7.1.3 Nutritional and Economic Benefits

While many upcycled products are currently discretionary foods like snacks, CMC enables the creation of healthier, more nutritious options by improving texture and retaining the beneficial properties of by-products. For instance, CMC has been used in biscuits made from upcycled defatted sunflower seed flour, offering sustainable and economically viable alternatives [83]. The upcycled food market is projected to grow substantially, with a market value of $46.7 billion and a compound annual growth rate of 5% over the next 6 years [84].

The use of CMC in upcycled food systems has significant environmental benefits, including reducing greenhouse gas emissions and mitigating resource depletion. By transforming food scraps into high-value products, the food industry can contribute to more sustainable practices and minimize environmental harm [85].

7.2 Biomedical Applications of Carboxymethylcellulose

CMC is a versatile biopolymer widely recognized for its unique properties, including biocompatibility, biodegradability, and tunable mechanical characteristics (Table 3). These attributes make it an ideal material for a variety of biomedical applications, such as drug delivery systems, wound healing, and tissue engineering scaffolds. Its ability to form hydrogels that mimic the extracellular matrix (ECM) enables it to create conducive environments for cell growth, proliferation, and tissue regeneration [86, 87].

Table 3. Biomedical applications of carboxymethylcellulose

Application	Key functions	Enhancements and modifications
Drug delivery systems	Controlled drug release Encapsulation of therapeutic agents Mucoadhesive properties	CMC hydrogels retain biological substances for sustained release Used in transdermal microneedle arrays for enhanced drug delivery Integrated into mucoadhesive films for localized drug release, e.g., imiquimod for skin conditions
Wound healing	Retains moisture, promoting epithelialization and tissue regeneration Minimizes irritation, allowing long-term use	Antimicrobial CMC hydrogels with silver nanoparticles prevent infections CMC-nanocellulose composites enhance mechanical strength and swelling properties for improved wound management
Tissue Engineering	Supports cell attachment, proliferation, and differentiation Mechanical properties can be tuned for specific tissue applications	Higher CMC concentration—Increased structural integrity (e.g., for bone tissue engineering) Methacrylate-modified CMC enables photo-crosslinking for customizable hydrogels
Advanced biomedical materials	Enhances bioactivity and antimicrobial properties	CMC-PVA composites improve mechanical properties in wound dressings Nanoparticle-infused CMC (silver, zinc ferrite) provides bioactive functionalities

7.3 Environmental Applications of Carboxymethylcellulose

CMC offers significant environmental benefits due to its biodegradability, renewable origin, and potential to reduce reliance on non-renewable resources (Table 4). Its use aligns with circular economy principles, emphasizing the reuse and recycling of materials to minimize waste. CMC is increasingly explored as a sustainable alternative in various industries, addressing environmental challenges such as plastic waste and pollutant removal.

Table 4. Environmental applications of carboxymethylcellulose

Application	Key functions and examples	Enhancements and modifications
Biodegradability and circular economy	Reduces reliance on fossil fuels by replacing traditional plastics in biodegradable films and food packaging	CMC-based materials provide moisture and gas barriers while decomposing naturally
Pollutant removal	Used in wastewater treatment to adsorb industrial dye pollutants and heavy metals, helping purify industrial effluents	Cross-linked CMC hydrogels improve adsorption efficiency by encapsulating contaminants
Composite materials	Enhances mechanical strength in eco-friendly composites for construction materials and sustainable engineering components	CMC-cellulose nanocomposites improve structural integrity for durability and sustainability
Innovative packaging solutions	Provides a biodegradable alternative to plastic packaging, ensuring reduced environmental impact in food and product packaging	CMC blends with cellulose derivatives enhance moisture and gas barrier properties for extended shelf life

7.4 Textile Applications

The textile industry generates substantial cellulose-rich waste, including cotton scraps and worn garments, which can be upcycled into CMC for use in a wide array of applications such as biodegradable coatings, hydrogels, and food additives (Table 5). Valorizing these waste streams reduces the need for virgin cellulose sources, supports sustainable resource use, and aligns with global waste reduction goals. This process offers innovative solutions for sustainable fabric production:

Table 5. Applications of upcycled CMC in textile enhancements

Application	Functionality
Finishing agents	CMC-based coatings improve textile durability, water resistance, and wrinkle resistance while maintaining environmental compatibility
Printing thickeners	Upcycled CMC ensures uniform dye distribution and reduces wastewater generation in textile printing processes
Antimicrobial finishes	Combining upcycled CMC with antimicrobial agents creates textiles with added functionalities, such as microbial resistance for hospital linens and protective clothing
Color stripping agents	Upcycled CMC can be used in processes like ozone-assisted color stripping of dyed textiles, reducing chemical usage and environmental impact
Wastewater treatment	CMC derived from textile waste can assist in the removal of dyes and contaminants from industrial wastewater

8 Conclusion

The upcycling of agricultural and industrial waste into high-value bioproducts, exemplified by the production and versatile applications of carboxymethylcellulose (CMC), represents a paradigm shift toward sustainable development and circular economy principles. This review underscores the immense potential of CMC, not only as a renewable and biodegradable alternative to synthetic polymers but also as a bridge between waste valorization and advanced material science.

The bioconversion of waste materials into CMC showcases the effectiveness of leveraging biological, chemical, and thermal methodologies to transform cellulose-rich byproducts into valuable resources. Agricultural residues such as corn husks, sugarcane bagasse, and cotton waste are demonstrated to be cost-effective, abundant, and renewable feedstocks for CMC production. Similarly, the upcycling of wastepaper through innovative recycling techniques reveals opportunities for sustainable cellulose extraction, minimizing environmental burdens associated with waste disposal.

Beyond its production, the applications of CMC across various domains highlight its multifaceted nature. In food systems, CMC contributes to improved texture, stability, and shelf life, enabling the development of healthier and eco-friendly products. Its role in advanced biomedical applications, including drug delivery systems and wound healing, emphasizes its compatibility with human health and its potential in regenerative medicine. Furthermore, the use of CMC in environmental remediation, such as pollutant adsorption and wastewater treatment, solidifies its position as an eco-friendly solution for addressing pressing environmental challenges.

In conclusion, the advancement of upcycling technologies and the strategic utilization of CMC as a sustainable biomaterial can profoundly impact industries and ecosystems. Continued research on optimizing production methods, enhancing functionality, and expanding application scopes will further position CMC as a cornerstone for sustainable innovation, aligning global industrial practices with environmental stewardship and economic viability.

References

1. Gennaro, M.: Design of carboxymethylcellulose/poloxamer-based bioformulation embedding trichoderma harzianum for agricultural applications. Langmuir. **40**(23), 12159–12166 (2024). https://doi.org/10.1021/acs.langmuir.4c01038
2. Taşkın, Ö., Hubble, D., Zhu, T., Liu, G.: Biomass-derived polymeric binders in silicon anodes for battery energy storage applications. Green Chem. **23**(20), 7890–7901 (2021). https://doi.org/10.1039/d1gc01814k
3. A study of properties of fibers extracted from karkade (Hibiscus sabdariffa l.) agricultural waste as potential reinforcing material. Int. J. Eng. Sci. Res. Technol. **9**(7), 207–213 (2020). https://doi.org/10.29121/ijesrt.v9.i7.2020.21
4. Obi, F., Ugwuishiwu, B., Nwakaire, J.: Agricultural waste concept, generation, utilization, and management. Niger. J. Technol. **35**(4), 957 (2016). https://doi.org/10.4314/njt.v35i4.34
5. Açıkbaş, B., Yaşasin, A., Gürbüz, M., Candar, S., Çınar, G.: The effect of grape marc and olive pomace organic wastes on nutrition of kalecikkarası and narince (vitis vinifera l.) grape varieties. Tekirdag Viticulture Research Institute (2022). https://doi.org/10.52001/vis.2022.8.25.32
6. Chen, S.: Effect evaluation of the agricultural waste on improving growth of potted sunflower (Helianthus annuus). Int. J. Hortic. (2023). https://doi.org/10.5376/ijh.2023.13.0012
7. Firmansyah, A.: Unraveling the impact of social innovation based on biocycle farming: the path to sustainable development. E3s Web Conf. **454**, 02012 (2023). https://doi.org/10.1051/e3sconf/202345402012
8. Sholichah, H., Wahyuni, A.: Pemanfaatanlimbahorganikberbasis bio cyclo farming (studikasus di desatelagamurni). J. Environ. Eng. Waste Manag. **4**(1), 21 (2019). https://doi.org/10.33021/jenv.v4i1.622
9. Yildirim, M., Şumnu, G., Şahin, S.: Rheology, particle-size distribution, and stability of low-fat mayonnaise produced via double emulsions. Food Sci. Biotechnol. **25**(6), 1613–1618 (2016). https://doi.org/10.1007/s10068-016-0248-7
10. Townsend, K., Laffan, J., Hayman, G.: Carboxymethylcellulose excipient allergy: a case report. J Med Case Reports. **15**(1), 565 (2021). https://doi.org/10.1186/s13256-021-03180-y
11. Kowalska, M., Woźniak, M., Żbikowska, A., Ivanišová, E., Molik, A.: Quality of emulsions containing fat blends modified by enzymatic catalysis. Catalysts. **11**(4), 453 (2021). https://doi.org/10.3390/catal11040453
12. Yue, Y., Pang, S., Li, N., Tong, L., Wang, L., Fan, B., et al.: Interactions between pea protein isolate and carboxymethylcellulose in neutral and acid aqueous systems. Food Secur. **10**(7), 1560 (2021). https://doi.org/10.3390/foods10071560
13. Wang, S., Lu, Y., Tsai, M.: Formulation and evaluation of chitosan/NaCl/maltodextrin microparticles as a saltiness enhancer: study on the optimization of excipients for the spray-drying process. Polymers. **13**(24), 4302 (2021). https://doi.org/10.3390/polym13244302
14. Rutz, J., Borges, C., Zambiazi, R., Rosa, C., Silva, M.: Elaboration of microparticles of carotenoids from natural and synthetic sources for applications in food. Food Chem. **202**, 324–333 (2016). https://doi.org/10.1016/j.foodchem.2016.01.140
15. Verardi, A.: Enhancing carotenoids' efficacy by using chitosan-based delivery systems. Forum Nutr. **3**(3), 451–480 (2023). https://doi.org/10.3390/nutraceuticals3030033

16. Zainol, M., Rasid, N., Asmadi, M., Amin, N.: Carboxymethyl cellulose synthesis from treated oil palm empty fruit bunch using ionic liquid and hydrogen peroxide. ASEAN Eng. J. **11**(4), 80–88 (2021). https://doi.org/10.11113/aej.v11.17866
17. Bahlool, S., Kamel, S.: Utilization of Egyptian cotton waste fibers for production of carboxymethyl cellulose (CMC). J. Textiles Color. Polym. Sci. **20**, 83 (2023). https://doi.org/10.21608/jtcps.2023.186804.1158
18. Rojsitthisak, P., Khunthon, S., Noomun, K., Limpanart, S.: Response surface method to optimize the preparation of carboxymethyl cellulose from corn peel agricultural waste. Sci. Asia. **43**(1), 8 (2017). https://doi.org/10.2306/scienceasia1513-1874.2017.43.008
19. Almlöf, H., Kreutz, B., Jardeby, K., Germgård, U.: The influence of extended mercerization on some properties of carboxymethyl cellulose (CMC). Holzforschung. **66** (2012). https://doi.org/10.1515/hf.2011.131
20. Abdullaevich, Y., Abdukhalilovich, S.: Low and highly substituted carboxymethyl cellulose synthesis from powdered and microcrystalline cellulose. Am. J. Appl. Sci. **2**(10), 83–96 (2020). https://doi.org/10.37547/tajas/volume02issue10-13
21. Mokatse, K., Mhlanga, H., Wyk, J.: Relative saccharification and initial degradation rates of different waste paper materials by cellulase from Trichoderma viride. J. Appl. Biosci. **105**(1), 10183 (2016). https://doi.org/10.4314/jab.v105i1.14
22. Hivechi, A., Bahrami, S., Arami, M., Karimi, A.: Ultrasonic mediated production of carboxymethyl cellulose: optimization of conditions using response surface methodology. Carbohydr. Polym. **134**, 278–284 (2015). https://doi.org/10.1016/j.carbpol.2015.07.045
23. Rodríguez, A.: Special issue "lignocellulosic biomass II". Molecules. **28**(17), 6230 (2023). https://doi.org/10.3390/molecules28176230
24. Capolupo, L., Faraco, V.: Green methods of lignocellulose pretreatment for biorefinery development. Appl. Microbiol. Biotechnol. **100**(22), 9451–9467 (2016). https://doi.org/10.1007/s00253-016-7884-y
25. Sritapunya, T., Sommanas, G., Surachaijarin, H.: The effect of carboxymethyl cellulose from various agriculture wastes on the viscosity and physical properties of low-concentration solutions of surfactants. Key Eng. Mater. **751**, 745–750 (2017). https://doi.org/10.4028/www.scientific.net/kem.751.745
26. Rachtanapun, P., Luangkamin, S., Tanprasert, K., Suriyatem, R.: Carboxymethyl cellulose film from durian rind. LWT. **48**(1), 52–58 (2012). https://doi.org/10.1016/j.lwt.2012.02.029
27. Landolsi, I., Rjiba, N., Hamdaoui, M., Harzallah, O., Boudokhane, C.: Homogeneous microwave-assisted carboxymethylation from totally chlorine-free bleached olive tree pruning residues pulp. J. Serb. Chem. Soc. **87**(2), 247–261 (2022). https://doi.org/10.2298/jsc210401046l
28. Churam, T.: Sustainable production of carboxymethyl cellulose: a biopolymer alternative from sugarcane (Saccharum officinarum L.) leaves. Sustain. For. **16**(6), 2352 (2024). https://doi.org/10.3390/su16062352
29. Suriyatem, R., Noikang, N., Kankam, T., Jantanasakulwong, K., Leksawasdi, N., Phimolsiripoletal..: Physical properties of carboxymethyl cellulose from palm bunch and bagasse agricultural wastes: effect of delignification with hydrogen peroxide. Polymers. **12**(7), 1505 (2020). https://doi.org/10.3390/polym12071505
30. Yaradoddi, J., Banapurmath, N., Ganachari, S., Soudagar, M., Mubarak, N., Halladetal: Biodegradable carboxymethyl cellulose-based material for sustainable packaging application. Sci. Rep. **10**(1), 21960 (2020). https://doi.org/10.1038/s41598-020-78912-z
31. Hidayati, S., Maulidia, U., Satyajaya, W., Hadi, S.: Effect of glycerol concentration and carboxymethyl cellulose on biodegradable film characteristics of seaweed waste. Heliyon. **7**(8), e07799 (2021). https://doi.org/10.1016/j.heliyon.2021.e07799

32. Fang, C., Hu, J., Zhang, Y., Yu, R., Zhang, M.: The status and progress of waste paper deinking. Adv. Mat. Res. **174**, 341–345 (2010). https://doi.org/10.4028/www.scientific.net/amr.174.341
33. Marchenko, R.: Use of a knifeless grinding plant for waste paper recycling. E3s Web Conf. **390**, 05006 (2023). https://doi.org/10.1051/e3sconf/202339005006
34. Ye, Y., Yu, L., Lizundia, E., Zhu, Y., Jiang, F.: Cellulose-based ionic conductor: an emerging material toward sustainable devices. Chem. Rev. **123**(15), 9204–9264 (2023). https://doi.org/10.1021/acs.chemrev.2c00618
35. Dai, T.: Assessing the ecological cost of material flow in China's waste paper recycling system. Sustain. For. **16**(4), 1610 (2024). https://doi.org/10.3390/su16041610
36. Zambrano, F., Márquez, R., Vera, R., Jameel, H., Venditti, R., Gonzalez, R.: Developing alternative, high-absorbency brown fibers: tissue paper from upcycled corrugated packaging waste to meet new consumer trends. ACS Sustain. Chem. Eng. **10**(40), 13343–13356 (2022). https://doi.org/10.1021/acssuschemeng.2c03280
37. Caldera, S., Jayasinghe, R., Desha, C., Dawes, L., Ferguson, S.: Evaluating barriers, enablers, and opportunities for closing the loop through 'waste upcycling': a systematic literature review. J. Sustain. Dev. Energy Water Environ. Syst. **10**(1), 1–20 (2022). https://doi.org/10.13044/j.sdewes.d8.0367
38. Le, L., Xu, L., Ding, C., Yuan, S., Zhang, Z., Chen, Y., et al.: Ultrasonic-assisted enzymatic extraction and antioxidant activity of polysaccharides from Setariaviridis. Sep. Sci. Technol. **51**(11), 1798–1805 (2016). https://doi.org/10.1080/01496395.2016.1178287
39. Brienzo, M., Carvalho, W., Milagres, A.: Xylooligosaccharides production from alkali-pretreated sugarcane bagasse using xylanases from Thermoascusaurantiacus. Appl. Biochem. Biotechnol. **162**(4), 1195–1205 (2010). https://doi.org/10.1007/s12010-009-8892-5
40. Yan, P., Li, Y., Cao, L.: Optimization of enzymatic hydrolytic extraction conditions for high-yield polysaccharide manufacturing from liquid fermented materials of Hypsizigus-marmoreus. Appl. Mech. Mater. **145**, 154–158 (2011). https://doi.org/10.4028/www.scientific.net/amm.145.154
41. Cano-Ramírez, C., Santiago-Hernández, A., Rivera-Orduña, F., Pineda-Mendoza, R., Zúñiga, G., Hidalgo-Lara, M.: One-step zymogram method for the simultaneous detection of cellulase/xylanase activity and molecular weight estimation of the enzyme. Electrophoresis. **38**(3–4), 447–451 (2016). https://doi.org/10.1002/elps.201600347
42. Andlar, M., Rezić, T., Marđetko, N., Kracher, D., Ludwig, R., Šantek, B.: Lignocellulose degradation: an overview of fungi and fungal enzymes involved in lignocellulose degradation. Eng. Life Sci. **18**(11), 768–778 (2018). https://doi.org/10.1002/elsc.201800039
43. Abdullaevich, Y., Yunusov, K., Abdukhalilovich, S., Gogli, G.: Synthesis and characterization of sodium-carboxymethylcellulose from cotton, powder, microcrystalline, and nanocellulose. Polym. Eng. Sci. **62**(3), 677–686 (2021). https://doi.org/10.1002/pen.25874
44. Köhler, S., Liebert, T., Heinze, T.: Ammonium-based cellulose solvents suitable for homogeneous etherification. Macromol. Biosci. **9**(9), 836–841 (2009). https://doi.org/10.1002/mabi.200900156
45. Hu, Q., Liebert, T., Meister, F., Heinze, T.: Homogeneous carboxymethylation of cellulose in the NaOH/urea aqueous solution. React. Funct. Polym. **69**(10), 779–784 (2009). https://doi.org/10.1016/j.reactfunctpolym.2009.06.007
46. Liebert, T., Heinze, T.: Exploitation of reactivity and selectivity in cellulose functionalization using unconventional media for the design of products showing new superstructures. Biomacromolecules. **2**(4), 1124–1132 (2001). https://doi.org/10.1021/bm010068m
47. Moreau, C., Falourd, X., Talantikite, M., Cathala, B., Villares, A.: Bifunctionalization of cellulose fibers by one-step Williamson's etherification to obtain modified microfibrillated cellulose. ACS Sustain. Chem. Eng. **10**(40), 13415–13423 (2022). https://doi.org/10.1021/acssuschemeng.2c03754

48. Kaldéus, T., Larsson, P., Boujemaoui, A., Malmström, E.: One-pot preparation of bi-functional cellulose nanofibrils. Cellul. **25**(12), 7031–7042 (2018). https://doi.org/10.1007/s10570-018-2066-y
49. Stigsson, V., Kloow, G., Germgård, U.: The influence of the solvent system used during the manufacturing of CMC. Cellul. **13**(6), 705–712 (2006). https://doi.org/10.1007/s10570-006-9083-y
50. Khullar, R., Varshney, V., Naithani, S., Heinze, T., Soni, P.: Carboxymethylation of cellulosic material (average degree of polymerization 2600) isolated from cotton (Gossypium) linters with respect to degree of substitution and rheological behavior. J. Appl. Polym. Sci. **96**(4), 1477–1482 (2005). https://doi.org/10.1002/app.21645
51. Bayarri, S., Costell, E.: Flow properties of carboxymethylcellulose dairy systems with different fat content. J. Food Process Eng. **34**(6), 1903–1914 (2011). https://doi.org/10.1111/j.1745-4530.2009.00383.x
52. Goodman-Smith, F., Bhatt, S., Moore, R., Mirosa, M., Ye, H., Deutsch, J., et al.: Retail potential for upcycled foods: evidence from New Zealand. Sustain. For. **13**(5), 2624 (2021). https://doi.org/10.3390/su13052624
53. Mohamed, D.: Potential impacts of edible coatings fortified by moringa and/or cedar extracts on quality and shelf life of chilled Turkey fillets. Benha Vet. Med. J. **45**(2), 209–214 (2023). https://doi.org/10.21608/bvmj.2023.252362.1758
54. Tongdeesoontorn, W., Mauer, L., Wongruong, S., Sriburi, P., Rachtanapun, P.: Effect of carboxymethyl cellulose concentration on physical properties of biodegradable cassava starch-based films. Chem. Cent. J. **5**(1), 6 (2011). https://doi.org/10.1186/1752-153x-5-6
55. Han, J., Shin, S., Park, K., Kim, K.: Characterization of physical, mechanical, and antioxidant properties of soy protein-based bioplastic films containing carboxymethylcellulose and catechin. Food Sci. Biotechnol. **24**(3), 939–945 (2015). https://doi.org/10.1007/s10068-015-0121-0
56. Goodman-Smith, F.: Consumer acceptance of upcycled craft beer: a New Zealand case study. Front. Nutr. **10** (2023). https://doi.org/10.3389/fnut.2023.1235137
57. Moshtaghian, H., Bolton, K., Rousta, K.: Public preferences for nutritional, environmental, and food safety characteristics of upcycled foods in Sweden. Int. J. Food Sci. Technol. **58**(10), 5616–5625 (2023). https://doi.org/10.1111/ijfs.16588
58. Thorsen, M., Skeaff, S., Goodman-Smith, F., et al.: Upcycled foods: a nudge toward nutrition. Front. Nutr. **9**, 1071829 (2022). https://doi.org/10.3389/fnut.2022.1071829
59. Ghazanfar, S., Abdullah, M., Ummar, R., Shabbir, R., Saqib, S.: Effect of sustainability claim on willingness to pay for upcycled food in the digital era. Front. Environ. Sci. **10**, 870401 (2022). https://doi.org/10.3389/fenvs.2022.870401
60. Grasso, S., et al.: The use of upcycled defatted sunflower seed flour as a functional ingredient in biscuits. Food Secur. **8**(8), 305 (2019). https://doi.org/10.3390/foods8080305
61. Voblikova, T., Laricheva, K.: Bifidobacteria encapsulation and viability of probiotic culture during oral delivery in a milk drink matrix. Int. J. Food Sci. **2023**, 1–8 (2023). https://doi.org/10.1155/2023/8484835
62. Jain, S., Gualandris, J.: When does upcycling mitigate climate change? The case of wet spent grains and fruit and vegetable residues in Canada. J. Ind. Ecol. **27**(2), 522–534 (2023). https://doi.org/10.1111/jiec.13373
63. Varma, D., et al.: Injectable, redox-polymerized carboxymethylcellulose hydrogels promote nucleus pulposus-like extracellular matrix elaboration by human MSCs. J. Biomater. Appl. **33**(4), 576–589 (2018). https://doi.org/10.1177/0885328218805216
64. Lin, H., et al.: Lower crosslinking density enhances functional nucleus pulposus-like matrix elaboration by human mesenchymal stem cells in carboxymethylcellulose hydrogels. J. Biomed. Mater. Res. A. **104**(1), 165–177 (2015). https://doi.org/10.1002/jbm.a.35552

65. Reeves, R., et al.: Synthesis and characterization of carboxymethylcellulose-methacrylate hydrogel cell scaffolds. Polymers. **2**(3), 252–264 (2010). https://doi.org/10.3390/polym2030252
66. Soullard, L.: Methacrylated cellulose nanocrystals as fillers for the development of photo-cross-linkable cytocompatiblebiosourced formulations targeting 3D printing. Biomacromolecules. **24**(12), 6009–6024 (2023). https://doi.org/10.1021/acs.biomac.3c01090
67. Ghosh, M., et al.: Injectable alginate-peptide composite hydrogel as a scaffold for bone tissue regeneration. Nanomaterials. **9**(4), 497 (2019). https://doi.org/10.3390/nano9040497
68. García-Lizarribar, A., et al.: Composite biomaterials as long-lasting scaffolds for 3D bioprinting of highly aligned muscle tissue. Macromol. Biosci. **18**(10), e1800167 (2018). https://doi.org/10.1002/mabi.201800167
69. Park, J., et al.: Advanced growth factor delivery systems in wound management and skin regeneration. Molecules. **22**(8), 1259 (2017). https://doi.org/10.3390/molecules22081259
70. Neacşu, I., et al.: Collagen-carboxymethylcellulose biocomposite wound-dressings with antimicrobial activity. Materials. **14**(5), 1153 (2021). https://doi.org/10.3390/ma14051153
71. Leau, S., et al.: Study of wound-dressing materials based on collagen, sodium carboxymethylcellulose, and silver nanoparticles. ICAMS. (2018). https://doi.org/10.24264/icams-2018.i.18
72. Lima, G., et al.: Nanocellulose as reinforcement in carboxymethylcellulose superabsorbent nanocomposite hydrogels. Macromol. Symp. **394**(1) (2020). https://doi.org/10.1002/masy.202000126
73. Tarawneh, O., et al.: Characterization of chlorhexidine-impregnated cellulose-based hydrogel films intended for the treatment of periodontitis. Biomed. Res. Int. **2021**, 9853977 (2021). https://doi.org/10.1155/2021/9853977
74. Oliveira, A., et al.: The development study of a prototype of sodium carboxymethylcellulose and poly(vinyl alcohol) wound dressings using a mixer torque rheometer. Adv. Mat. Res. **1171**, 25–47 (2022). https://doi.org/10.4028/p-h118we
75. Macieja, S., et al.: Bioactive carboxymethylcellulose (CMC)-based films modified with melanin and silver nanoparticles (AgNPs). Int. J. Mol. Sci. **23**(24), 15560 (2022). https://doi.org/10.3390/ijms232415560
76. Ganjali, F., et al.: A novel nanocomposite containing zinc ferrite nanoparticles embedded in carboxymethylcellulose hydrogel plus carbon nitride nanosheets with multifunctional bioactivity. RSC Adv. **13**(32), 21873–21881 (2023). https://doi.org/10.1039/d3ra02822d
77. Paiva, A., Borsagli, F.: Ecofriendly multiphase aqueous colloidal based on carboxymethylcellulose nanoconjugates for potential bioimaging cancer cells. J. Polym. Environ. **28**(12), 3076–3096 (2020). https://doi.org/10.1007/s10924-020-01825-5
78. Maisani, M., et al.: A new composite hydrogel combining the biological properties of collagen with the mechanical properties of a supramolecular scaffold for bone tissue engineering. J. Tissue Eng. Regen. Med. **12**(3), e1489–e1500 (2017). https://doi.org/10.1002/term.2569
79. Aranda, F.: Polymers in circular economy: a comprehensive approach to sustainability. An overview. J. Chil. Chem. Soc. **68**(3), 5950–5962 (2023). https://doi.org/10.4067/s0717-97072023000305950
80. Rendón-Villalobos, R., et al.: The Role of Biopolymers in Obtaining Environmentally Friendly Materials. IntechOpen, London (2016). https://doi.org/10.5772/65265
81. Nepal, A., et al.: Advances in haemostatic sponges: characteristics and the underlying mechanisms for rapid haemostasis. Bioact. Mater. **27**, 231–256 (2023). https://doi.org/10.1016/j.bioactmat.2023.04.008
82. Gil, M., Rudy, M.: Innovations in the packaging of meat and meat products—a review. Coatings. **13**(2), 333 (2023). https://doi.org/10.3390/coatings13020333

83. Capanema, N., et al.: Eco-friendly and biocompatible cross-linked carboxymethylcellulose hydrogels as adsorbents for the removal of organic dye pollutants for environmental applications. Environ. Technol. **39**(22), 2856–2872 (2017). https://doi.org/10.1080/09593330.2017.1367845
84. Wang, B., et al.: Aligned bioinspired cellulose nanocrystal-based nanocomposites with synergetic mechanical properties and improved hygro mechanical performance. ACS Appl. Mater. Interfaces. **7**(8), 4595–4607 (2015). https://doi.org/10.1021/am507726t
85. Prathapan, R., et al.: Recent progress in cellulose nanocrystal alignment and its applications. ACS Appl. BioMater. **3**(4), 1828–1844 (2020). https://doi.org/10.1021/acsabm.0c00104
86. Rahimi, S., et al.: Textile waste valorization via Carboxymethylation for biopolymer production. J. Clean. Prod. **271**, 122422 (2020). https://doi.org/10.1016/j.jclepro.2020.122422
87. Abdelrahman, S.H., Al-Sayed, W.: Sustainable chemistry in textile processes. In: Green Chemistry for Sustainable Textiles. Elsevier, Amsterdam (2021)

Experimental Analysis of Spiral Finned Tube Heat Exchanger Using PCM as Thermal Energy Storage

R. Anish[1(✉)], Albin Laji[1], Alan Antony[1], Aibel Joseph[1], Abhijith Biju[1], Mathew J. Joseph[1], Amal Chummar[1], and Jijo Jose[2]

[1] Department of Mechanical Engineering, Amal Jyothi College of Engineering (Autonomous), Kanjirappally, Kerala, India
ranish@amaljyothi.ac.in

[2] Department of Metallurgical and Materials Engineering, Amal Jyothi College of Engineering (Autonomous), Kanjirappally, Kerala, India

Abstract. The experimental study focused on spiral finned tube heat exchangers utilising phase change materials (PCM) for efficient heat storage. The research shows how the thickness of the fins can significantly influence heat transfer rates and shows the advantages of using twisted fins over traditional straight fins, which can enhance PCM storage rates. To understand the system's performance, a detailed parametric analysis is conducted, examining how variations in flow rates and temperatures affect the charging time of the PCM. The study aims to identify the energy storage capacity of spiral finned tube configurations and the performance of the fin in real-world applications. This study examines the performance of heat exchangers across different operating conditions, focusing heat transfer fluid (HTF) flow rate such as 20 LPM, 24 LPM and 28 LPM, as well as inlet temperatures such as 65 °C and 75 °C. According to the results, the PCM's melting time decreased from 120 min to 90 min when the input temperature was changed from 65 °C to 75 °C, indicating a 25% increase in charging efficiency. Additionally, an increase in HTF flow rate from 20 LPM to 24 LPM resulted in a 15% decrease in melting time. The findings show the potential of spiral-finned tube heat exchangers in improving thermal energy management, making them suitable for various applications.

Keywords: Phase Change Materials · Heat Transfer Fluid · Thermal Energy Storage · Melt Fraction

1 Introduction

Energy storage is an essential part of modern energy management, which enables the storage of energy produced to be used after, and this service is particularly relevant to the challenge of balancing energy consumption and production rates, as renewable energy sources generate energy sporadically [1, 2]. Also, growing awareness of environmental pollution has driven scientists to explore eco-friendly replacements for synthetic

R. Rajasekharan et al. (Eds.): MatZone 2024, SPPHY 316, pp. 20–34, 2025.
https://doi.org/10.1007/978-981-96-6541-9_2

options [3]. Energy systems are evolving globally, and demand for energy storage is stimulating innovation providing efficient and sustainable systems [4]. Energy storage systems exist in many forms, depending on the characteristics and applications of the systems. Mechanical energy storage could be in the form of pressurised water storage or pressurised air energy storage, which channels electrical energy into the storage system, converting it to mechanical energy and enabling the energy to be available when required [5]. During off-peak hours, pumped hydro storage transfers water from the bottom reservoir to an elevated reservoir using electricity to generate power. The water is redirected to the lower reservoir during peak hours [6].

Conversely, batteries are a more familiar type of energy storage system with comparatively high-efficiency rates. Batteries utilise chemical energy stored inside the system and release it as electrical energy as the chemical process occurs. However, environmental concerns over sourcing lithium and cobalt used in battery parts are prominent in the literature. The process of mining these materials poses concerns related to deforestation, water contamination, and human rights abuses [7, 8]. Once a battery is 'spent' or loses its intended purpose, environmental concerns associated with what happens to the battery once its intention has been exhausted are also notable, as concerns over potential pollutant effects can re-emerge. ICE convert and store chemical energy to thermal energy; however, within the ICE, the heat lost due to chemical to-mechanical energy conversion dissipated as a heat loss occurs has a high energy loss. ICEs take chemical energy and create mechanical energy with a machine that creates energy by producing electricity or driving a vehicle forward [9]. Even when the finishing output produces motion or energy, the dissipated energy and overall system output result in notable loss. Heat storage has been viewed as an effective solution to address this issue. Thermal energy storage involves storing energy as heat or cold that can be used to create electricity, heat or cool [10]. The experiment presented in this research focuses on combining sensible and latent heat storage using an experimental apparatus for a thermal energy management system [5].

Sensible heat storage is the direct heating or cooling of a material such as water or air in order to store energy. This storage would be applicable to buildings or industrial processes, where energy is stored as hot or cold water or air [7, 11]. Energy is stored in latent heat storage using phase change materials (PCMs). The PCMs absorb or release energy as they change states to or from solid, liquid, or gas [12, 13]. The experimentation of this work relied on the effective insulation of the temperature-controlled experimental apparatus to minimise heat loss, particularly during incubation and experimentation. This effective insulation will be critical in maximising the thermal energy when arisen since heat will be lost to the environment without it [14]. Additionally, the PCM used in this work is a critical indicator of how or if the heat absorbed/released by phase change is effective.

Paraffin wax is the selected PCM due to its thermal properties for the temperature range that it will be utilised and other environmentally friendly factors. Paraffin wax is a natural, non- toxic, biodegradable material that can easily be found from petroleum-based or vegetable- based sources, has a high latent heat of fusion that will make it effective at storing energy, and it has a fairly low melting point [15]. Each of these features makes paraffin wax a suitable PCM for a variety of applications [16]. The experimental apparatus for this research includes a heat exchanger, pump, flow control valve, and rotameter. The heat exchanger is the main apparatus component where the HTF flows through inner tube, while it transfers heat to the PCM contained in the shell. The heat exchanger also has a spiral finned design, enabling improved heat transfer due to modified surface area, increasing turbulence in the fluid flow, and maximising mixing of the flow within the HTF. The pump circulates the HTF throughout the system at a constant flow rate for effective heat transfer. The flow control valve controls the HTF flow rate, enabling the user to adjust HTF flow rate according to experimental needs for the heat transfer rate. The Function of the rotameter is to measure the flow rate for the HTF, in which the information is essential for monitoring and controlling the test system's performance and the experimental concern's proper function [6].

The combined usage of sensible and latent heat storage methods in the thermal systems is a solution for better thermal energy management. Using paraffin wax as PCM maximises the latent storage capacity, and energy release during the phase change absorption process applies to latent heat transfer [15, 17]. Integrating PCMs in heat storage and thermal systems significantly enhances overall energy efficiency by allowing for excellent heat absorption during charging. The experimental apparatus provides an arrangement for both horizontally and vertically oriented thermal energy storage for ongoing research. This allows the apparatus to study the performance of thermal energy systems for latent and sensible heat storage systems without concern about what is being utilised to maximise production with transient performing guidelines. Modern energy management includes energy storage, which combines a sensible heat reservoir with energy systems to achieve thermal heating performance, addressing energy storage issues. It also permits any auxiliary flow with the test system outside of its performance, maximizing both systems as a single unit. Optimising the design of heat exchangers can lead to a substantial improvement in storage systems performance [18].

2 Experimental Setup—Components and Procedure

2.1 Experimental Setup

Figure 1a depicts a schematic diagram of the experimental setup, whereas Fig. 1b shows a pictographic perspective of the same. The components used are heater with an auto-transformer, a constant temperature water bath of 30 L, a 0.5HP electric hot water pump, a three-way plug valve with flow control, a rotameter leading HTF to spiral finned heat exchanger with PCM paraffin wax (RT 50) connected with K-type thermocouple and data logger. It is enhanced with an inner spiral-finned copper tube. A 230 V power supply is provided to an autotransformer that regulates power to an electric heater in the water bath, maintaining a constant water temperature during the charging phase. This ensures that the HTF has a consistent temperature input for accurate system performance analysis. The HTF path is initiated through a 0.5HP electric hot water pump at constant or variable rates. The pump's purpose is to aid with steady flow, assisting thermal energy transfer. A three-way plug valve is positioned at the pump's output, [15] with one pathway leading to the flow control valve to control flow into the heat exchanger and the other returning the HTF directly to the HTF storage tank, allowing precise control over circulation.

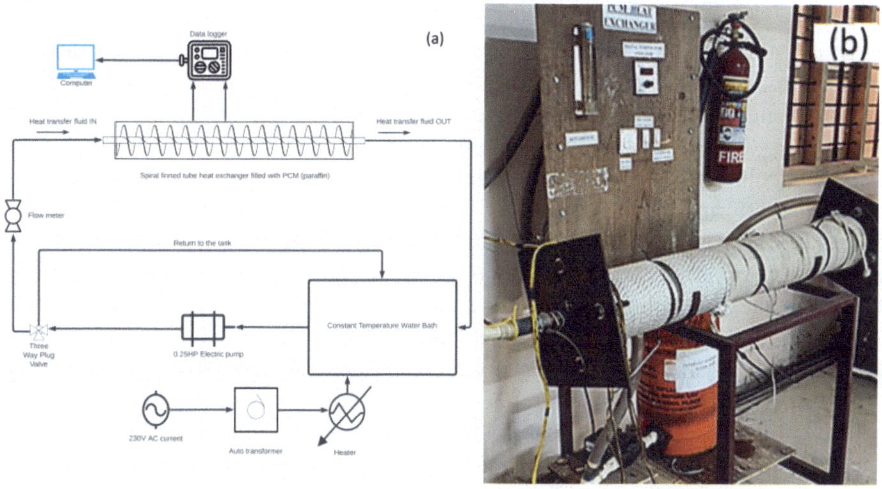

Fig. 1. (a) Schematic Diagram and (b) Experimental Setup

After the three-way plug valve, the HTF passes through a flow control valve, allowing further precision to the temperature flow to investigate rates. It is necessary to adjust the HTF circulation rate in order to investigate the melting as well as solidifying of PCM at various HTF circulation rates [18]. Along with the HTF path, a rotameter is connected to closely measure and monitor flow rates. The rotameter consistently gives flow information and retains a valuable observation of flow, whether the HTF has infusion circulation or matches the retention flow for energy output in the experiment. The HTF then enters the spiral-finned inner copper tube, which receives energy when the system first charges the accounting PCM energy. Subsequently, the HTF can exhibit discharge savings when the PCM returns the energy to the space or environment. Paraffin wax is chosen as the PCM for its acceptable thermophysical properties, including proper melting point and high latent heat capacity [18].

In the charging direction, as the hot HTF runs through the inner tube, it transfers energy to the PCM, enabling it to melt within and store thermal energy into the PCM storage media. During the discharging direction, the solid PCM transfers its stored heat content back to the cooler HTF that runs through the tube. PCM surrounds the spiral finned inner copper tube and is located entirely within the shell, 4 mm thick and 5 inch diameter GI pipe of the heat exchanger fully insulated to prevent heat loss by convection. The inner tube contains the spiral fin design, which is a critical aspect of this experimental setup. The fins increase the effective surface area for heat transfer from the HTF to the PCM. The presence of the fins allowed for the properly induced turbulence into the fluid flow and the mixing within the HTF, thus improving heat transfer rates [10]. Because of the spiral fin design, PCMs melting time should be reduced and the efficiency over traditional straight fins should be enhanced, is observed from the literature review. The entire heat exchanger system is well insulated to prevent heat loss to the environment so that most of the energy is accumulated in the PCM, or transferred back to the HTF.

K-type temperature sensors are placed at the two end of the storage unit to observe the temperature variation of HTF. Also 6 K type thermocouples are placed radially at 3 different positions optimally as shown in Fig. 1, so that relevant data can be obtained regarding temperature gradients and progression of the melting or solidification heights in the PCM. Connections to different heat sources, including electric heater or simulated solar collectors, are available to emulate operational conditions [15]. Therefore, the system performance with a range of input temperature and heat flux can be studied to emulate real scenarios, including renewable energy systems or industrial waste heat recovery. The experimental apparatus has incorporated data acquisition systems to constantly document the temperature and flow rate. This holistic data compilation will allow a detailed interrogation of the physics underpinning the system's performance, such as heat transfer rates, stored thermal energy capacity, and comprehensive cycle efficiency.

For safety purposes, pressure relief valves and emergency shut-off equipment are included in the experimental devices, with the understanding that experiments will be conducted under high-temperature and high-pressure conditions. An experimental apparatus of this type has been carefully composed to address fundamental, applied, and useful research on PCM-based thermal energy storage systems. By changing the initial conditions of experimental variables, HTF flow rate, input temperature and cycle operation, researchers are then able to make minor to drastic changes to the aspects of resulting system performance [15]. This enables experiments to characterise PCM heat exchangers and trends that could lead to new designs and operational objectives for thermal energy storage system performance. With this capability, a wide variety of studies are possible, including but not limited to the fundamental work on assessing heat transfer pathways in PCM materials, or to applied and useful work assessing new options for the thermal efficiency of renewable energy systems [8]. This experimental process ultimately bridges the gap between theoretical and practical utility, which is important to the advancement of thermal energy storage technologies as a practical response for efficient energy management and sustainability.

2.2 Heat Transfer Fluid and PCM

The properties of paraffin wax utilised as a PCM are presented in Table 1. Water is utilised as the HTF due its thermal properties, low cost, and its abundance, The HTF will be heated to a particular temperature range, which is typically in the range of 60 °C–75 °C. Temperature control is achieved using a constant temperature water bath for the precision required while conducting experiments. Heating the HTF temperature at levels above the melting point of the PCM will achieve a more efficient energy transfer method, allowing the PCM to absorb significant quantities of thermal energy while in the charging phase, thus maximising amounts of thermal energy storage.

2.3 Latent Heat Storage Prototype

The spiral-finned tube latent heat storage prototype is illustrated in Fig. 2. The prototype consists of a shell with dimensions 1 m in length and 5-inch diameter. The shell in which the PCM is filled is generally constructed of stainless steel. The shell must possess resistance to corrosion and good structural integrity. Furthermore, the shell must be

Table 1. Thermo physical properties of PCM material (paraffin wax)

Properties	Value
Latent heat of fusion	200 kJ/kg
Melting point	58 °C
Thermal conductivity:	0.2 to 0.25 W/m K
Density	850 kg/m^3
Specific heat:	
a. Solid state	2.1 kJ/kg K
b. Liquid state	2.5 kJ/kg K
Dynamic viscosity	0.03 kg/m s
Thermal expansion coefficient	0.0005 to 0.001 K^{-1}

designed to withstand several thermal shock cycles, maintaining its inherent capability over time. The inner tubes have a length of 1.12 m and an outer diameter of 0.025 m. The inner tubes are typically manufactured by copper and Copper has a good mechanical and thermal properties, including good thermal conductivity.

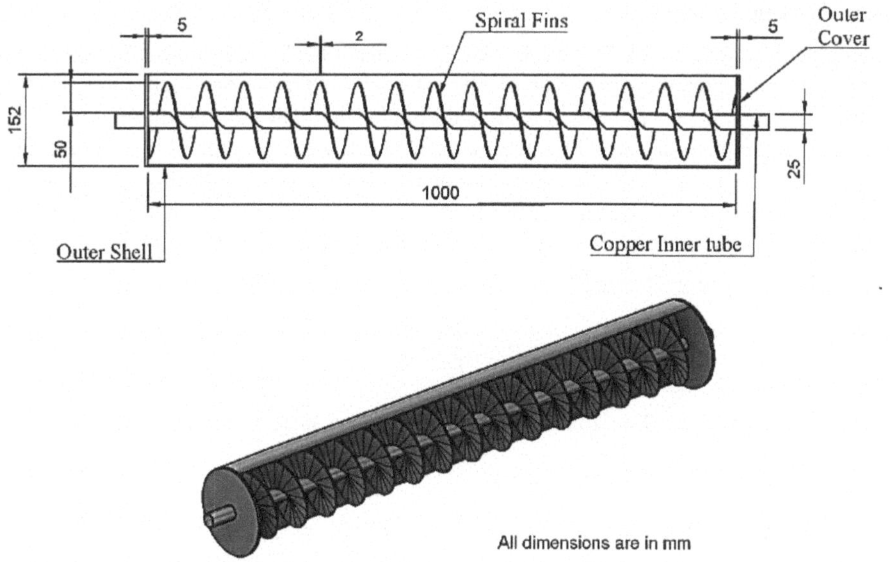

Fig. 2. Details of spiral finned tube PCM heat exchanger

Heat transfer rates are improved by adding conductive fillers, which raise the PCMs thermal conductivity [10]. The effectiveness of heat transmission between the HTF and the PCM is largely dependent on thermal conductivity, hence material choice is crucial. The heat transmission capabilities of the solid PCM and the HTF used significantly impact how well the phase change process works. The heat exchange characteristics were further enhanced using spiral fins attached to the outer surface of the inner tubes. The spiral fins have a height of 0.02 m, width of 0.005 m and a pitch of 0.05 m at a 6° angle. The fins are positioned to create turbulence and increase the surface area available for heat transmission. The spiral fins are extended along the whole length of the inner tubes. The fins' diameters are the same as the inner tubes' outer diameter.

T1 is the inlet temperature, T2 is the outlet temperature, T3, T4,T5, T6, T7 and T8 are the other K- type thermocouple. T3 and T4 is thermocouple is kept 3 cm towards the inside of the outer shell also T5 and T6 is kept 2 cm towards the inside. T7 and T8 are kept 2.5 cm towards the inside of the outer shell. Thus, the melting and solidifying rates of the PCM would be improved by the effectiveness of the spiral fins. The whole assembly's outer walls are coated to enhance the thermal transfer area. The PCM adapts the melting and solidifying work substance based on the fin diameter and thickness. In nearly all prototype designs, paraffin wax has been generally applied as the PCM owing to its significant thermo-physical properties, representing a low melting point and high latent heat. The prototype is well insulated by attaching a glass wool layer measuring 50 mm in thickness. The glass wool decreased the heat lost by the fabric in both charging and discharging phases.

2.4 Performance Parameter

2.4.1 Melting Fraction (θ)

Phase change materials (PCMs), particularly mixtures of salts, have a specific temperature range over which they melt. The mushy zone refers to the temperature range where a material simultaneously exists in solid and liquid states. The solidus temperature (T_s) is the highest temperature at which the material remains fully solid, while the liquidus temperature (T_l) is the lowest temperature at which it becomes fully liquid. The melt fraction, a dimensionless quantity, represents the proportion of the liquid phase in the mushy zone. The melt fraction is given by Eqs. 1 and 2 [19, 20]. 'T' stated in Eq. (1) is the temperature that the temperature sensors in the PCM area measured.

$$\theta = \frac{T - T_s}{T_l - T_s} \tag{1}$$

$$\theta = \{0 \quad for\, T < T_s\; 0-1 \quad for\, T_s \le T \le T_l\; 1 \quad for\, T > T_l\,\} \tag{2}$$

2.4.2 Charging Time

A crucial factor in the functioning of thermal systems that use PCM is the charging time parameter, which shows how quickly the system can store thermal energy. Experimental investigations with the latent heat storage prototype indicated that raising the input temperature of the HTF had a dramatic impact on the charging time; for example, raising 65 °C to 75 °C of temperature substantially reduced the charging time of the PCM from 120 min to 90 min. Increasing the HTF flow rate from 20 LPM to 28 LPM caused a reduction in melting time of 15%, emphasising the importance of having efficient heat transfer. Gaining knowledge about how to maximize mechanical operating parameters may enhance thermal energy storage system performance, which in turn can help with improved energy management and sustainability initiatives across a range of applications [14].

2.4.3 Energy Stored

In the experimental evaluation of the PCMs in a thermal energy storage system, the total amount of energy stored is an important parameter including both sensible and latent heat energy. The total energy stored indicates the system's total thermal capacity to return heat if needed. Sensible energy is defined as the energy that the PCM absorbs while it is being heated prior to melting, using the equation $Q_{sensible} = m \cdot C_p \cdot \Delta T$. During the experiment, it was noted that the PCM was effective at storing a large amount of sensible energy due to the heating process, thus contributing to the entire thermal capacity of the system. When the PCM reached the melting point, the next form of energy storage would be latent heat energy storage, PCM absorbs energy but does not change in temperature when geometrically changing form from solid to liquid [21]. The latent heat energy stored, evaluated by Q latent $= mL_f\,\theta$, where L_f is the latent heat of fusion, made the system capable of large amounts of energy storage due to its efficiency [9, 17]. The experimental results indicated that storage of both sensible and latent heat energy storage systems was synergistic, offering an effective energy management solution for stable and reliable energy performance [15].

Sensible heat,

$$Q_s = mC_{ps}(T_m - T_{ini}) + C_{pl}(T - T_m) \tag{3}$$

Latent heat,

$$Q_l = mL_f\theta \tag{4}$$

Total energy stored,

$$Qc = Q_s + Q_l \tag{5}$$

3 Results and Discussion

Analysis and research were done on the LHS prototype's performance during the energy storage or charging process. Results of a study on melting temperature, temperature change, melt fraction, flow rate effects and intake temperature, stoarage time, and energy storage rate are given below. In addition to providing opportunities for future study on increased thermal efficiency and sustainability, this study has confirmed the feasibility of employing PCM-based thermal energy storage devices. Thus, according to the studies, the HTF input temperature and flow rate had an impact on the phase change material's (PCM) charging time. By changing the input temperature from 60 °C to 80 °C, the PCM's melting time decreased from 120 to 90 min. Such a reduction infers that thermal gradients improve the effectiveness of heat transfer. Similarly, an increment of HTF from a flow rate of 0.01 m^3/s to 0.02 m^3/s has decreased the melting time by almost 15%. The storage systems efficiency can be significantly improved by optimising the flow rates of heat transfer fluids [21]. Temperature measurements of thermocouples T3 to T8 demonstrate an almost uniform profile across the entire PCM, thereby clearly indicating that this is one of the effective heat sinks with relatively uniform temperature distribution during charging. Synergism between sensible and latent storage has been demonstrated, wherein both components play their part in the total storage of the system. The efficiency of storage units utilising PCMs is significantly influenced by heat storage capacity and thermal conductivity [21]. Some observations of the mushy zone during which the PCM is in both solid and liquid phases are carried out concerning the melting process. A comparative analysis of the temperature profile indicates that this optimisation technique of HTF temperature and flow rate may indeed improve performance for better energy management [22]. Important relationships like temperature vs. time, melt fraction vs. time, and energy storage vs. time are therefore generated and show the heating patter, phase change efficiency, and heat storage during the process. The results have indicated the suitability of using PCM for future applications related to heat storage enhancement efficiency.

3.1 Temperature Variation

3.1.1 Temperature vs. Time–24 LPM and 65 °C

The six curves of the graph representing temperatures of PCM at different fixed locations T3 (black), T4 (Red), T5 (Blue), T6 (green), T7 (purple), and T8 (yellow). These curves track temperature changes over time. Each curve starts at a similar temperature (around 30 °C to 35 °C) at the beginning (time $\approx$ 0) and then gradually increases. The slopes of the temperature curves suggest the rate of heating varies slightly between the different temperature points. For example, the blue curve (T5) reaches the highest temperature (around 65 °C) by the end, while the yellow curve (T8) increases the slowest, reaching around 50 °C. All temperatures steadily increase over time, indicating that the system is heating gradually. The temperature differences (especially around time $\approx$ 250 onwards) suggest varying thermal gradients, indicating different heat absorption properties or varying distances from the heat source. Small step-like fluctuations are visible in each temperature curve, which could represent periodic adjustments in the heating process, temperature measurement intervals, or changes in environmental conditions.

3.1.2 Inlet and Outlet Temperature of HTF vs Time (at 65 °C)

The inlet temperature (T_1) fluctuates between approximately 63 °C and 67 °C over time. These fluctuations are relatively minor, indicating that the inlet temperature remains relatively stable with only small variations throughout the observed period. The red line represents the outlet temperature (T_2). It fluctuates between approximately 54 °C and 58 °C, which is consistently lower than the inlet temperature. Similar to the inlet temperature, the outlet temperature also shows fluctuations but remains lower than T1 because the paraffin wax absorbed the heat from the inlet HTF and underwent a phase transition. The change in the input (T_1) and outlet (T_2) temperatures is a result of heat transfer, likely due to the paraffin wax absorbing heat and undergoing a phase transition.

The paraffin absorbs heat from the fluid, causing the outlet temperature (T_2) to be lower than the inlet temperature (T_1), reflecting the heat loss. This behavior is typical in systems where a heat sink (like paraffin wax) absorbs heat from the fluid as it passes through, maintaining the temperature difference. The small fluctuations suggest variations in heat absorption or external influences on the system [23].

3.1.3 Average Temperature (T_{avg}) vs Time

The curve shows a steady increase in temperature as the process progresses. The curve has some fluctuations or minor dips, but the overall trend is upward. After reaching approximately 60, the curve seems to stabilise, indicating that the average temperature approaches a maximum or equilibrium point near the end of the process. Initial Stage: The temperature increases smoothly from around 30 °C to 50 °C during the early stages, which could represent the heating phase of the process. Middle Stage: There are some fluctuations or uneven increases in temperature after around 150 on the x-axis. These could represent temporary fluctuations or changes in heating rates. Final stage: Near the end (around 300–350 on the x-axis), the temperature continues to rise more gradually, suggesting that the system is approaching a stable temperature or has reached a thermal equilibrium [10]. The graph indicates a general heating process where the average temperature increases steadily and stabilises over time. The minor fluctuations in the middle suggest that there might be changes in heat input, thermal properties, or other factors influencing the rate of temperature change [15]. Ultimately, the system reaches an equilibrium near 60 on the y-axis.

3.1.4 Temperature vs. Time—24 LPM and 75 °C

Figure 3 shows the temperature changes at 6 different regions of the tube (T3, T4, T5, T6, T7, T8) over time. All groups start with temperatures between 30 °C and 40 °C and exhibit a steady increase as time progresses. Group T5 reaches the highest temperature, peaking near to 75 °C by the end experiment, while T8 remains the lowest, with a final temperature just below 50 °C. T4 and T6 also show significant temperature increases, though they stay slightly lower than T5. As time passes, the temperature differences between the groups become more distinct, indicating the varying heating rates, with T5 heating the fastest and reaching the highest temperature while T8 rising the slowest. This suggests that the regions closer to the inlet heat up faster and to higher temperatures than others.

3.1.5 Inlet and Outlet Temperature of HTF vs Time (at 75 °C)

The temperature of the fluid entering the system stays consistent over time. The represents T2 Outlet Temperature, which starts lower, around 64 units, and gradually increases, reaching about 68 units by the end. Paraffin wax which absorbs heat from the fluid [24]. This process consumes a significant amount of energy (latent heat), which explains why the outlet temperature (T2) is lower than the inlet temperature (T1) despite some heat transfer occurring. Over time, as the paraffin wax absorbs more heat, T2 gradually increases, indicating the progressive melting of the wax. The outlet temperature does

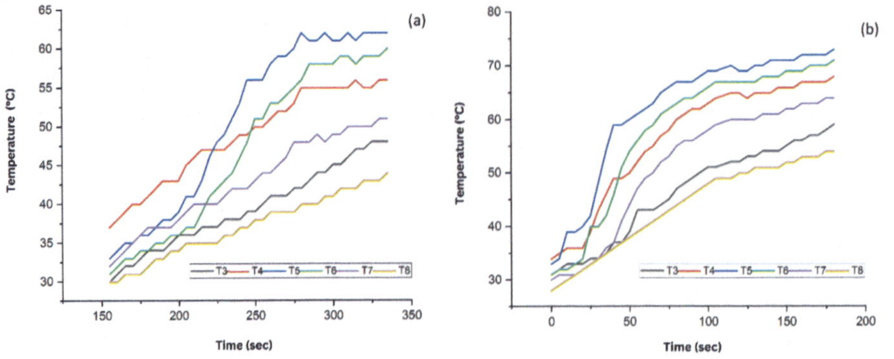

Fig. 3. (a) Temperature vs. Time (at 24 LPM, 65 °C) and (b) Temperature vs Time (at 24 LPM, 75 °C

not rise rapidly because the phase transition absorbs much of the energy, maintaining a slower temperature rise. The stable T1 represents consistent heat input. The increasing T2 reflects the wax gradually absorbing heat and undergoing a phase change, which moderates the temperature rise at the outlet. As the paraffin wax absorbs this heat, it transitions from solid to liquid.

3.1.6 T_{avg} (Average Temperature) vs Time

The average temperature starts at around 30 units. As time progresses, the average temperature increases steadily, showing a near-linear rise up to around 60 units by the end of the period. Towards the graph's later stages, the temperature increase rate slows slightly, suggesting the system is approaching a steady state. This increase in the average temperature over time likely reflects the heat absorption process involving the paraffin wax. In the early stages, the system warms up as heat is transferred, and the paraffin wax absorbs energy to transition from solid to liquid. Figure 4b shows the gradual increase in temperature, which means more energy is required as the phase transition progresses. Towards the end, the slowing rate of temperature rise suggests that the wax may be nearing the completion of its phase change, with less heat being absorbed for melting.

3.1.7 Energy Stored vs Time for 65 °C and 75 °C at Constant Flow Rate 24 LPM

Figure 5a shows the relationship between energy storage and time at two different temperatures, 65 °C and 75 °C. The red line, representing energy stored at 75 °C, shows a steeper rise compared to the black line, which represents energy stored at 65 °C. This suggests that energy is stored more rapidly and efficiently at higher temperatures. The step-like increases in the graph likely correspond to phase change events, where the material absorbs latent heat, temporarily pausing the temperature rise while energy is absorbed to overcome the phase transition.

At 75 °C, the system can store energy faster due to the higher thermal driving force, which increases the heat transfer rate into the material. At elevated temperatures, the phase change material (PCM) can more easily transition between phases (such as solid

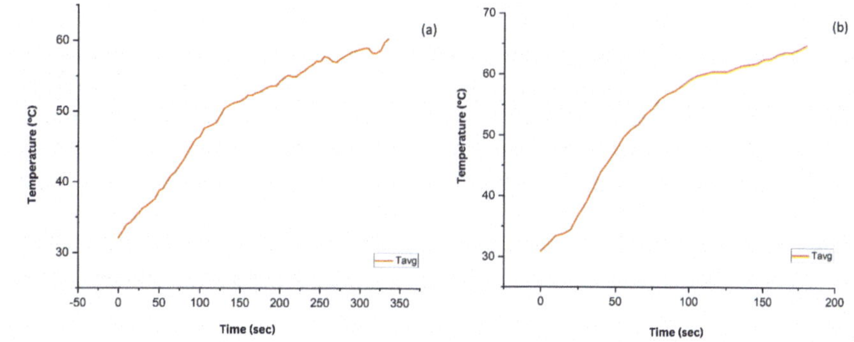

Fig. 4. (a) Average Temperature vs Time (at 24 LPM, 65 °C) and (b) Average Temperature vs. Time of HTF (at 24 LPM, 75 °C)

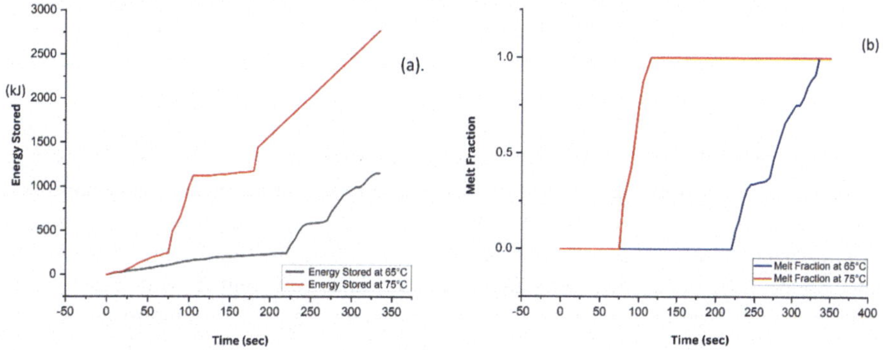

Fig. 5. (a) Energy Stored vs. Time (at 24 LPM, 65 °C and 75 °C) and (b) Melt Fraction vs. Time (at 24 LPM, 65 °C and 75 °C)

to liquid), allowing for quicker absorption of latent heat. The energy storage at 65 °C, represented by the black line, progresses more slowly, indicating that lower temperatures result in a slower rate of energy input due to reduced heat transfer efficiency and a slower phase change process [14]. Overall, the graph highlights the importance of operating temperature in enhancing energy storage efficiency, particularly in systems utilising PCM for thermal energy management.

3.1.8 Melt Fraction vs Time

Figure 5b shows the melt fraction of a material as a function of time at 65 °C and 75 °C. The red line, representing the melt fraction at 75 °C, shows a much steeper rise, indicating that the material reaches full melting (melt fraction = 1) in a significantly shorter time compared to the blue line at 65 °C. The graph demonstrates that the phase change material (PCM) undergoes the melting process at higher temperatures more rapidly. At 75 °C, the higher thermal energy provided accelerates the molecular motion within the PCM, enabling it to overcome the latent heat barrier more efficiently, thus reaching full melt in less time.

The steep slope of the red line suggests that the phase transition occurs quickly once the material reaches its melting point. In contrast, the blue line at 65 °C shows a more gradual slope, indicating that the PCM takes longer to absorb sufficient heat to complete the phase change. This slower process can be attributed to the reduced thermal gradient at lower temperatures, which decreases the energy transfer rate into the material and prolongs the melting process. The difference in melt fraction behaviour between the two temperatures emphasises the critical role of operating temperature in phase change efficiency [21]. Higher temperatures speed up the phase change and ensure a faster release of latent heat, which is essential for optimising thermal energy storage systems utilising PCM [2].

4 Conclusion

The PCM-based storage system's thermal performance is greatly enhanced by the spiral-finned tube heat exchanger. This enhancement is evident as increasing the heat transfer fluid (HTF) inlet temperature from 65 °C to 75 °C reduced the PCM's melting time from 120 min to 90 min. Additionally, a flow rate increase to 28 LPM cut the melting time by 15% compared to a 20 LPM flow rate. After 60 min at an inlet temperature of 75 °C and a flow rate of 28 LPM, the PCM achieved an 80% melt fraction and demonstrated an energy storage capacity of approximately 150 kJ/kg. The temperature of the PCM increased steadily, reaching 75 °C after 60 min, with the temperature profile showing a significant initial rise followed by a constant as the PCM neared its melting point. Furthermore, using copper fins in the heat exchanger resulted in a 30% higher heat transfer rate than other fins, such as aluminium, which are mentioned in the literature because of copper's superior thermal conductivity. Overall, the system's thermal efficiency was 45%, reflecting effective heat transfer and energy storage capabilities. The findings show the potential of spiral-finned tube heat exchangers in improving thermal energy management, making them suitable for various applications.

References

1. Arun, B.S., Mariappan, V., Anish, R., Nimisha, P.: Performance investigation of regenerative evaporative cooled air-to-air heat recovery ventilator under different humidity conditions. Mater. Today Proc. **46**, 9901–9907 (2021). https://doi.org/10.1016/J.MATPR.2021.02.099
2. Rahman, M.M., Khan, I., Alameh, K.: The role of energy storage technologies for sustainability in developing countries. In: Renewable Energy and Sustainability, pp. 347–376. Elsevier, Amsterdam (2022). https://doi.org/10.1016/B978-0-323-88668-0.00004-8
3. Vishnu Nandan, H., et al.: Experimental investigations on the mechanical properties and thermal analysis of coir-paper hybrid composite carry bag material. Mater. Lett. **353**, 135252 (2023). https://doi.org/10.1016/J.MATLET.2023.135252
4. Ribezzo, A., Falciani, G., Bergamasco, L., Fasano, M., Chiavazzo, E.: An overview on the use of additives and preparation procedure in phase change materials for thermal energy storage with a focus on long term applications. J. Energy Storage. **53**, 105140 (2022). https://doi.org/10.1016/j.est.2022.105140

5. Joshi, G.K., Rongali, B., Biswal, M.: A review on mechanical energy storage technology. In: 2022 International Conference on Intelligent Controller and Computing for Smart Power (ICICCSP), pp. 1–5. IEEE, Piscataway, NJ (2022). https://doi.org/10.1109/ICICCSP53532.2022.9862382
6. Harby, A., Sauterleute, J., Korpås, M., Killingtveit, Å., Solvang, E., Nielsen, T.: Pumped storage hydropower. In: Transition to Renewable Energy Systems, pp. 597–618. Wiley, Hoboken, NJ (2013). https://doi.org/10.1002/9783527673872.ch29
7. Sayed, E., et al.: Renewable energy and energy storage systems. Energies (Basel). **16**(3), 1415 (2023). https://doi.org/10.3390/en16031415
8. Ang, T.-Z., Salem, M., Kamarol, M., Das, H.S., Nazari, M.A., Prabaharan, N.: A comprehensive study of renewable energy sources: classifications, challenges and suggestions. Energ. Strat. Rev. **43**, 100939 (2022). https://doi.org/10.1016/j.esr.2022.100939
9. Wazeer, A., Das, A., Abeykoon, C., Sinha, A., Karmakar, A.: Phase change materials for battery thermal management of electric and hybrid vehicles: a review. Energy Nexus. **7**, 100131 (2022). https://doi.org/10.1016/j.nexus.2022.100131
10. Chinnasamy, V., Palaniappan, S.K., Raj, M.K.A., Rajendran, M., Cho, H.: Thermal energy storage and its applications. In: Materials for Solar Energy Conversion, pp. 353–377. Wiley, Hoboken, NJ (2021). https://doi.org/10.1002/9781119752202.ch13
11. Anish, R., Karale, S.R.: Experimental analysis of cold thermal energy storage system for cooling applications. Int. J. Adv. Res. Sci. Eng. (IJARSE). **2**(6), 45–55 (2013)
12. Anish, R., Mariappan, V., Suresh, S.: Experimental investigation on melting and solidification behaviour of erythritol in a vertical double spiral coil thermal energy storage system. Sustain. Cities Soc. **44**, 253–264 (2019). https://doi.org/10.1016/J.SCS.2018.10.012
13. Anish, R., Mariappan, V., Mastani Joybari, M.: Experimental investigation on the melting and solidification behavior of erythritol in a horizontal shell and multi-finned tube latent heat storage unit. Appl. Therm. Eng. **161**, 114194 (2019). https://doi.org/10.1016/J.APPLTHERMALENG.2019.114194
14. Borri, E., Zsembinszki, G., Cabeza, L.F.: Recent developments of thermal energy storage applications in the built environment: a bibliometric analysis and systematic review. Appl. Therm. Eng. **189**, 116666 (2021). https://doi.org/10.1016/j.applthermaleng.2021.116666
15. Imran Khan, M., Asfand, F., Al-Ghamdi, S.G.: Progress in research and development of phase change materials for thermal energy storage in concentrated solar power. Appl. Therm. Eng. **219**, 119546 (2023). https://doi.org/10.1016/j.applthermaleng.2022.119546
16. Tawalbeh, M., Khan, H.A., Al-Othman, A., Almomani, F., Ajith, S.: A comprehensive review on the recent advances in materials for thermal energy storage applications. Int. J. Thermofluids. **18**, 100326 (2023). https://doi.org/10.1016/j.ijft.2023.100326
17. Chen, Q., et al.: Recent advances in thermal-conductive insulating polymer composites with various fillers. Compos. Part A Appl. Sci. Manuf. **178**, 107998 (2024). https://doi.org/10.1016/j.compositesa.2023.107998
18. Kharaji, S.: Heat exchanger design and optimization. In: Heat Exchangers. IntechOpen, London (2022). https://doi.org/10.5772/intechopen.100450
19. Anish, R., Joybari, M.M., Seddegh, S., Mariappan, V., Haghighat, F., Yuan, Y.: Sensitivity analysis of design parameters for erythritol melting in a horizontal shell and multi-finned tube system: numerical investigation. Renew. Energy. **163**, 423–436 (2021). https://doi.org/10.1016/J.RENENE.2020.08.153
20. Anish, R., Mariappan, V., Suresh, S., Joybari, M.M., Abdulateef, A.M.: Experimental investigation on the energy storage/discharge performance of xylitol in a compact spiral coil heat exchanger. Int. J. Therm. Sci. **159**, 106633 (2021). https://doi.org/10.1016/J.IJTHERMALSCI.2020.106633

21. Parhizi, M., Jain, A.: The impact of thermal properties on performance of phase change based energy storage systems. Appl. Therm. Eng. **162**, 114154 (2019). https://doi.org/10.1016/j.applthermaleng.2019.114154
22. Elalfy, D.A., Gouda, E., Kotb, M.F., Bureš, V., Sedhom, B.E.: Comprehensive review of energy storage systems technologies, objectives, challenges, and future trends. Energ. Strat. Rev. **54**, 101482 (2024). https://doi.org/10.1016/j.esr.2024.101482
23. De Carne, G., et al.: The role of energy storage systems for a secure energy supply: a comprehensive review of system needs and technology solutions. Electr. Pow. Syst. Res. **236**, 110963 (2024). https://doi.org/10.1016/j.epsr.2024.110963
24. Jia, Z., Hu, C., Zhang, Y., Zhang, S., Tang, B.: Exploring electro-thermal conversion in phase change materials: a review. Compos. Part A Appl. Sci. Manuf. **175**, 107809 (2023). https://doi.org/10.1016/j.compositesa.2023.107809

Redesigning Diphenylalanine Conjugated Push-Pull Chromophores for Nonlinear Optical Applications Through a DFT Study

Aswathy Soman[1](✉) and R. Renjith[2]

[1] Department of Physics, University College, Affiliated to University of Kerala, Thiruvananthapuram 695034, Kerala, India
aswathysoman19590@gmail.com
[2] Department of Physics, Government College Kasaragod, Kasaragod 671123, Kerala, India

Abstract. Nonlinear Optics (NLO) is significant in scientific research due to its foundational role in optical communication, ultrafast laser pulse generation, optical switching and signal processing, frequency conversion, nonlinear microscopy, Quantum optics and photonic devices. Peptide based molecular materials are very significant in nonlinear optics due to their electronic and luminous properties. Diphenylalanine (FF) is a short peptide that has garnered promising attention due to its unique properties like self-assembly, optical transparency, optical stability and biocompatibility. Diphenylalanine provides a variety of applications in nonlinear optics (NLO) such as second harmonic generation (SHG), third harmonic generation (THG), optical waveguides, nonlinear optical switching, optical data storage and processing. In the present work we design a series of diphenylalanine (FF) based push-pull chromophores in order to check the influence of a set of various electron donors. Here diphenylalanine plays the role of π- spacer and cyano (-CN-) group is the common electron acceptor. Density functional theory (DFT) and time dependent density functional theory (TD-DFT) methods at B3LYP/6-31+G(d) level of theory is used to calculate nonlinear and linear optical properties of designed compounds. The computational study shows that by choosing suitable donor moieties, the designed compounds can be effectively tuned to a better nonlinear optical property.

Keywords: Peptides · nonlinear optical properties · diphenyl alanine · DFT

1 Introduction

Nonlinear optics (NLO) is a wide area of study that describes the nonlinear interaction between light and matter. NLO describes the analysis of behaviour of the nonlinear interaction of the medium when light is incident on the material system. The special effects in which the optical properties of a material system change due to the electric and magnetic fields associated with the incident light are called nonlinear effects. The materials which exhibit such effects are called nonlinear materials [1]. Nonlinear optical

R. Rajasekharan et al. (Eds.): MatZone 2024, SPPHY 316, pp. 35–46, 2025.
https://doi.org/10.1007/978-981-96-6541-9_3

materials provide a significant role in modern technologies including telecommunication, data storage, super-resolution lithography, signal processing etc. [2].The excellent properties of organic compounds make them potential candidates in nonlinear optics (NLO) [3].

Organic molecules with a D-π-A structure having a strong electron donating group, electron accepting group and a polarizable π- bridge (hence known as push-pull chromophores) are well known among researchers due to their strong NLO behaviour. Intramolecular charge transfer (ICT) occurs in such push-pull chromophores from the donor to the acceptor through the π- spacer [4]. The nature of the electron donors and acceptors as well as the length and nature of the π- bridge highly influence the nonlinear optical properties of push-pull chromophores. The strong electron donor and acceptor groups increase the difference in electron density across the molecule. Which results in an efficient intramolecular charge transfer (ICT) across the molecule. As a result, the polarizability and hyperpolarizability (strong indicators of NLO response) increase due to the strength of the donor and acceptor groups. The NLO behaviour of molecules are also affected by the length of the π-spacer. The extent of electron delocalisation depends on the length of the π- bridge. Longer π-conjugated bridge enhances the extent of electron delocalisation and increases the molecule's polarizability. The NLO behaviour of the system can also be influenced by the nature of π-bridge. Certain studies reported that NLO responses of the push-pull systems can be enhanced by using aromatic or hetero aromatic π- bridges because they improve electron delocalisation and enhance the conjugation and charge stability [5]. Such types of structural tuning of D-π-A systems allow researchers to design new category of compounds with high NLO responses. The D-π-A architecture has predominantly featured organic molecules due to their tunable electronic properties and ease of functionalization. However, in recent years, biomolecules have gained much attention as promising π-spacers, especially in optoelectronic studies [6].Biomolecules such as amino acids peptides, and nucleotides offer unique benefits like biocompatibility and environmental friendliness, self-assembly properties, tunable optoelectronic properties. Thus, we selected peptide conjugated push-pull chromophores for the NLO study.

Diphenylalanine (Phe-Phe or FF) is a dipeptide composed of two amino acids named phenylalanine. It has a considerable interest in research due to their potential applications in nanotechnology and materials science. Their interesting properties include self-assembly, mechanical stability, thermal stability, optical properties, piezoelectricity, biocompatibility and biodegradability. Thus, FF provides an excellent platform for photonics and nonlinear optics [7]. This inspired us to design a set of push-pull chromophores with diphenylalanine (FF) as the π-bridge with a common acceptor unit. Inorder to check the influence of various types of electron donors, we selected a set of different types of donors to the D-π-A architecture, and made the investigated compounds as potential candidates for nonlinear optical applications.

2 Computational Design

Density functional theory (DFT) is used to calculate nonlinear optical properties of the designed derivatives. DFT is a powerful approach for obtaining insights into their electronic and optical behaviour. Geometry optimizations of the investigated derivatives were done by DFT at B3LYP functional with 6-31+G(d) basis set. The choice of functional strictly influences the nonlinear optical properties of the designed compounds [8]. B3LYP (Becke, 3-parameter, Lee-Yang-Parr) is indeed one of the most widely used hybrid functionals in density functional theory (DFT) for quantum chemical calculations. It is popular because it balances computational cost and accuracy for a broad range of molecular systems. The frequency calculations confirmed that the optimized geometries are at the global minima of the potential energy curve. Which ensures that the optimized geometries are stable and not transition states. The static and dynamic polarizabilities (α) and hyperpolarizabilities (β) were estimated using DFT at B3LYP/6-31+G(d) basis set at an incident wavelength 1907 nm.

The linear polarizability α can be calculated by Eq. (1) [9–11].

$$\alpha = \left(\alpha_{xx} + \alpha_{yy} + \alpha_{zz}\right) \tag{1}$$

The first order hyperpolarizability β can be computed using Eq. (2) [9–11].

$$\beta = \left(\beta_x{}^2 + \beta_y{}^2 + \beta_z{}^2\right)^{1/2} \tag{2}$$

Where

$\beta_x = \beta_{xxx} + \beta_{xyy} + \beta_{xzz}$
$\beta_y = \beta_{yyy} + \beta_{xxy} + \beta_{yzz}$
$\beta_z = \beta_{zzz} + \beta_{xxz} + \beta_{yyz}$

Time-dependent density functional theory (TD-DFT) is used to record the UV-Vis absorption spectra of the designed compounds at B3LYP/6-31+G(d) level of theory. Gaussian 16 program package is used to execute DFT and TD-DFT calculations. The multifunctional wavefunction analyser named Multiwfn version 3.5(dev) were used to estimate the components of linear polarizabilities α and first order hyperpolarizabilities β [12]. All the calculations were executed in the gaseous phase [13]. The components of polarizability and hyperpolarizability were computed by multiwfn software [12].

3 Results and Discussion

3.1 Design of Diphenylalanine (FF) Derivatives

The study mainly focused into the designing of a set of D-π-A structures for enhancing NLO response of the reference compound (parent compound) diphenylalanine. Because diphenyl alanine has a wide range of applications in nonlinear optics including second and third harmonic generation (SHG & THG), nonlinear optical switching, optical waveguides and optical signal processing [14]. It is quite effective to tune the NLO response of organic molecules using donor-π-acceptor (D-π-A) structures. These structures enhance charge transfer which is crucial for enhancing NLO responses [15].Thus, structural modification is done by altering the strength of donors or acceptors, by increasing the length of π- conjugated bridge and varying the nature of the π-bridge [16, 17]. In the current study, D-π-A structures were designed by incorporating diphenylalanine as the π-conjugated peptide bridge. Cyano group (-CN) is treated as the common acceptor moiety in all the designed derivatives. The triple bond between the carbon and nitrogen in the cyano group provide a strong electron withdrawing effect. The reference (parent) compound is abbreviated as DA. Furan, pyrrole, aniline, pyran, thiophene, diphenylamine (DPA) and triphenylamine (TPA) are the various donor moieties included in the diphenylalanine conjugated D-π-A (push-pull) structures. These derivatives are labelled as DA1, DA2, DA3, DA4, DA5, DA6 and DA7 respectively. All these donors have π-electrons within the ring and are able to create electron density difference between the donors and acceptor. Their molecular structures are illustrated in the Fig. 1.

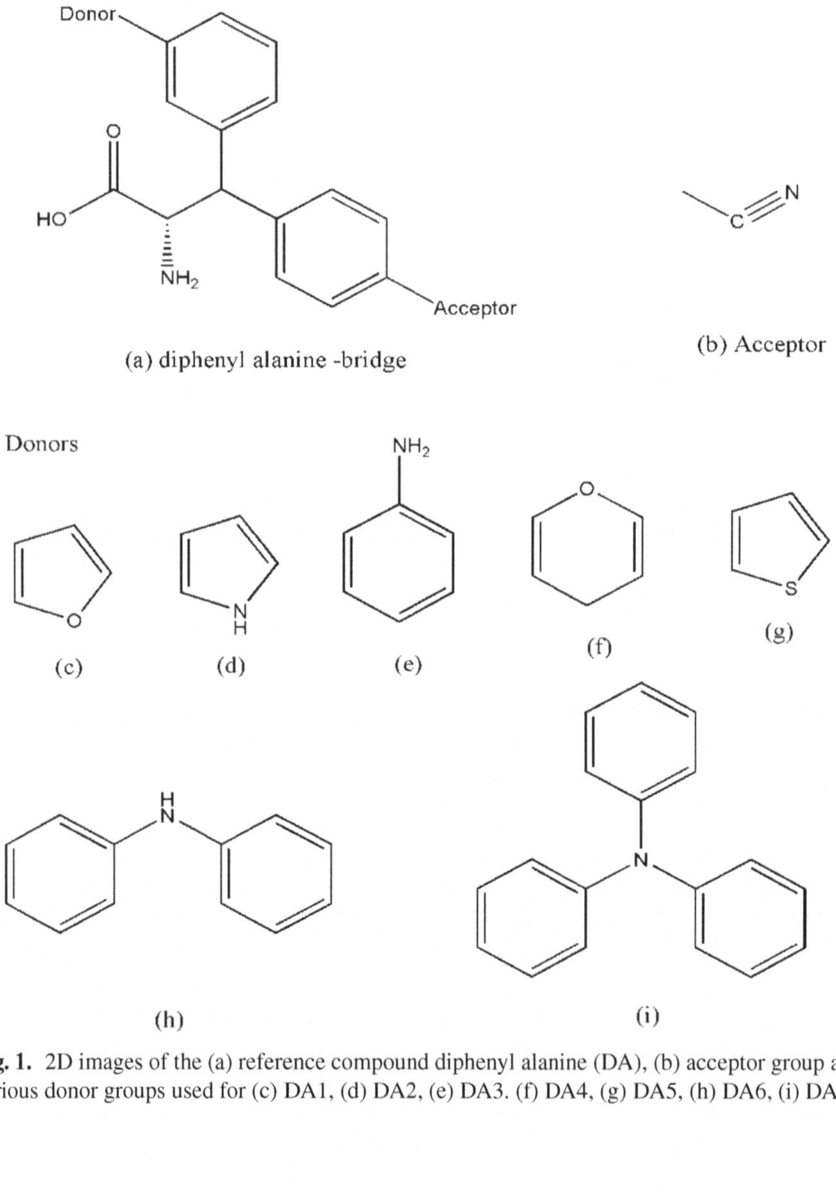

Fig. 1. 2D images of the (a) reference compound diphenyl alanine (DA), (b) acceptor group and various donor groups used for (c) DA1, (d) DA2, (e) DA3. (f) DA4, (g) DA5, (h) DA6, (i) DA7.

3.2 Electronic Properties

3.2.1 FMO Analysis

Frontier molecular orbital analysis is a key tool in computational calculations for understanding electronic properties especially, the behaviour of highest occupied molecular orbital (HOMO) and lowest unoccupied molecular orbital (LUMO) in the molecule [18]. The energy gap between HOMO and LUMO is calculated as

$$E_g = E_{LUMO} - E_{HOMO} \tag{3}$$

Energy gap is the strong indication of molecule's stability and reactivity. The Table 1 given below shows the energy gap values of the reference and derivatives.

Table 1. HOMO-LUMO gap for reference compound DA and designed compounds DA1-DA7

Compounds	Energy gap (ev)
DA	4.81
DA 1	4.32
DA 2	4.7
DA 3	4.62
DA 4	4.41
DA 5	4.27
DA 6	4.17
DA 7	3.86

From the table, it is clear that, all the derivatives have lower energy gap values compared to the reference compound DA. Among the investigated compounds DA 7 has the lowest energy gap value of 3.86 eV. Lower the energy gap more reactive is the compound. Hence it actively participates in electronic transitions between the energy levels. The Fig. 2 shows the the distribution of HOMO and LUMO in the molecules. In the case of DA 6 and DA 7 HOMO is exactly distributed in the donor region and LUMO in the region of bridge to the acceptor. This is due to the efficient intra molecular charge transfer (ICT) from the donor moiety to the acceptor.

Figure 2 shows the HOMO-LUMO behaviour of the reference compound and designed derivatives. In the HOMO-LUMO diagram the color indicates the different phase of the wavefunction. The positive phase of the wavefunction is indicated by red color. Green indicates the negative phase of the wavefunction.

3.2.2 UV-Visible Absorption Spectra

UV-Visible absorption spectra are an efficient tool for material characterization especially, for obtaining linear optical properties of molecules.When the molecule is illuminated by light electron is excited from ground state to higher electronic state. Due to this

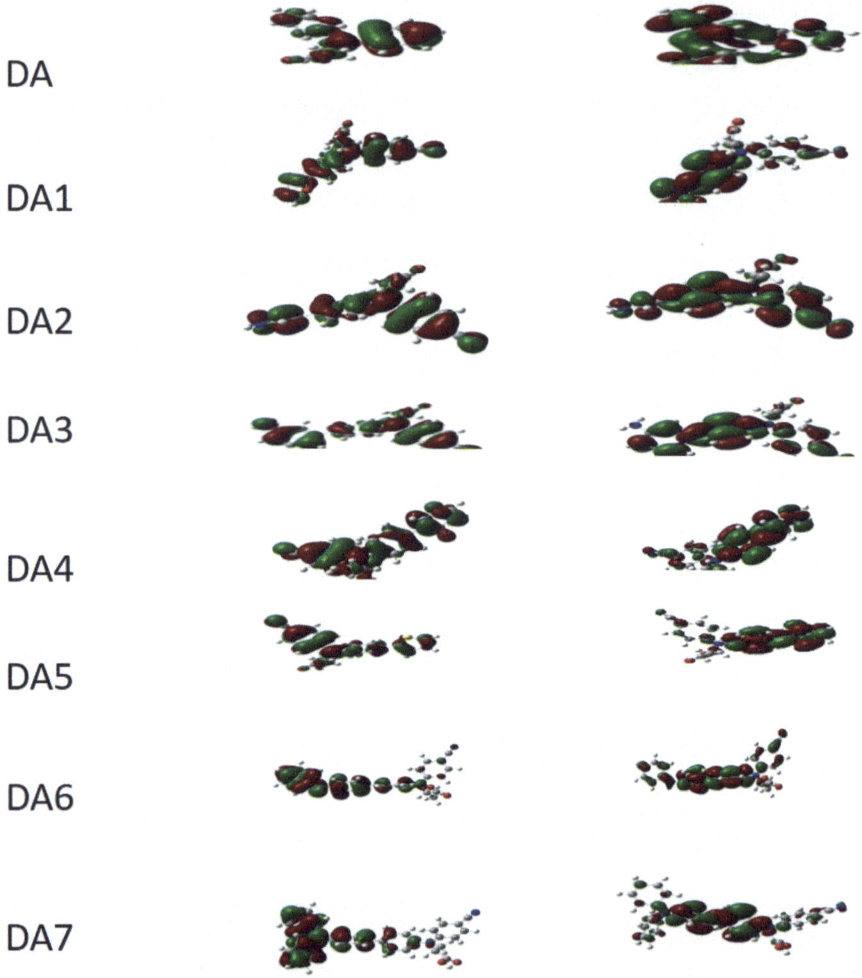

Fig. 2. HOMO-LUMO diagrams of the reference compound DA and designed derivatives DA1-DA7

electronic transition peaks are obtained in the uv-vis absorption spectra corresponding to the wavelengths where absorption takes place. It helps to understand conjugation and interactions between donor and acceptors in push-pull chromophores. The electronic transition energies and absorption characteristics are modulated by structural tuning by incorporating different donor moieties [18]. Figure 3 shows UV-Vis absorption spectra of the parent compound and the designed compounds. The absorption spectra are recorded in the gaseous phase at TD-B3LYP/6-31+G(d) level of theory. All the designed compounds have larger absorption maxima (λ_{max}) and lower excitation energy compared to the parent compound DA (Table 2).

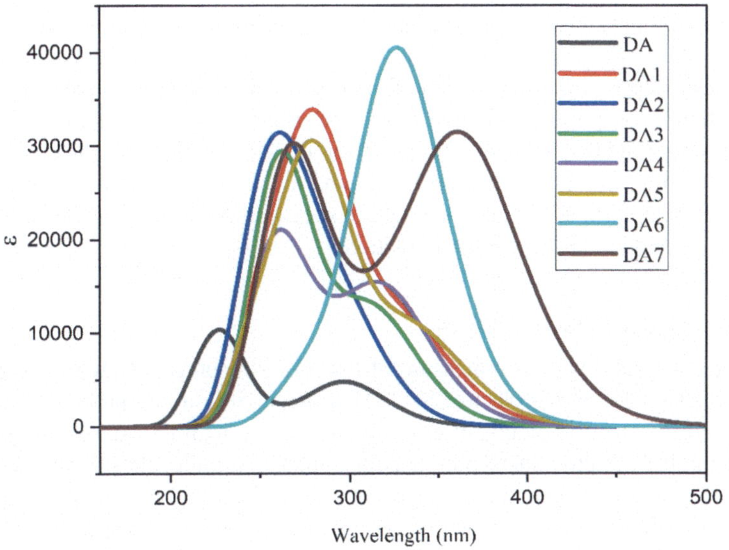

Fig. 3. The UV-visible spectra of the reference compound DA and designed derivatives DA1-DA7 in the gaseous phase

Table 2. Excitation energy (eV), oscillator strength and maximum absorption wavelength (nm) of parent compound (DA) and designed compounds (DA 1-DA 7) in gaseous phase using TD-B3LYP/6-31+G(d) level of theory

Compounds	Excitation energy (eV)	Oscillator strength, f_{osc}	Max. absorption wavelength, λ_{max} (nm)
DA	5.35	0.1893	231.82
DA 1	4.36	0.6333	284.09
DA 2	4.89	0.2986	253.54
DA 3	4.62	0.3009	268.03
DA 4	4.85	0.3475	255.81
DA 5	4.35	0.5618	284.55
DA 6	3.77	0.9307	328.67
DA 7	3.39	0.7214	364.95

Here the absorption maxima range from 231.82 nm and 364.95 nm. The graph shows the highest absorption maximum is obtained for DA 7 compound. Table 2 shows that the highest absorption maximum is obtained for compound DA7. It is due to the efficient π–π* electronic transition from the triphenyl amine (TPA) donor to the cyano acceptor. The maximum absorption for DA 7 is 364.95 nm. Even though the strongest absorption of light occurs for the compound DA 6, Since the molar extinction coefficient (ε) and oscillator strength (f) for DA 6 is the largest value compared to the other designed compounds. Because, molar absorptivity or molar extinction coefficient is a strong indication of how strongly a compound absorbs light at a particular wavelength. Thus, we can interpret that the structural modification by the introduction of various donors shift the absorption maximum towards the longer wavelength region compared to the parent compound DA. The experimentally reported maximum absorption wavelength is around 265 nm [19]. The excitation energy and maximum absorption wavelength are inversely related. In the UV-visible absorption spectrum peaks in the lower wavelength region corresponds to higher energy transition.

3.2.3 First Order Polarizability (α)

The linear polarizability, α is a measure of first order response of the dipole moment with respect to the external electric field. That means it is a key property explaining the behaviour of electron clouds of atoms or molecules with an external electric field [20]. In complex systems linear polarizability impacts a molecule's spectroscopy and reactivity. It provides a fundamental explanation for electronic transitions between the energy levels. Which results in better designing of compounds with specific optical and electronic properties.

The Table 3 shows that the investigated compounds (DA1-DA7) have larger linear optical properties compared to the parent compound DA. However, DA6 and DA7 possess largest linear optical properties due to the effect of their donor moieties diphenyl amine (DPA) and triphenyl amine (TPA).

Table 3. Static and dynamic (at 1907 nm) linear polarizabilities of parent compound DA and designed compounds DA1-DA7

Compound	Static ($\times 10^{-24}$ esu)	Dynamic ($\times 10^{-24}$ esu)
DA	24.27	24.37
DA1	24.37	37.24
DA2	36.43	36.65
DA3	40.57	40.84
DA4	37.64	37.89
DA5	38.82	39.10
DA6	53.77	54.25
DA7	71.29	71.94

3.3 Nonlinear Optical Properties

The second order nonlinear optical (NLO) properties of molecules can be characterized by the theoretical HRS (Hyper-Rayleigh Scattering) hyperpolarizability (β_{HRS}). β_{HRS} corresponds to first hyperpolarizability. For the study of materials for applications such as second harmonic generation (SHG) and electro-optic modulation, β_{HRS} is a significant parameter for the evaluation. The strength of second order nonlinear optical response of molecules can be determined by the magnitude of β_{HRS}. Larger β_{HRS} values signify a material is highly nonlinear active and become potential candidates for photonic applications like optical switching, frequency conversion and modulation for telecommunication. The pre-screening of molecules before synthesis is possible through the prediction of β_{HRS} values. Which saves time and resources. In push-pull chromophores β_{HRS} provides an insight into the influence of donor-acceptor groups and charge transfer characteristics.

The Table 4 shows that parent compound DA has a low value of HRS hyperpolarizability. The structural tuning provides an increase in β_{HRS} values due to the effect of various donor groups connected to the diphenylalanine spacer. Among the designed compounds DA7 has the largest β_{HRS} value. One of the strongest donors triphenyl amine (TPA) is used in DA7.This stronger donor increased the electron density difference between the donor and the acceptor groups. Which enhanced the charge transfer from the donor moiety to the acceptor. Thus, β_{HRS} value increased 16 times compared to the parent compound DA. Other parameters which describe NLO response of molecules are depolarisation ratio (DR) and anisotropy parameter (ρ). They provide insights into symmetry and hyperpolarizability. Asymmetric molecules (high DR) have a stronger nonlinear optical (NLO) property due to the broken Centro symmetry That results in variations in dipole moments during optical transitions.

Table 4. HRS hyperpolarizability β_{HRS}($\times$ 10^{-30} esu), depolarization ratio (DR), nonlinear anisotropy parameter (ρ), dipolar and octupolar contributions in percentage to first order hyperpolarizability of parent compound DA and designed compounds DA1-DA7 at DFT-B3LYP/6-31+G(d) level of theory using dispersion wavelength 1907 nm

Compound	β_{HRS} ($\times 10^{-30}$ esu)	DR	ρ	Φ ($\beta_{j=1}$)	Φ ($\beta_{j=3}$)
DA	1.897	1.870	3.351	23.0	77.0
DA1	7.008	5.098	0.795	55.7	44.3
DA2	8.272	5.639	0.688	59.2	40.8
DA3	6.465	3.585	1.231	44.8	55.2
DA4	6.726	4.859	0.848	54.1	45.9
DA5	5.220	3.021	1.515	39.8	60.2
DA6	21.055	5.807	0.658	60.3	39.7
DA7	30.359	4.321	0.984	50.4	49.6

The polar nature of HRS hyperpolarizability can be analysed by dipolar and octupolar interactions of the molecules in the parent compound DA and designed compounds DA1-DA7. Table 4 show the value of dipolar and octupolar contributions of the investigated compounds. Figure 4 shows the values of dipolar and octupolar contributions of the investigated compounds. In the parent compound DA 77% of octupolar contribution is estimated. The structural modification implies a transformation of octupolar contribution to dipolar contribution for the designed derivatives DA1-D7. Octupolar contribution signifies that the system exhibits a symmetric charge distribution. Whereas an asymmetric charge distribution is associated with a dipolar system. The DA6 derivative shows strong dipolar interactions. The DA6 and DA7 compounds produce strong HRS hyperpolarizability due to their strong dipolar interactions. Dipolar systems are more efficient in generating second order nonlinear optical (NLO) response due to their asymmetric charge distribution and ability to produce large β values.

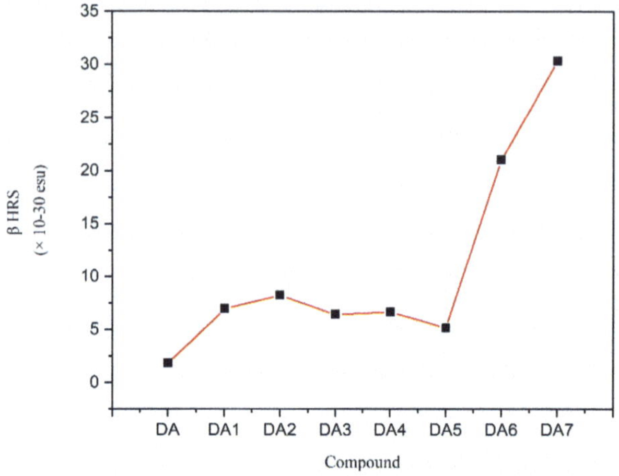

Fig. 4. HRS hyperpolarizability $\beta_{HRS}(\times 10^{-30}$ esu) of parent compound DA and designed compounds DA1-DA7.3.4 Dipolar and Octopolar interactions

4 Conclusion

A set of diphenyl alanine conjugated push-pull chromophores (DA1-DA7) were designed and analysed electronic and NLO responses using DFT and TD-DFT. The FMO analysis described that all the designed compounds have lower energy gap ranges between 4.32–3.86 eV compared to the parent (reference) compound DA (4.81 eV). The smaller energy gap is obtained for compounds DA6 and DA7. The HOMO-LUMO diagram indicates efficient ICT occurs in compounds where HOMO is distributed on the donor part and LUMO on the acceptor part. Compounds with smaller HOMO-LUMO gap attributes easier electronic excitation. The UV-Visible absorption spectra provided detailed information regarding the excitation energy and maximum absorption wavelength. Which showed that all the designed compounds (DA1-DA7) have lower excitatuion energy and higher maximum absorption compared to the parent compound. Maximum absorption wavelength occurs for DA6 and DA7 with lowest excitation energy. Nonlinear optical properties revealed that all the designed compounds have larger first order hyperpolarizability compared to the parent compound implies the designed compounds possess good second order nonlinear optical properties especially DA6 and DA7 compounds.

Thus, it can be concluded that among the designed compounds DA6 and DA7 possess better electronic and second order nonlinear optical properties due to the presence of strong donors, diphenyl amine (DPA) and triphenyl amine (TPA). Also, the transformation of octupolar to dipolar contribution from the parent to the investigated compounds during structural tuning confirmed good second order nonlinear optical response for the proposed compounds DA6 and DA7. Thus, the study concluded that further structural modifications on DA6 and DA7 will provide pronouncing applications in optical switching.

References

1. Ye, J.T., Qiu, Y.Q.: The Inspiration and challenge for through-space charge transfer architecture: from thermally activated delayed fluorescence to nonlinear optical properties. J. Phys. Chem. Chem. Phys. **23**, 15881 (2021)
2. Sergey, S., Li, X., Duan, Y., Rasing, T.: J. Adv. Opt. Mater. **9**(23), 2100327 (2021)
3. Pascal, D., Santos, H.F.D.: J. Orgel. **28**, 111–117 (2016)
4. Vineetha, P.K., Aswathy, A., Shiju, E., Chandrashekharan, K., Manoj, N.: Structure property correlations on nonlinear optical properties of a few bipodal D-π-A molecules-experimental and theoretical approach. New J. Chem. **44**, 6142–6150 (2020)
5. Albert, I.D.L., Marks, T.J., Ratner, M.A.: Large molecular hyperpolarizabilities. Quantitative analysis of aromaticity, and auxiliary donor-acceptor effects. J. Am. Chem. Soc. **120**(43), 11174–11181 (1998)
6. Boddula, R., Singh, S.P.: Peptide-based novel small molecules and polymers: unexplored optoelectronic materials. J. Mater. Chem. C. **9**, 12462 (2021)
7. Khanra, S., Vassiliades, S.V., Alves, W.A., Yang, K., Classer, R., Ghosh, K., Bhattacharya, P., Yu, P., Guha, S.: Enhanced piezo response and nonlinear optical properties of fluorinated self-assembled nanotubes. J. Adv. **9**, 115202 (2019)
8. Wang, H., Chen, F.: Theoretical perspective on two-state "ON=OFF NLO" switch of Rh (III)-Azobenzene complex with considerable first hyperpolarizability. J. Phys. Chem. A. **127**, 5030–5038 (2023)
9. Avci, D., Basoglu, A., Atalay, Y.: Ab initio HF and DFT calculations on an organic nonlinear optical material. J. Struct. Chem. **21**(1), 213–219 (2010)
10. Avci, D.: Second and third order nonlinear optical properties and molecular parameters of azo chromophores; semiempirical. J. Spectrochim. Acta A. **82**, 37–43 (2011)
11. Abdel-Rahman, L.H., Abu-Dief, A.M., Moustafa, H., Hassan Abdel-Mawgoud, A.A.: Design and nonlinear optical properties (NLO) using DFT approach of new Cr (III), Vo (II), and Ni (II) chelates incorporating tri-dentate imine ligand for DNA interaction, antimicrobial, anticancer activities and molecular docking studies. Arab. J. Chem. **13**(1), 649–670 (2020)
12. Tian, L., Chen, F.: Multiwfn: a multifunctional wavefunction analyzer. J. Comput. Chem. **33**, 580–592 (2012)
13. Khalid, M., Usman Khan, M., Shafiq, I., Hussain, R., Mahmood, K., Hussain, A., Jawaria, R., Hussain, A., Imran, M., Assiri, M.A., Ali, A., Fayyaz ur Rehmon, M., Sun, K., Li, Y.: NLO potential exploration of D-π-A heterocyclic organic compounds by incorporation of various π-linkers and acceptor units. Arab. J. Chem. **14**(8), 103295 (2021)
14. Khanra, S., Vassiliades, S.V., Alves, W.A., Yang, K., Glaser, R., Ghosh, K., Bhattacharta, P., Yu, P., Guha, S.: Enhanced piezoresponse and nonlinear optical properties of fluorinated self-assembled peptide nanotubes. J. Adv. **9**, 115202 (2019)

15. Khalid, M., Khan, M., Shafiq, I., Mahmood, K., Akhtar, M.N., Iqbal, J., Al-Sadoon, M.K., Zaman, W., Carmo Brage, A.B.: Role of donors in trigerring second order nonlinear optical properties of non-Fullerene FCO-2FRA based derivatives: a theoretical perspective. J. Heliyon. **9**(2), e13033 (2023)
16. Atalay, Y., Avci, D., Basoglu, A.: Linear and nonlinear optical properties of some donor-acceptor oxadiazoles by ab initio Hartree-Fock calculations. J. Struct. Chem. **19**, 239–246 (2008)
17. Mahmood, R., Janjua, M.R.S.A., Jamil, S.S.: DFT molecular simulation for design and effect of core-bridging acceptors (BA) on NLO response: first theoretical framework to enhance nonlinearity through BA. J. Clust. Sci. **28**, 3178–3183 (2017)
18. Mohammed, N., Wiles, A.A., Belsley, M., Fernandus, S.M., Carielli, M., Rotello, V.M., Rapson, M.M.M., Cooke, G.: Synthesis and characterisation of push-pull flavin dyes with efficient second harmonic generation (SHG) properties. J. RSC Adv. **7**, 24462 (2017)
19. Catalini, S., Bagni, F., Cicchi, S., Di Donato, M., Lagatti, A., Lapini, A., Foggi, P., Petrillo, C., Dimichele, A., Paolantoni, M., Schiro, G., Comez, L., Paciaroni, A.: Multiple length-scale control of BCC-protected diphenyl alanine aggregates through solvent composition. J. Mater. Adv. **5**, 3802 (2024)
20. Champagne, B.: Jean-Mary Andre.: towards the calculations of polarizabilities of stereoregular polymers. In: Advances in Quantum Chemistry. Elsevier, Amsterdam (2019)

Influence of Explosophoric Groups on Energetic Performance of 1,2,4-Oxadiazoles: A DFT Study

Hridya Rajan[1] and R. Renjith[1,2(✉)]

[1] Department of Physics, University College, Thiruvananthapuram 695034, Kerala, India
renjithkadavoor@gmail.com

[2] Department of Physics, Government College, Kasaragod 671123, Kerala, India

Abstract. High energy compounds rich in nitrogen have accepted significant attention at present for their potential applications as explosives. Pentagonal heterocycles rich in nitrogen, such as oxadiazoles are preferred as High Energy Density Materials (HEDMs) because of the imine and nitroxyl bonds in their ring enhance both oxygen balance and heat of formation (HOF). Computational screening is an appealing approach for discovering new materials, as it provides a cost-effective means of evaluating compounds before synthesis. The present study involved the design and optimization of energetic compounds by incorporating different energetic substituent groups to 1,2,4-oxadiazole backbone. The designed compounds are evaluated for HOF, density, detonation performance, electronic structures and impact sensitivity. The detonation performance of optimized compounds was calculated using Kamlet-Jacobs equations. The impact sensitivity of HEDMs is evaluated in terms of impact energy (*EI*). The introduction of nitro and azido group to the oxadiazole compound has resulted in energetic materials with greater detonation performance and lower sensitivity than RDX and HMX. Four compounds may be selected as potential candidates for HEDMs.

Keywords: Oxadiazoles · Detonation performance · Oxygen balance · Kamlet-Jacobs equation

1 Introduction

Energetic materials, categorized into pyrotechnics, explosives, and propellants have wide-ranging applications in both civilian and military sectors [1–6]. Research on energetic materials demand materials with higher density, positive HOF, superior detonation performance and lower sensitivity to impact [7–10]. The choice of a heterocyclic backbone is the favourable choice in designing energetic materials, since it not only increases the heats of formation (HOFs) but also enhance its detonation properties during the decomposition process [11–13]. Azoles are a significant class of heterocyclic compounds, whose incorporation enhances both the nitrogen content and the stability of energetic materials, while also improving detonation performance [14, 15]. Azoles can be classified based on the number of oxygen and nitrogen atoms in the ring structure as pyrrole, pyrazole, imidazole, triazole, tetrazole, and oxadiazole. The presence of reactive

R. Rajasekharan et al. (Eds.): MatZone 2024, SPPHY 316, pp. 47–59, 2025.
https://doi.org/10.1007/978-981-96-6541-9_4

oxygen atoms in oxadiazoles, make them a unique class of energetic units. Recently, 1,2,4-oxadiazole heterocycles have garnered increasing attention as a crucial pillar for nitrogen-rich energetic compounds. This structure has shown significant potential in the development of new high explosives that offer a balance between energy output and safety [16, 17].

A series of compounds were designed here, from basic structural unit of 1,2,4-oxadiazole with its 3, 5 position occupied by ethanimine to form (1,2,4-oxadiazole-3,5-diyl) dimethanimine (Fig. 1). Further modification is performed by introduction of energetic groups (N_3, NH-NO_2, NO_2, O-NO_2, NH_2, CH-$(NO_2)_2$, OH, C-$(NO_2)_3$) at R_1 position in scheme A and scheme B shown in Fig. 2.

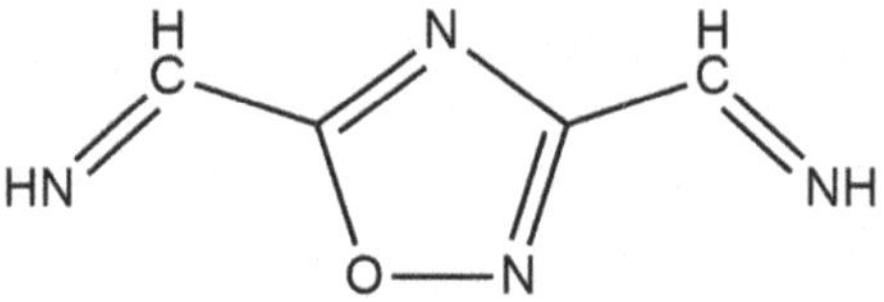

(1,2,4-oxadiazole-3,5-diyl)dimethanimine

Fig. 1. Basic structural unit of designed compounds

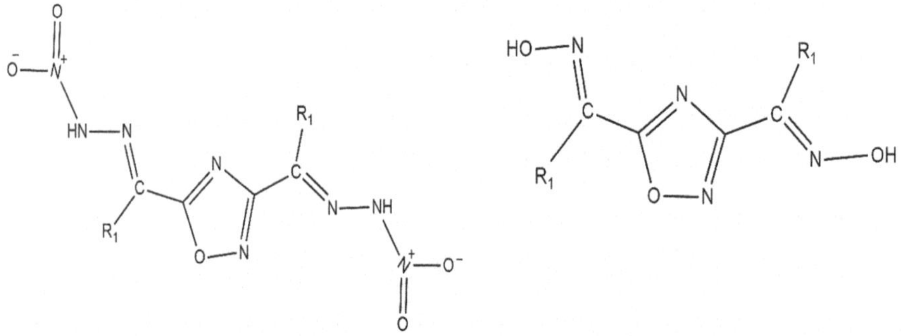

R_1=N_3, NH-NO_2, NO_2, O-NO_2, NH_2, CH-$(NO_2)_2$, OH, C$(NO_2)_3$

Scheme A **Scheme B**

Fig. 2. Scheme A and Scheme B of the designed compounds

The optimized compounds were evaluated in terms of Oxygen Balance (%), HOF (kJ/mol), Density (g/cm^3), Detonation Performance (Detonation velocity (km/s) and Detonation Pressure (GPa), Impact Energy (J) and Frontier Molecular Analysis were investigated. The results will help chemist to synthesis potential energetic materials rather than consume time and resources in blind synthesis.

2 Computational Details

The geometrical structures of the designed compounds were optimized by density functional theory (DFT) at the B3PW91/6-31G (d,p) level. The vibrational analysis was performed to confirm that the optimized compounds are in local energy minima. The free hydrogen (N-H) of molecule in Fig. 1 is substituted with Nitramine and free hydrogen (C-H) of the same is substituted with different explosophoric group (scheme 1). If free hydrogen in N-H is substituted with hydroxyl group we get scheme 2. Different compounds designed are briefed in Table 1.

Table 1. List of Designed compounds

Compound name	Substituent group for H in N-H	Substituent group for H in C-H (R_1 position)
A_1	$NH\text{-}NO_2$	$\mathbf{N_3}$
B_1	OH	
A_2	$NH\text{-}NO_2$	$\mathbf{NH\text{-}NO_2}$
B_2	OH	
A_3	$NH\text{-}NO_2$	$\mathbf{NH_2}$
B_3	OH	
A_4	$NH\text{-}NO_2$	$\mathbf{NO_2}$
B_4	OH	
A_5	$NH\text{-}NO_2$	$\mathbf{O\text{-}NO_2}$
B_5	OH	
A_6	$NH\text{-}NO_2$	$\mathbf{CH\text{-}(NO_2)_2}$
B_6	OH	
A_7	$NH\text{-}NO_2$	$\mathbf{C(NO_2)_3}$
B_7	OH	
A_8	$NH\text{-}NO_2$	**OH**
B_8	OH	

The gas phase HOF [18] ($\Delta H_{f,\,gas}$) were calculated by equation

$$\Delta H_{298} = \sum \Delta H_{f,P} - \sum \Delta H_{f,R} \tag{1}$$

where $\Delta H_{f,\,P}$, $\Delta H_{f,\,R}$ are HOF of products and reactants at 298 K

$$\Delta H_{298} = \Delta E_{298} + \Delta PV = \Delta E_0 + \Delta ZPE + \Delta H_T + \Delta nRT \tag{2}$$

where ΔE_0 is the energy changes between product and reactant, ΔZPE is the zero point energy difference between products and reactants, ΔH_T is the thermal correction to enthalpy and Δn is the change in the number of energetic groups.

Solid state heat of formation $\Delta H_{f,\,solid}$ is calculated by Hess's Law [19] as

$$\Delta H_{f,solid} = \Delta H_{f,gas} - \Delta H_{sub} \tag{3}$$

Enthalpy of sublimation [20] can be evaluated using the relation (4).

$$\Delta H_{sub} = aA^2 + b\sqrt{v\sigma^2_{total}} + c \tag{4}$$

Where A is the isosurface area at 0.001 electrons/bohr^3 density, υ is the degree of balance between positive and negative surface potentials and σ^2_{total} reflects the variability of the electrostatic potential. The coefficients a = 2.670*10^{-4} kcal mol^{-1}A^{-4}, b = 1.650 kcal mol^{-1} and c = 2.966 kcal mol^{-1} calculated at B3PW91 / 6-31G (d, p) level.

The energetic compounds that fulfill the condition $2a + \frac{b}{2} > c > \frac{b}{2}$for explosives with the formula $C_aH_bN_dO_c$ are characterized by N, the number of moles of detonation gases per gram of the compound, $\overline{M}$ the average molecular weight of the gas products, and Q the maximum detonation heat per unit mass. These parameters can be determined using the following equations.

$$N = \frac{(b + 2c + 2d)}{(4 * M_w)}$$

$$\overline{M} = \frac{(56d + 88c - 8b)}{(b + c + 2d)}$$

$$Q = \frac{28.9b + 94.05\left(\frac{c}{2} - \frac{b}{4}\right) + 0.239 * \Delta H_f}{M_w}$$

The detonation velocity (D), and pressure (P) of the compounds were calculated using the Kamlet-Jacobs equations [21].

$$D = 1.01 * \sqrt{\left(N * \sqrt{\overline{M}} * \sqrt{Q}\right)} * (1 + 1.30 * \rho) \tag{5}$$

$$P = 1.558 * \rho^2 * N * \sqrt{\overline{M}} * \sqrt{Q} \tag{6}$$

where D represents the detonation velocity (km/s), P denotes the detonation pressure (GPa), N is the number of moles of gaseous products per gram of the explosive, $\overline{M}$ is the average molecular weight of these gases, Q is the heat of detonation (J/g) and ρ is the crystal density of explosives (g/cm^3).

Accurate densities were found by an upgraded equation by Politzer et al. [22]

$$\rho = \alpha * \left(\frac{M}{V(0.001)}\right) + \beta * v * \sigma^2_{total} + \gamma \tag{7}$$

where the coefficients $\alpha = 0.9183$, $\beta = 0.0028$ *and* $\gamma = 0.0443$, M was the molecular mass (g/mol) and V is the volume of a molecule.

High energy density materials release high heat when C and H are fully oxidized to CO_2 and H_2O. The oxygen balance (%) for explosives with molecular mass M and formula $C_aH_bN_cO_d$ is calculated as

$$OB(\%) = \frac{\left(d - 2a - \frac{b}{2}\right)^*}{M} 1600 \tag{8}$$

The bond dissociation energies (BDE) of the optimized compounds were found using the equation.

$$BDE_0(A - B) = E_0\left(A^*\right) + E_0\left(B^*\right) - E_0(A - B) \tag{9}$$

$BDE_0(A-B)$ is BDE of A–B at 0 K, $E_0(A^*)$, $E_0(B^*)$ and $E_0(A-B)$ are total energy of product radicals A*, B* and neutral molecule (A–B).

3 Result and Discussion

A total of 16 compounds were designed as given as Table 1 and their optimized structure is in Fig. 3. The positive values of all frequencies of the designed structure confirms that they are in global minimum.

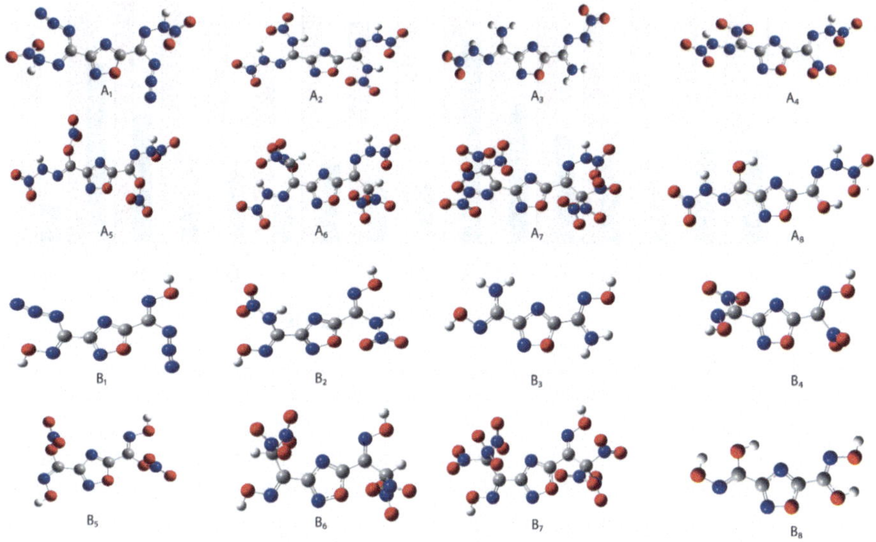

Fig. 3. Optimized structure of designed compounds

3.1 Energetic Properties

The HOF indicates the energy content of a molecule. All compounds showed positive HOF value, which is ideal for HEDM's. Fig. 4 indicates the variation of HOF solid and gas phase with introduction of different substituent group. HOF of gas phase ranged from 199 to 380 kJ/mol for compounds in scheme A and scheme B, HOF_{gas} ranged from 139 to 315 kJ/mol. Trinitomethyl group substituted at R1 position in scheme A and scheme B showed greater HOF value. This can be due to higher NO2 group in these compounds [23]. Each substituent group in both series has different effect and their contribution of HOF is in the order -C-(NO2)3> -CH-(NO2)2> O- NO2> NH-NO2> N3> NO2 > NH2> OH.

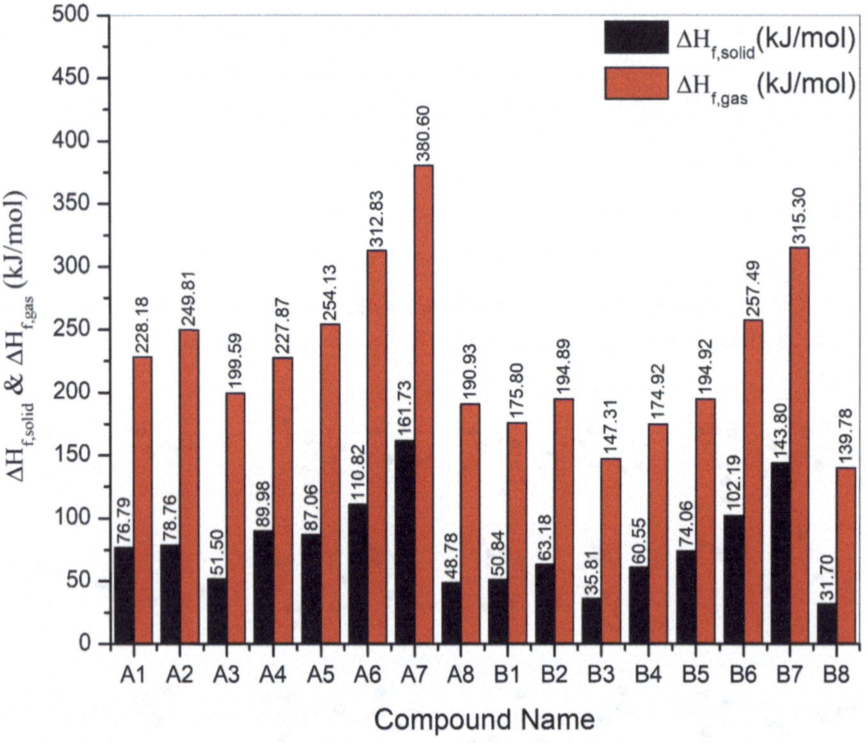

Fig. 4. HOF solid and gas phase of the designed compounds

The term Oxygen Balance of HEDM's indicate the extent to which all the carbon and hydrogen atoms present in the compound are oxidized to CO_2 and H_2O respectively. Figure 5 indicates the variation of Oxygen Balance (%) with introduction of different substituent group. Most of the designed compounds showed negative oxygen balance, indicating higher relative hydrogen content, as the case of A_3 (−35%), A_8 (−17%), B_3 (−69%) and B_8 (−43%). In compounds belonging to scheme B (R = OH) and R_1 varied with NH-NO_2, NO_2, CH-$(NO_2)_2$ group resulted in less negative oxygen balance than RDX and HMX. In compounds belonging to scheme A (R = NH-NO_2) and varied with NH-NO_2 and CH-$(NO_2)_2$ resulted in near zero negative oxygen balance. When R_1 was occupied by O-NO_2 and C-$(NO_2)_3$ in both schemes resulted in zero or positive oxygen balance (<12%).

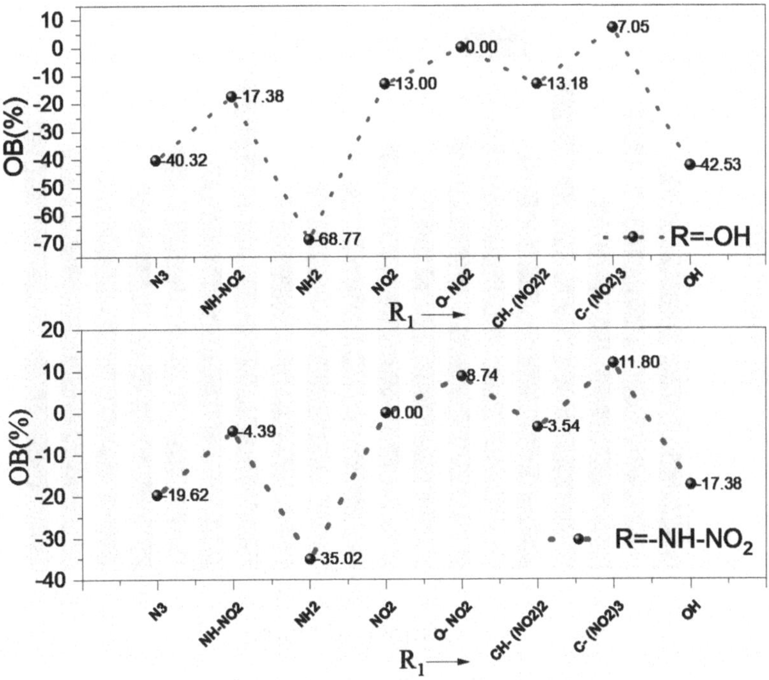

Fig. 5. Oxygen Balance (%) of the designed compounds

According to the K-J equation, the detonation velocity and detonation pressure of an energetic material are proportional to its ρ and ρ^2 respectively. Additionally, the heat of detonation (Q), which is derived from the enthalpy of formation (ΔH_f), is also a key factor in the equation. The density (ρ) was predicted using a Monte Carlo method by calculating the molecular volume within an electron density contour of 0.0001 e $Bohr^{-1}$ [24–26]. The variation of density with different substituent group in R_1 position in scheme A and B is shown in Fig. 6. There is a good agreement of calculated density of traditional explosives RDX, HMX and CL-20 with the values reported in literature [27]. For Scheme A, density ranges from 1.67 g/cm^3 for NH_2 group to 2.21 g/cm^3 for NO_2 group in the order NO_2 > NH-NO_2 > CH-$(NO_2)_2$ > C-$(NO_2)_3$ > -O-NO_2 > N_3 > OH > NH_2. For Scheme B, density ranges from 1.71 g/cm^3 for NO_2 group to 2.21 g/cm^3 for -O-NO_2 group in the order O- NO_2 > N_3 > C-$(NO_2)_3$ > CH-$(NO_2)_2$ > OH > NH_2 > NH-NO_2 > NO_2.

Heat of detonation (Q) is another parameter which effects detonation performance of energetic materials. Figure 7 shows the variation of Q with different substituent group in R_1 position in scheme A and B. Among the designed structures, heat of detonation of scheme B compounds (1116–1873 cal/g) are higher than scheme A compounds (1458–810 cal/g).

The variation of D and P with different substituent groups in R_1 position in scheme A and B is shown in Fig. 8. There is a decent agreement of calculated detonation velocity and detonation pressure of traditional explosives such as RDX, HMX and CL-20 with its values reported in literature [27]. For Scheme A, detonation velocity ranges from

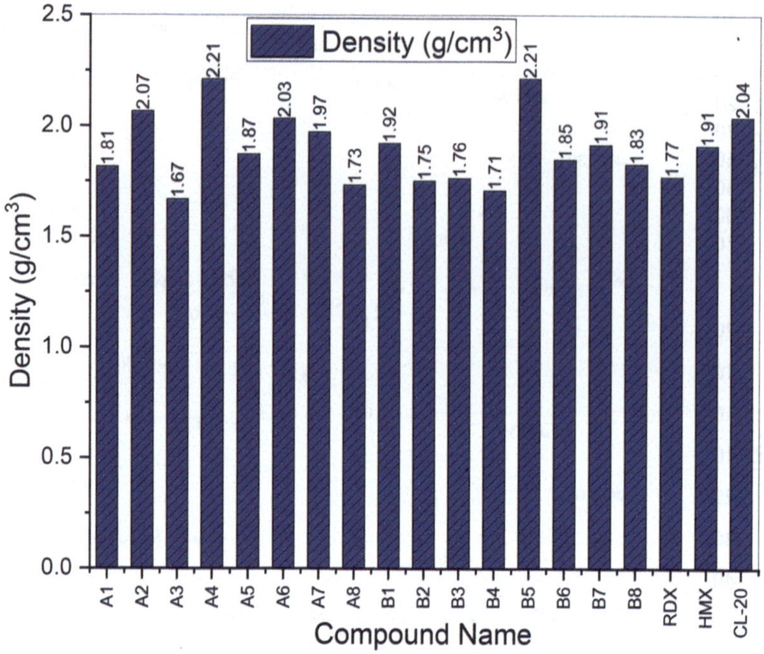

Fig. 6. Density of the designed compounds

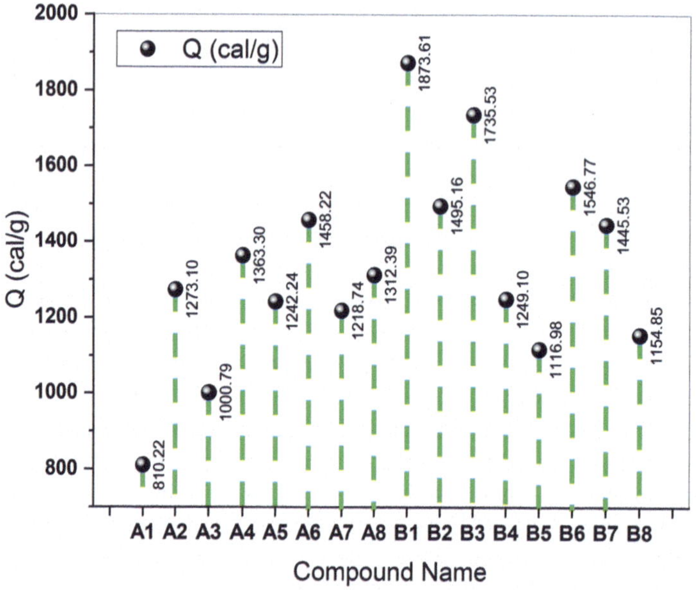

Fig. 7. Heat of detonation (Q) of the designed compounds

7.35 km/s for NH_2 group to 9.88 km/s for NO_2 in the order NO_2 > CH-$(NO_2)_2$ > NH-NO_2 > C-$(NO_2)_3$ > -O-NO_2 > N_3 > OH > NH_2 and detonation pressure ranges from 22.88 GPa for NH_2 group to 48.45 for NO_2. The variation trend of D and P are same. The variation trend of density and detonation performance are similar except that for substituent CH-$(NO_2)_2$ the compound showed higher detonation velocity and lower density than compound with substituent NH-NO_2. This may be due to higher heat of detonation value of compound with CH-$(NO_2)_2$ substituent than NH-NO_2. For Scheme B, detonation velocity ranges from 7.77 km/s for NO_2 and OH group to 9.62 km/s for N_3 group in the order N_3 > O- NO_2 > C-$(NO_2)_3$ > NH_2 > CH-$(NO_2)_2$ > NH-NO_2 > NO_2 = OH and detonation pressure ranges from 25.93 GPa for NO_2 group to 42.69 for N_3. The variation trend of D and P are almost similar. The variation trend of ρ and detonation performance are different. This is due to higher Q value of compound with N_3 and NH_2 leading to their better detonation performance and lower value of compound with OH substituent leading to inferior detonation performance.

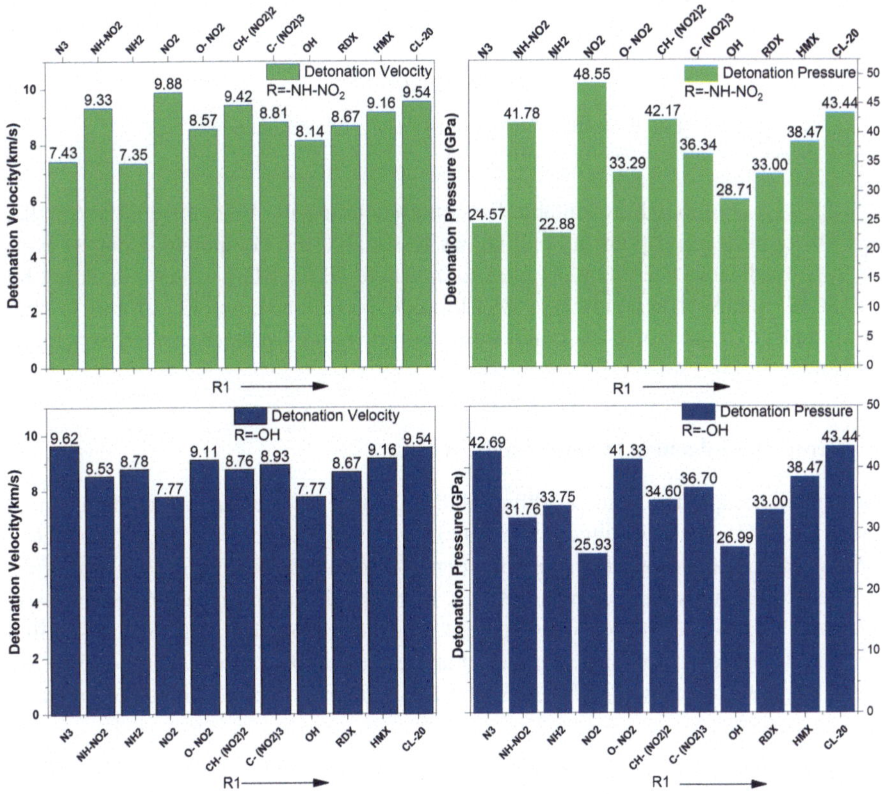

Fig. 8. Detonation Velocity and Pressure of designed compounds

Impact sensitivity is frequently used for evaluating the sensitivity of HEDM's. Impact sensitivity is generally reported as vertical distance (h_{50} (cm)) at which there is 50% probability of detonation or as impact energy (EI (J)). Figure 9 shows the variation of EI with substituent group at R_1 position in scheme A and B compounds. Maximum Impact energy is observed for R_1 = OH and minimum for R_1 = C-$(NO_2)_3$ in both schemes.

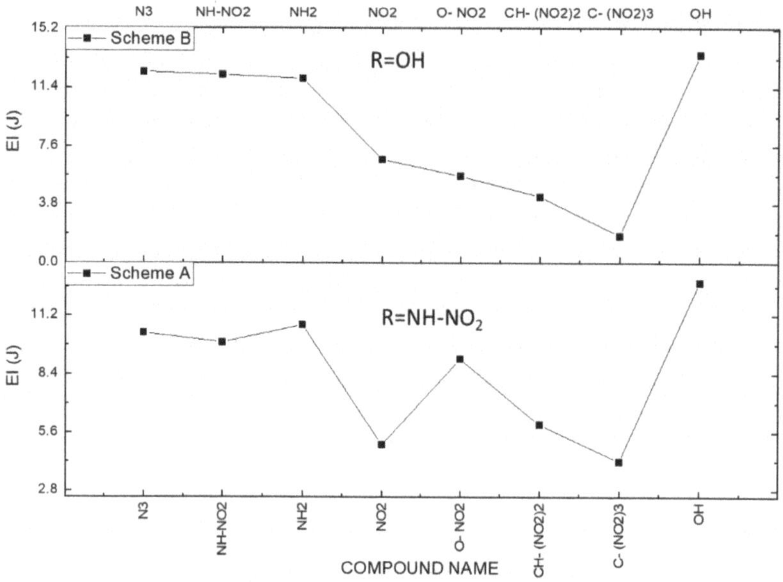

Fig. 9. Impact Energy of the designed compounds

BDE of a compound are suitable for comprehending its thermal stability. Smaller the BDE required to rupture a bond shows less stable the compound. Longest bond in 1,2,4- oxadiazole derivatives is used to calculate BDE. BDE values calculated for A_2 (156 kJ/mol, N-NO_2 bond), A_4 (188 kJ/mol, C-NO_2 bond), A_6 (137 kJ/mol, C-NO_2 bond) and B_1 (323 kJ/mol, C-N_3 bond) showed that these compounds are stable energetic molecule.

3.2 Frontier Molecular Orbital Analysis

The highest occupied molecular orbital (HOMO) and the lowest unoccupied molecular orbital (LUMO) of the designed compounds were analysed to investigate their electronic structures and properties in greater detail. The calculated $\Delta E_{(LUMO\text{-}HOMO)}$ of the designed compounds lies between 3.83 eV for OH substituent group in scheme A to 5 eV for NO_2 substituent group in scheme B. The band gap values suggest that stability of all the designed compounds is greater than traditional explosive RDX (3.4 eV) [28]. Figure 10 shows the HOMO-LUMO distribution and energy gap of designed compounds.

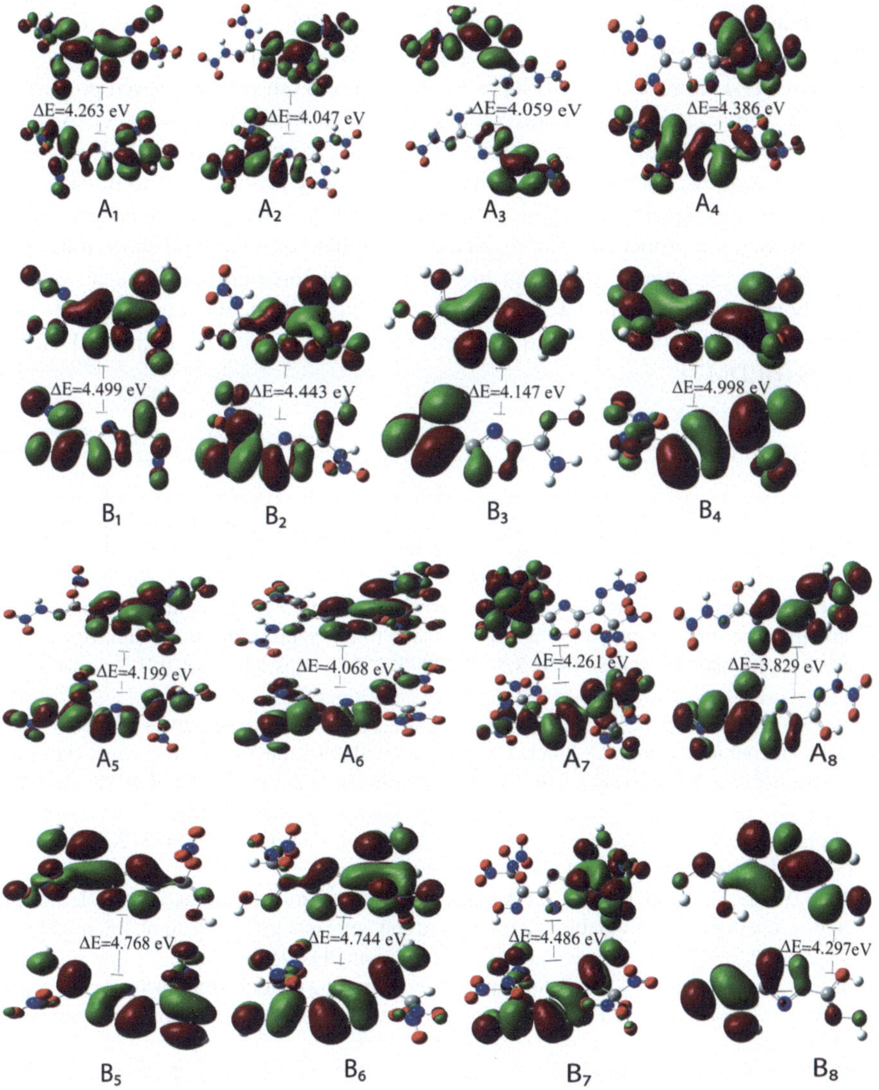

Fig. 10. Frontier Molecular orbital analysis of designed compounds

4 Conclusion

A series of 16 derivatives of 1,2,4-oxadiazole were designed by incorporation of set of eight different energetic groups in schemes. The designed compounds were then evaluated for their energetic properties, electronic structure and impact sensitivities by DFT at B3PW91/6-31G (d, p) level. All the designed compound showed positive HOF value, with trinitromethyl substituent contributing to highest ΔH_f (solid) value. When the substituent group was O-NO_2 and C-$(NO_2)_3$ in both scheme resulted in zero or positive oxygen balance (%). Compounds A_2, A_4, A_6, A_7, B_1, B_5, B_7 showed density value greater than that of well known explosive HMX. Results of detonation performance showed that among the 16 designed compounds, four of them (when R_1 = NH-NO_2, NO_2, CH-$(NO_2)_2$ in scheme A and R_1 = N_3 in scheme B) showed improved properties than HMX. Impact sensitivity of the compounds revealed that compounds A_1, A_3, A_8, B_1, B_2, B_3, B_8 displayed impact energies of the value better than that of HMX. This shows that the above 7 designed compounds are relatively less sensitive to impact. The bond dissociation energies are calculated for A_2, A_4, A_6 and B_1 compounds due to their superior detonation properties. The results revealed that B_1 compound showed highest BDE value of 296 kJ/mol. Based on the calculated detonation performance, impact sensitivity and BDE of designed compounds, one can conclude that compound B_1 (R_1 = N_3, Scheme B) satisfies all the requirements needed for an ideal High Energy Density Materials (HEDM's).

References

1. Cao, W., Dong, W., Zhang, J.G., et al.: Construction of coplanar bicyclic backbones for 1,2,4-triazole-1,2,4-oxadiazole-derived energetic materials. Chemistry. **2755**, 13807–13818 (2021). https://doi.org/10.1002/chem.202101884
2. Cao, W., Wang, T., Zhang, J.G., et al.: Access to green pyrotechnic compositions via constructing coordination polymers: a new approach to the application of 3,4-dinitropyrazole. ACS Appl. Mater. Interfaces. **1428**, 32084–32095 (2022). https://doi.org/10.1021/acsami.2c07758
3. Kuang, B.L., Wang, T.W., Zhang, J.G., et al.: Synthesis and properties of novel high nitrogen energetic complex of nickel perchlorate with tetrazole5formylhydrazine. Hanneng Cailiao/Chin. J. Energ. Mater. **316**, 531e8 (2023). https://doi.org/10.11943/CJEM2023041
4. Wang, H., Liu, Y., Liu, Q., et al.: Hydrogen bonding distribution and its effect on sensitivity of planar tricyclic polyazole energetic materials. Chem. Eng. J. **433**, 134479 (2022). https://doi.org/10.1016/j.cej.2021.134479
5. Tang, Y., He, C., Imler, G.H., et al.: Aminonitro groups surrounding a fused pyrazolotriazine ring: a superior thermally stable and insensitive energetic material. ACS Appl. Energy Mater. **23**, 2263–2267 (2019). https://doi.org/10.1021/acsaem.9b00049
6. Ma, Q., Zhang, Z., Fan, G., et al.: Strategies for constructing melt-castable energetic materials: a critical review. Energ. Mater. Front. **21**, 69e85–69e85 (2021). https://doi.org/10.1016/j.enmf.2021.01.006
7. Jin, X.H., Xiao, M.H., Wang, C.Y., Zhang, C., Zhou, J.H., Hu, B.C.: Eur. J. Org. Chem. **2019**, 988–994 (2019)
8. Fischer, D., Gottfried, J.L., Klapötke, T.M., Karaghiosoff, K., Stierstorfer, J., Witkowski, T.G.: Angew. Chem. Int. Ed. **55**(4), 16132–16135 (2016)
9. Ling, Y., Ren, X.L., Lai, W.P., Luo, J.: 4,4,8,8-Tetranitroadamantane-2,6-diyl dinitrate: a high-density energetic material. Eur. J. Org. Chem. **2015**, 1541–1547 (2015)
10. Jin, X.H., Xiao, M.H., Zhou, G.W., Zhou, J.H., Hu, B.C.: Molecule design and properties of bridged 2,2-bi(1,3,4-oxadiazole) energetic derivatives. RSC Adv. **9**(7), 5417–5430 (2019)
11. Dalinger, I.L., Serushkina, O.V., Muravyev, N.V., Meerov, D.B., Miroshnichenko, E.A., Konkova, T.S., Suponitsky, K.Y., Vener, M.V., Sheremetev, A.B.: J. Mater. Chem. A. **6**, 18669–18676 (2018)
12. Tsyshevsky, R., Smirnov, A.S., Kuklja, M.M.: Comprehensive end-to-end design of novel high energy density materials: III. Fused heterocyclic energetic compounds. J. Phys. Chem. C. **123**, 8688–8698 (2019)

13. Ma, C.M., Pan, Y., Jiang, J.C., Liu, Z.L., Yao, Q.Z.: Synthesis and thermal behavior of a fused, tricyclic pyridine-based energetic material: 4-amino-5-nitro-[1,2,5]oxadiazolo[3,4-e]tetra-zolo[1,5-a]pyridine-3-oxide. New J. Chem. **42**, 11259–11263 (2018)
14. Hervé, G., Roussel, C., Graindorge, H.: Selective preparation of 3,4,5-Trinitro-1*H*-Pyrazole: a stable all-carbon-nitrated arene. Angew. Chem. Int. Ed. **49**, 3177–3181 (2010)
15. Yu, Q., Yin, P., Zhang, J., He, C., Imler, G.H., Parrish, D.A., Shreeve, J.M.: Pushing the limits of oxygen balance in 1,3,4-oxadiazoles. J. Am. Chem. Soc. **139**, 8816–8819 (2017)
16. Yan, C., Wang, K., Liu, T., Yang, H., Cheng, G., Zhang, Q.: Exploiting the energetic potential of 1,2,4-oxadiazole derivatives: combining the benefits of a 1,2,4-oxadiazole framework with various energetic functionalities. Dalton Trans. **46**(41), 14210–14218 (2017). https://doi.org/10.1039/C7DT03320F
17. Cao, W.-L., Tariq, Q.-N., Li, Z.-M., Yang, J.-Q., Zhang, J.-G.: Recent advances on the nitrogen-rich 1,2,4-oxadiazole-azoles-based energetic materials. Defence Technol. **18**(3), 344–367 (2022). https://doi.org/10.1016/j.dt.2021.12.002
18. Elioff, M.S., Hoy, J., Bumpus, J.A.: Calculating heat of formation values of energetic compounds: a comparative study. Adv. Phys. Chem. **2016**, 1–11 (2016). https://doi.org/10.1155/2016/5082084
19. Xiao, M., Jin, X., Zhou, J., Hu, B.: 1,2,5-Oxadiazole-1,2,3,4-tetrazole-based high-energy materials: molecular design and screening. Struct. Chem. **32**(4), 1619–1628 (2021). https://doi.org/10.1007/s11224-021-01740-y
20. Byrd, E.F.C., Rice, B.M.: Improved prediction of heats of formation of energetic materials using quantum mechanical calculations. Chem. A Eur. J. **110**(3), 1005–1013 (2006). https://doi.org/10.1021/jp0536192
21. Kamlet, M.J., Jacobs, S.J.: Chemistry of detonations. I. A simple method for calculating detonation properties of C–H–N–O explosives. J. Chem. Phys. **48**(1), 23–35 (1968). https://doi.org/10.1063/1.1667908
22. Politzer, P., Martinez, J., Murray, J.S., Concha, M.C., Toro-Labbé, A.: An electrostatic interaction correction for improved crystal density prediction. Mol. Phys. **107**(19), 2095–2101 (2009). https://doi.org/10.1080/00268970903156306
23. Fasila, P.M., Rahana, A., Biju, A.R.: Theoretical investigation of energetic performance and impact sensitivities of nitro and trinitromethyl substituted ozonides of ethylene and cyclopentene. Comput. Theor. Chem. **1205**, 113425 (2021). https://doi.org/10.1016/j.comptc.2021.113425
24. Zhang, J.Y., Chen, G.L., Dong, J., Gong, X.D.: Effects of electronic delocalization and hydrostatic compression on structure and properties of cage compound 4-trini troethyl-2, 6, 8, 10, 12-pentanitrohexaazaisowurtzitane. Chem. Sel. **4**, 1567–1575 (2019)
25. Zhang, J.Y., Chen, G.L., Gong, X.D.: Screening for potential energetic C-N cages with high energy and good stability: a theoretical comparative study. Mol. Simulat. **45**, 129–136 (2018)
26. Zhang, J.Y., Gong, X.D.: The effect of nitro groups on the structures and energetic properties of the derivatives composed of TATB and cubane. Struct. Chem. **28**, 645–654 (2017)
27. Talawar, M.B., Sivabalan, R., Mukundan, T., Muthurajan, H., Sikder, A.K., Gandhe, B.R., Rao, A.S.: Environmentally compatible next generation green energetic materials (GEMs). J. Hazard. Mater. **161**, 589–607 (2009)
28. Ansbert, C., Pogrebnoi, A., Pogrebnaya, T.: High energy density materials based on fluorinated bridged trinitromethyl azo triazole derivatives: a quantum chemical study of thermodynamic and energetic properties. SN Appl. Sci. **2**, 1843 (2020)

Performance Analysis of Double Slope Single Basin Solar Still Using Various Absorbing Materials: An Experimental Study

Safeeruddin Khan[1] and Sharmas Vali Shaik[2](✉)

[1] Department of Mechanical Engineering, Aditya University, Surampalem 533437, Andhra Pradesh, India
[2] Mechanical Engineering Department, Srinivasa Ramanujan Institute of Technology (Autonomous), Ananthapuramu 515701, Andhra Pradesh, India
sharmasvali.nitk@gmail.com, drsharmasvalishaik.nitk@gmail.com

Abstract. The consumption of conventional energy sources like fossil fuels, fire wood and oil are rapidly depleting day by day and they are exhausting over a period of time. Therefore, the present work deals with conversion of brackish water into distilled water using one of the renewable energy resources called solar energy as heat input to the solar still. This work mainly investigates the ways to improve the productivity of water and still efficiency by using various materials as absorbents and also to minimize the total dissolved salts present in water. Materials used to improve the absorptivity of water for solar radiation are potassium permanganate ($KMnO_4$)and charcoal and these materials are added in various proportions to the water to analyze the productivity and still efficiency of the system. Experiments are conducted on solar still for water with different absorbing materials. The impact of using these absorbing materials on heat transfer through the still, productivity and efficiency is calculated. The total dissolved salts (TDS) present in water are found by TDS equipment and also the pH value of water is measured by using pH indicator papers. Experimental results revealed that the addition of charcoal resulted in 13.27% improvement in practical efficiency and 16.67% enhancement in practical productivity compared to baseline fluid water. Further, TDS presented in the water is reduced from 650 PPM to 55.5 PPM and pH value of water is 7.2.

Keywords: Solar still · Absorbing materials · productivity · efficiency of still · Total dissolved salts

1 Introduction

Many regions of the world are facing a severe shortage of clean drinking water. Producing fresh water has become a major challenge, especially in remote areas [1]. To address this, there is a growing need for affordable and efficient solutions. One promising method involves using non-conventional energy sources namely solar energy, and the solar Still device is a key technology being researched for this purpose. The solar still is commonly used in solar desalination, but its productivity is relatively low. Hence, several researchers made an attempt to increase productivity of still using various absorbing materials.

R. Rajasekharan et al. (Eds.): MatZone 2024, SPPHY 316, pp. 60–68, 2025.
https://doi.org/10.1007/978-981-96-6541-9_5

Researchers have investigated using natural materials like banana leaf powder, bamboo straw, sawdust, molasses and rice husk to enhance evaporation of seawater and condensation in solar stills [2]. The use of rice straw and sawdust led to a 62.88% increase in output compared to conventional solar stills. Suraparaju and co-worker (2021) reported that enhancement of 25.23% productivity was observed with luffa acutangula fibers [3]. The study by Sharshir et al. (2021) revealed that using several fibers and materials, including nanoscale carbon black particles, cotton fabrics and floating coal will enhance the performance in thermal of solar stills [4]. Ramzy et al. (2023) carried out experiments on two solar stills using several artificial and natural absorbing materials, including steel wool pads, fine steel wool, luffa and black luffa [5]. Results showed that steel wool pads significantly improved solar still productivity, yielding 4.384 L/m^2 with a thermal efficiency of 32.74%. Baskaran and Saravanane (2021) studied the impact of Spirulina algae on the productivity of solar stills [6]. They found that the inclusion of Spirulina algae increased the solar still's yield by 30.24% compared to conventional solar stills (CSS). Velmurugan and Sridhar (2011) reviewed both numerical and experimental studies on single-basin solar stills, highlighting that inclined solar stills are effective in increasing the evaporative heat transfer coefficient, thereby improving distillate output [7].

Panchal and Mohan (2017) presented that double-basin solar still with evacuated tubes is more productive than a standard double-basin solar still, as the evacuated tubes supply hot water to the lower basin, enhancing overall efficiency [8]. Elango et al. (2015) developed nanofluids with brine water as the base fluid to enhance the solar still's performance [9]. They used nanoparticles like Aluminium Oxide, Zinc Oxide, Fe_2O_3, and SnO_2, and discovered that the Aluminium Oxide nanofluid produced a yield rate 29.95% higher than the other nanoparticles. Akash et al. (1998) conducted experimental tests on a solar still consists of single basin, using various absorbing materials such as charcoal and rubber mats to enhance the absorption capacity of the basin still [10]. Madani and Zaki (1995) studied the performance of porous basin type solar stills. They found that when carbon powder (40–50 μm in size) was used as the basin material, the average yield ranged from 2.5 to 4 L per square meter per day [11]. Abdallah et al. (2009) found that maximum water collection was obtained for the uncoated sponge [12]. Panchel and Patel (2014) study reported that the charcoal particles serve as effective heat storing materials, resulting in a 15% more production compared to traditional solar stills [13]. Bhargva and Yadav (2021) revealed that for lower water depths in the basin, productivity was increased [14]. Ahmed et al. (2022) found that by employing single tracking parabolic trough heating for inclined and tubular solar technology can boost the production by 35.62% and generate the fresh water of 7.8 L/day [15].

The literature review clearly shows that the productivity and efficiency of solar stills are improved by using various absorbing materials. Hence, in this study an attempt is made to analyze the performance (productivity and efficiency) of double slope single basin solar still with charcoal, and potassium permanganate ($KMnO_4$) as heat absorbing materials.

2 Experimental Analysis of Still

The analysis of solar still involves the calculation of coefficients of convective heat transfer, radiative heat transfer, evaporative heat transfer and total heat transfer in the solar still. Moreover, productivity and still efficiency of the solar still are also computed. In this work, the pH value and TDS values of distilled water output are measured. The various correlations used are presented below [16, 17]. The physical dimensions of still are given in Table 1.

Table 1. Physical dimensions of still

Parameters	Values
Basin area (A_W)	0.63 m^2
Glass cover area (A_g)	0.77 m^2
Glass coven inclination	30°
Depth of basin	10 cm

Hourly production (yield) of still is given by

$$m_w = \frac{q_{ew} \times A_w}{L} \times 3600 \tag{1}$$

A_w = Area of water

$$q_{ew} = h_{ew} \times (T_w - T_g) \tag{2}$$

$$h_{ew} = 16.27 \times 10^{(-3)} \times h_{cw} \times \frac{(P_w - P_g)}{(T_w - T_g)} \tag{3}$$

$$h_{cw} = 0.884\left[T_w - T_g + \frac{(P_w - P_g)(T_w + 273)}{268.9 \times 10^3 - P_w}\right]^{(1/3)} \tag{4}$$

$$P_w = \exp\left(25.317 - \frac{5144}{T_w + 273}\right) \tag{5}$$

$$P_g = \exp\left(25.317 - \frac{5144}{T_g + 273}\right) \tag{6}$$

The correlation for latent heat of water evaporation (L) is given by

$$L = 3044205.5 - 1679.1109T_w - 1.14258{T_w}^2 \tag{7}$$

Theoretical efficiency of solar still is given by

$$\eta_{the} = \frac{q_{ew}}{I(t)} \tag{8}$$

$$I(t) = I_{sc}\left[1 + 0.333\cos\frac{360(n-2)}{365}\right] \tag{9}$$

$I_{sc} = 1353$ watt per square meter.

Practical efficiency of still is calculated as

$$\eta_{pra} = \frac{m_w \times L}{I(t) \times A_g \times \Delta t} \times 100 \tag{10}$$

Total heat transfer coefficient from water surface to condensing cover is given by

$$h_{total} = h_{cw} + h_{ew} + h_{rw} \tag{11}$$

$$h_{rw} = e_{effect} \times S[[(T_w + 273)^2 + (T_g + 273)^2](T_w + T_g + 546)] \tag{12}$$

$$\begin{aligned} e_{effect} &= \left(\tfrac{1}{e_g} + \tfrac{1}{e_w} - 1\right)^{(-1)} \\ e_g &= e_w = 0.9 \\ S &= 5.669 \times 10^{-8} W/m^2K^4 \end{aligned} \tag{13}$$

Experimental Observations for water and various considered absorbing materials are presented in Table 2.

Table 2. Mean temperatures of water basin and glass cover for every 5 days of each considered absorbing material

Material	T_W (°C)	T_g (°C)
Water	62	51.8
$KMnO_4$	53	44.54
Charcoal	60.4	48.2

3 Results and Discussion

3.1 Effect of Absorbing Materials on Theoretical Productivity of Still

Figure 1 shows the theoretical productivity of still using different absorbing material. From Fig. 1, it is observed that maximum theoretical productivity of still is obtained for charcoal (2.25 L/day) and it is 60.71% higher than that of water (1.4 L/day). Similarly, productivity of $KMnO_4$ is 50% higher than that of water.

3.2 Effect of Absorbing Materials on Practical Productivity of Still

Figure 2 shows the practical productivity of still using different absorbing material. From Fig. 2, it is observed that maximum enhancement in practical productivity of still is obtained for charcoal (1.40 L/day) and it is 16.67% higher than that of water (1.2 L/day). Similarly, productivity of $KMnO_4$ is 4.16% higher than that of water.

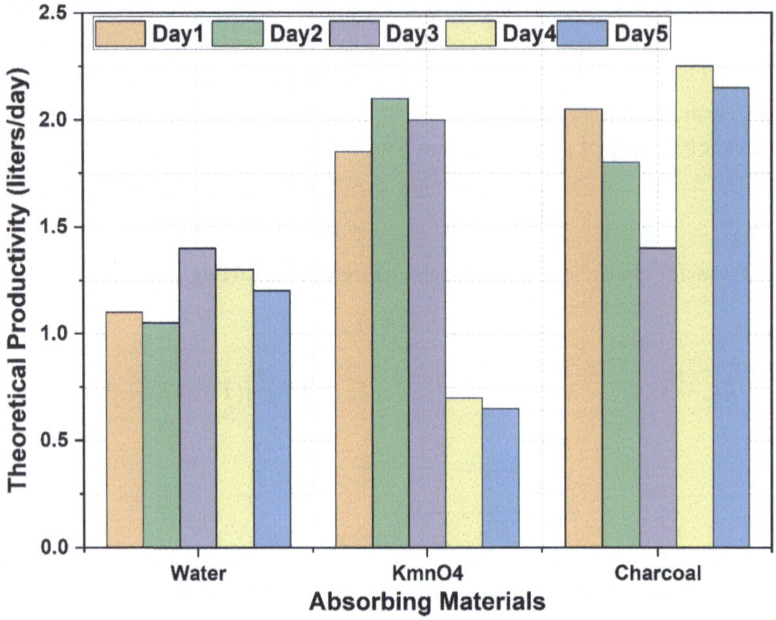

Fig. 1. Theoretical productivity of still using $KMnO_4$ and charcoal material

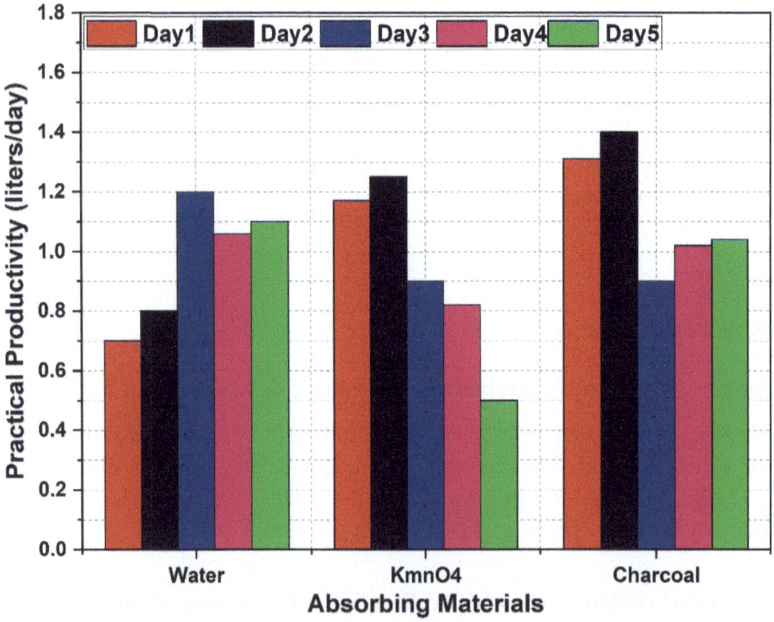

Fig. 2. Practical productivity of still using $KMnO_4$ and charcoal material

3.3 Effect of Absorbing Materials on Theoretical Efficiency of Still

Figure 3 shows the theoretical efficiency of still using different absorbing material. From Fig. 3, it is observed that maximum enhancement in theoretical efficiency of still is obtained for charcoal (26.87%) and it is 12.71% higher than that of water, whereas efficiency of $KMnO_4$ is 16.73% lower than that of water.

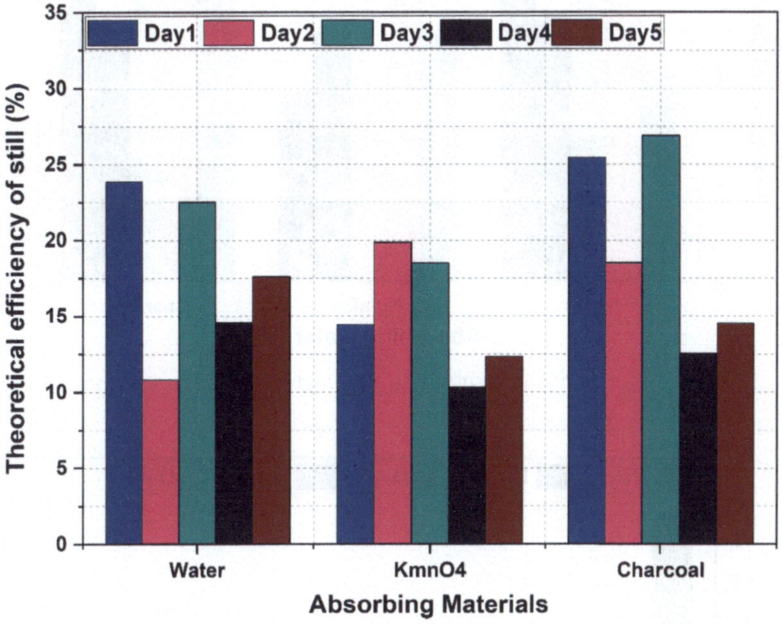

Fig. 3. Theoretical efficiency of still using $KMnO_4$ and charcoal material

3.4 Effect of Absorbing Materials on Practical Efficiency of Still

Figure 4 shows the practical efficiency of still using different absorbing material. From Fig. 4, it is observed that maximum enhancement in practical efficiency of still is obtained for charcoal (22.10%) and it is 13.27% higher than that of water, whereas efficiency of $KMnO_4$ is 23.57% lower than that of water.

3.5 Effect of Absorbing Materials on TDS of Distilled Water

Figure 5 shows the total dissolved salts (TDS) of distilled water using different absorbing material. From the Fig. 5, it is observed that TDS of all the investigated materials as well as base fluid water is reduced from 650 PPM to 6 PPM which is fair the usage of distilled water.

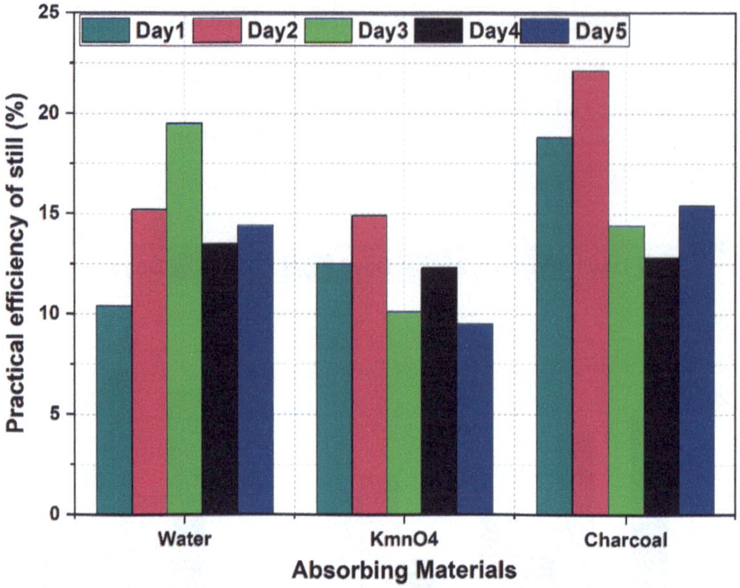

Fig. 4. Practical efficiency of still using $KMnO_4$ and charcoal material

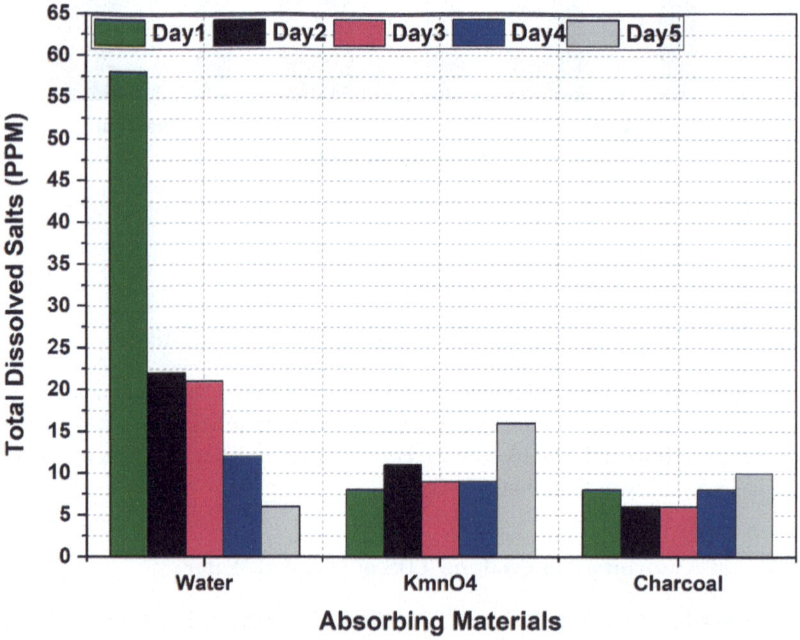

Fig. 5. Total dissolved salts of distilled water using $KMnO_4$ and charcoal material

3.6 Effect of Absorbing Materials on pH Value of Distilled Water

Figure 6 shows the pH values of distilled water using different absorbing material. From Fig. 6, it is observed that pH values of charcoal obtained is 7.5 and 7.2 on day 1 and day 3, respectively. These pH values of distilled water obtained from charcoal is within permissible limits of drinking water purpose. On the other side, pH values of $KMnO_4$ and base fluid water are ranging from 5.5 to 7.2.

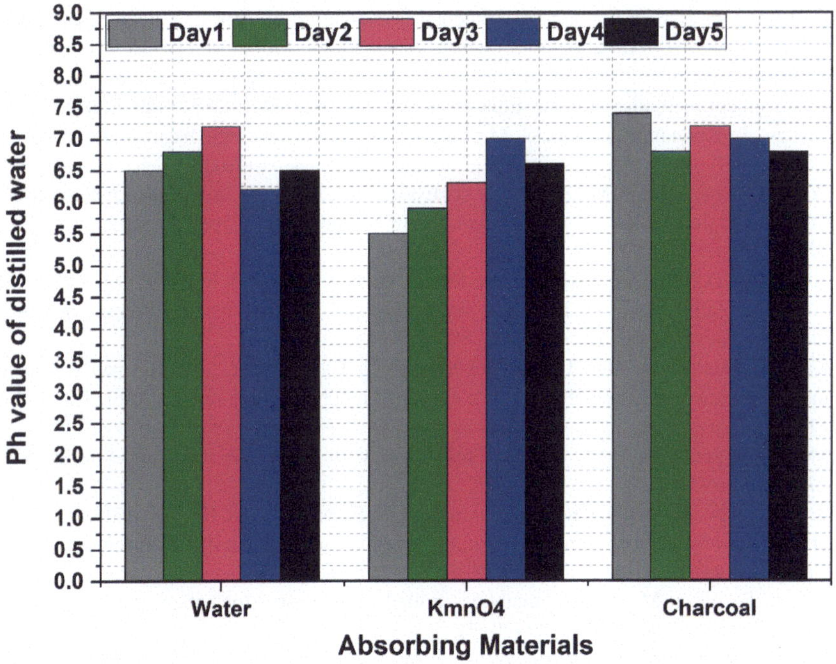

Fig. 6. pH values of distilled water using $KMnO_4$ and charcoal material

4 Conclusions

Conclusions drawn from the present study is given below.

- Theoretical and practical efficiency of double slope solar still obtained for charcoal is the highest among all the investigated materials. Exclusively, practical efficiency of charcoal was 13.27% higher than that of water.
- Practical productivity for charcoal is the highest among all the materials studied and it was 16.67% higher than that of water.
- TDS of all the investigated materials as well as base fluid water is reduced from 650 PPM to 6 PPM. pH values of distilled water for charcoal was ranging from 6.1 to 7.5.

References

1. Abdallah, S., Mazen, M.A., Omar, B.: Effect of various absorbing materials on the thermal performance of solar stills. Desalination. **242**, 128–137 (2009)
2. Mohamed, M.Z.A., Fuhaid, A., Ibrahim, A., Muapper, A., Mohamed, E.: A novel solar desalination system integrated inclined and tubular solar still with parabolic concentrator. Appl. Therm. Eng. **213**, 118665 (2022)
3. Bilal, A.A., Mousa, S.M., Omar, O., Yaser, E.: Experimental evaluation of a single basin solar still using different absorbing materials. Renew. Energy. **14**, 307–310 (1998)
4. Baskaran, V., Saravanane, R.: Experimental investigation on spirulina algae based thermal still for effective bio-desalination. Environ. Nanotechnol. Monit. Manag. **15**, 100434 (2021)
5. Mohit, B., Avadhesh, Y.: Factors affecting the performance of a solar still and productivity enhancement methods: a review. Environ. Sci. Pollut. Res. **28**, 54383–54402 (2021)
6. Elango, T., Kannan, A., Kalidasa, M.: Performance study on single basin single slope solar still with different water nanofluids. Desalination. **360**, 45–51 (2015)
7. Wissam, H.A., Suha, A.M., Hayder, A.D., Essac, F.A., Abdullah, A.S.: Performance analysis of a double-slope solar still with elevated basin comprehensive study. Desalin. Water Treat. **223**, 12–35 (2021)
8. Ali O.A.S., Amimul, A.,Basim A.R.A.,Mahir, M.H., Nik, N.N.D.,Idrus, S., Omer A.A., Elzbieta M., 'Zaher M.Y., 2022. Double-slope solar still productivity based on the number of rubber scraper motions. Energies,15, 7881
9. Kabeel, A.E., Arunkumar, T., Denkenberger, D.C., Ravishankar, S.: Performance enhancement of solar still through efficient heat exchange mechanism—a review. Appl. Therm. Eng. **114**, 815–836 (2017)
10. Madani, A.A., Zaki, G.M.: Yield of solar stills with porous basins. Appl. Energy. **52**(3), 273–281 (1995)
11. Sendhil, K.N., Subbarama, K.S., Rajvikram, M.E., Rishi, P., Eklas, H.: An experimental study on eco-friendly and cost-effective natural materials for productivity enhancement of single slope solar still. Environ. Sci. Pollut. Res. **29**, 1917–1936 (2022)
12. Hitesh, P., Indra, M.: Various methods applied to solar still for enhancement of distillate output. Desalination. **415**, 76–89 (2017)
13. Hitesh, N.P., Prathik, P.: Energy absorbing materials used in solar still for enhancement in distillate output: a review. Int. J. Ambient Energy. **37**, 528–540 (2014)
14. Khaled, R., Mohamed, A., Abd, E.K., Heba, M.: Performance of a single slope solar still using different porous absorbing materials: an experimental approach. Environ. Sci. Pollut. Res. **30**, 72398–72414 (2023)
15. Swellam, W.S., Ismail, M., Kandeal, A.W., Faisal, B.B., Ayman, E., Younes, M.M.: Improving thermal, economic, and environmental performance of solar still using floating coal, cotton fabric, and carbon black nanoparticles. Sustain Energy Technol Assess. **48**, 101563 (2021)
16. Subbarama, K.S., Sendhil, K.N.: Augmentation of freshwater productivity in single slope solar still using Luffa acutangula fibres. Water Sci. Technol. **84**, 2944–2957 (2021)
17. Velmurugan, V., Srithar, K.: Performance analysis of solar stills based on various factors affecting the productivity—a review. Renew. Sustain. Energy Rev. **15**, 1294–1304 (2011)

Myo-Electric Prosthetic Limb Fabrication Using Kenaf, Nickel and Polyester Based Composites

Aju Jo Sankarathil[1(✉)], Melvin Santhosh Thomas[1], K. A. Muhammed Easa[1], V. Neeraj Kumar[1], Paul J. Illickan[1], and Aswathy Ann Mathew[2]

[1] Department of Mechanical Engineering, Saintgits College of Engineering (Autonomous), Pathamuttom 686532, Kerala, India
aju.js@saintgits.org

[2] Department of Civil Engineering, National Institute of Technology, Calicut 673601, Kerala, India

Abstract. This research focuses on developing an affordable myo-electric prosthetic limb using Kenaf-Nickel Polyester composite, a material selected for its cost-effectiveness, tensile strength, and biodegradability. Myo-electric prosthetics, which convert muscle signals into movement, are often prohibitively expensive and lack essential sensory feedback, limiting accessibility for many users. The proposed design incorporates CAD modeling, finite element analysis, and additive manufacturing for precision and efficiency, while programming through Arduino IDE ensures system reliability and responsiveness. Preliminary testing indicates that the composite provides an excellent strength-to-weight ratio, significant cost savings, and improved durability compared to alternatives like Jute- Titanium-Epoxy and Flax-Carbon Fiber Epoxy. Early prototypes demonstrate promising functionality, including reliable movement and sensory feedback, though further refinement is required to enhance motion precision and user experience. This innovative approach highlights the potential for eco-friendly, accessible prosthetic solutions to meet the needs of a broader population.

Keywords: Prosthetic limb · Myo-electric Sensors · Kenaf-nickel Polyester

1 Introduction

The evolution of prosthetic limbs has made remarkable strides in recent years, aiming to restore functionality and enhance the quality of life for individuals with upper-limb amputations. Among these, myoelectric prosthetic limbs have emerged as an innovative solution, leveraging electrical signals from muscle contractions to deliver precise control and mimic natural limb movements. These devices utilize sensors placed on the residual limb to detect muscle-generated signals, which are processed by a microcontroller to enable specific hand movements such as gripping, finger manipulation, and wrist rotation [1]. However, despite these advancements, the high manufacturing costs of myoelectric prosthetics remain a significant barrier to accessibility, particularly in low- income regions. Additionally, many of these devices lack sensory feedback, an

R. Rajasekharan et al. (Eds.): MatZone 2024, SPPHY 316, pp. 69–78, 2025.
https://doi.org/10.1007/978-981-96-6541-9_6

essential feature for users to perceive limb strength and position effectively. Addressing these challenges is critical to making such devices more accessible and effective for a larger population.

The demand for affordable and functional prosthetic solutions is particularly pronounced in developing countries, where rates of limb loss due to trauma, illness, or congenital conditions are high [2]. Traditional prosthetics, whether mechanical or electrically powered, often fail to provide the intuitive control and functionality that myoelectric devices offer [3].This research seeks to bridge this gap by developing an affordable, functional myoelectric prosthetic limb utilizing a novel natural composite material, Kenaf-Nickel Polyester. This material promises to significantly reduce costs while ensuring the durability and strength necessary for daily use. By integrating sensory feedback capabilities, the research aims to create an effective and economical prosthetic limb tailored for users in resource- limited settings [4].

2 Design Approach

The research focuses on designing a prosthetic limb using Kenaf-Nickel Polyester, a sustainable and cost-effective natural composite material [5]. Kenaf, a renewable and biodegradable fiber, offers a lightweight and eco-friendly alternative to traditional materials, while the nickel component enhances tensile strength. Polyester resin serves as the binding matrix, providing durability and ease of processing. This unique combination results in a composite material capable of withstanding the physical demands of everyday use without compromising affordability.

The prosthetic system incorporates myoelectric sensors to detect electromyographic (EMG) signals from muscle contractions, which are then processed by a microcontroller to activate actuators for precise hand and wrist movements [1, 6]. DC and servo motors facilitate these movements, while a power source and vibration motors are integrated to enhance user feedback. The design process leverages advanced CAD software such as SolidWorks and CATIA for prototyping, while Finite Element Analysis (FEA) conducted in Ansys ensures material strength and durability under simulated conditions. Additive manufacturing techniques are employed to fabricate the prosthetic components [2] with precision, enabling seamless integration of the Kenaf-Nickel Polyester composite. Figure 1 demonstrates the effectiveness of the finely meshed model, with a 0.1 mm quad mesh ensuring precise stress distribution and accurate identification of localized stress concentrations.

3 Material Analysis

3.1 Characterization of the Material

The development of a myoelectric prosthetic limb necessitated the selection of a material that offers an optimal balance of mechanical strength, weight, cost, and sustainability [7]. Kenaf-Nickel Polyester composite was identified as the most suitable material due to its unique combination of properties. The composite consists of 30% kenaf fiber, 31% nickel powder, and 39% polyester resin.

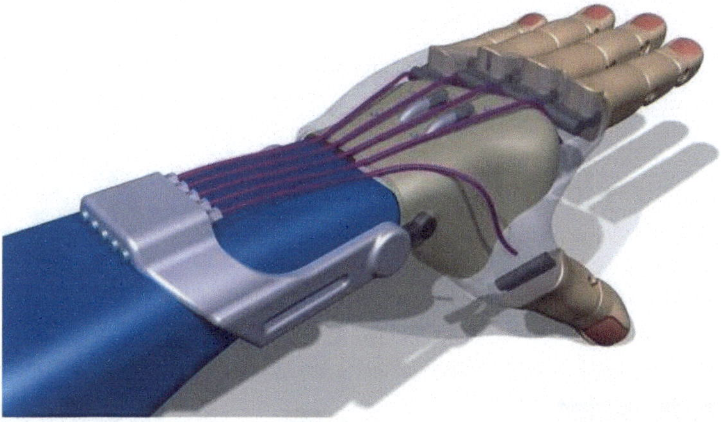

Fig. 1. 3-D Model Design of a Prosthetic Hand

Kenaf, a renewable and biodegradable natural fiber, was chosen for its lightweight nature, tensile strength, and environmental sustainability. To enhance the compatibility of kenaf fibers with the resin matrix, they underwent an alkali treatment using a 5% NaOH solution [8]. This treatment improved the surface roughness of the fibers, promoting better adhesion to the resin and thereby enhancing the overall mechanical properties of the composite. Nickel powder was incorporated to increase tensile strength and provide corrosion resistance, critical for ensuring durability in real-world conditions. Polyester resin served as the binding matrix, offering cost-effectiveness, ease of processing, and resistance to wear and tear. The combination of these materials resulted in a composite that addresses the dual needs of high mechanical performance and affordability, making it an ideal candidate for prosthetic applications, especially in resource-constrained environments [9].

3.2 Fabrication of the Composite

The fabrication of the composite was carried out using the Vacuum-Assisted Resin Transfer Molding (VARTM) technique, a widely used method for producing high-quality composites with excellent fiber-resin integration. In this process, the treated kenaf fibers and nickel powder were layered in a mold, and the polyester resin, mixed with a hardener in a 10:1 ratio, was infused into the mold under vacuum pressure (Table 1).

The vacuum ensured uniform resin distribution, eliminated air pockets, and maximized bonding strength throughout the composite. Once the resin was infused, the mold was sealed and subjected to controlled curing conditions to solidify the composite. The cured composite was then removed and machined into standard test specimens according to ASTM guidelines using water jet cutting. This method ensured precision in sample dimensions, which is critical for consistent and reliable testing results. The VARTM process not only ensured the structural integrity of the composite but also made it possible to produce lightweight and durable components efficiently, aligning with the project's goal of cost-effective manufacturing.

Table 1. Materials used for composite material selection

Fiber	Metallic Powder	Resin
Bamboo 	Titanium 	Vinyl Ester 
Jute 	Titanium 	Epoxy 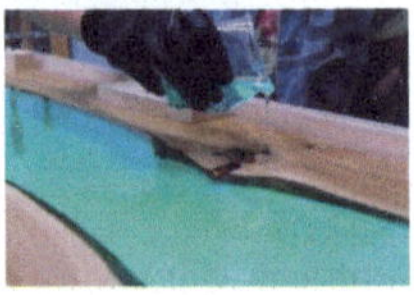
Flax 	Carbon Fiber 	Toughened Epoxy Resin 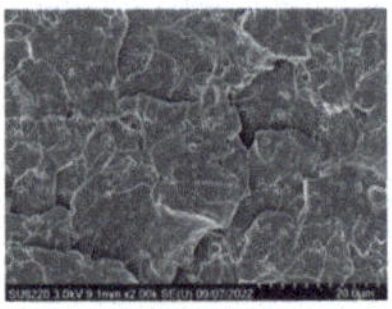
Kenaf 	Nickel 	Polyester 

4 Results and Discussion

4.1 Finite Element Analysis (FEA)

Finite Element Analysis (FEA) was employed using Ansys software to simulate and evaluate the mechanical performance of several composite materials under varying conditions. Given that the prosthetic would be subject to daily physical stresses, FEA was crucial for assessing materials for factors like stress distribution, strain, and deformation. This analysis helped to ensure that the prosthetic's chosen material could withstand the forces experienced during regular use without excessive wear or failure.

The research considered multiple natural and synthetic composite materials, including Kenaf-Nickel Polyester, Bamboo-Titanium Vinyl Ester, Jute-Titanium Epoxy, and

Flax-Carbon Fiber Toughened Epoxy [10]. Each material was tested in simulated 1 cm^3 samples for its density, tensile strength, stiffness, and durability. These simulations allowed the team to predict the performance of each material without needing to conduct extensive physical testing on full-scale prototypes, saving both time and resources. Kenaf-Nickel Polyester was selected as the primary material due to its high tensile strength (387.25 MPa), favourable density, and cost-effectiveness. The material's natural fibers offered weight savings and sustainability, while Nickel contributed additional tensile strength and corrosion resistance, meeting the project's dual goals of performance and affordability.

Selecting an appropriate material for the prosthetic limb involved rigorous testing of various natural composites to achieve an optimal balance between weight, strength, and sustainability [11]. Natural fibers were prioritized due to their eco-friendliness, renewability, and cost advantages over conventional materials like metals and synthetic fibers [12]. The material ultimately selected, Kenaf-Nickel Polyester, was chosen after extensive testing for its density, tensile strength, and overall durability.

The testing phase will involve evaluating the material for properties essential to daily prosthetic use, including stiffness, impact resistance, and corrosion resistance [13]. Kenaf, a lightweight natural fiber, will contribute significantly to weight reduction, while Nickel will enhance tensile strength and durability, allowing the prosthetic to withstand the mechanical demands of daily composite material, the team aims to develop a material that is lightweight yet strong enough to meet the user's needs, ensuring that the prosthetic will be both functional and accessible

Figure 2 represents the use of a 0.1 mm quad mesh for structural analysis of a 1 cm^2 component. This fine mesh enhances stress distribution accuracy, effectively identifies localized stress concentrations, and ensures smooth gradient transitions for more realistic results. It minimizes element distortion under large deformations, handles stress variations precisely, and promotes efficient numerical solution convergence. Despite higher computational demands, this approach ensures consistent modelling of structural behaviours, making it ideal for detailed and reliable analysis.

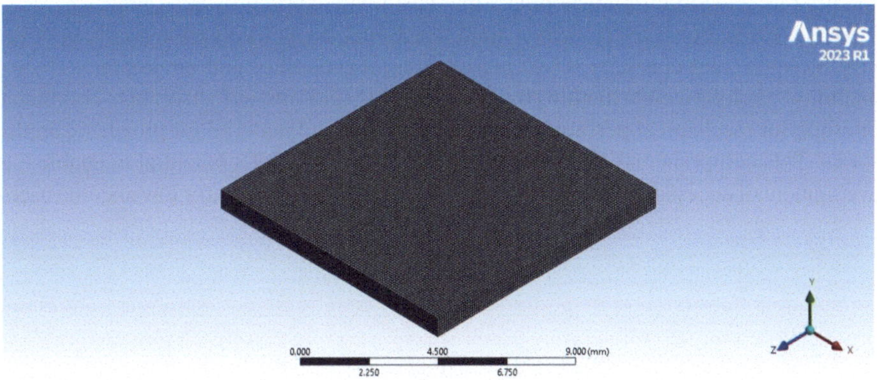

Fig. 2. Mesh structure

Figure 3a shows the maximum principal stress distribution in the Kenaf-Nickel Polyester composite from FEA analysis. The stress remains below the material's tensile strength of 387.25 MPa, ensuring structural integrity under operational loads. This confirms the composite's suitability for prosthetic applications, with stress concentrated in specific regions requiring optimization.

Figure 3b depicts the total deformation of the Kenaf-Nickel Polyester composite under applied loads as determined by FEA. The deformation remains minimal and within acceptable limits, confirming the composite's ability to maintain structural stability during use. This result highlights its suitability for prosthetic applications requiring durability and reliability.Figure 3c illustrates the maximum shear elastic strain distribution in the Kenaf-Nickel Polyester composite from FEA analysis. The strain values indicate the material's capacity to withstand shear forces without permanent deformation. The results confirm the composite's elasticity and durability, essential for the dynamic stresses encountered in prosthetic applications.

Figure 3d depicts the equivalent total strain distribution in the Kenaf-Nickel Polyester composite as obtained through FEA analysis. The strain values are well within the material's elastic limits, indicating its ability to endure operational stresses without permanent deformation. This confirms the composite's reliability and suitability for prosthetic applications.

Figure 3e shows the strain energy distribution in the Kenaf-Nickel Polyester composite from FEA analysis. The strain energy indicates the material's capacity to absorb and store energy under applied loads. The results confirm that the composite can efficiently handle mechanical stresses, making it ideal for durable prosthetic applications.

Figure 4 illustrates the safety factor distribution in the Kenaf-Nickel Polyester composite as determined by FEA analysis. The safety factor values exceed the required thresholds, indicating that the composite can withstand operational loads with a significant margin of safety. This ensures reliability and durability in prosthetic applications.

The analysis of the Kenaf-Nickel Polyester composite using Finite Element Analysis (FEA) highlights its superior performance across critical mechanical parameters, validating its selection for developing a low-cost myoelectric prosthetic limb.

These results collectively demonstrate that Kenaf-Nickel Polyester excels in durability, strength, and elasticity while maintaining affordability and environmental sustainability. Its lightweight nature, biodegradability, and cost-effectiveness, combined with superior mechanical performance, make it an optimal material for prosthetic applications. This comprehensive analysis underscores the material's potential to enable the development of accessible, robust, and sustainable prosthetic limbs for low-income users

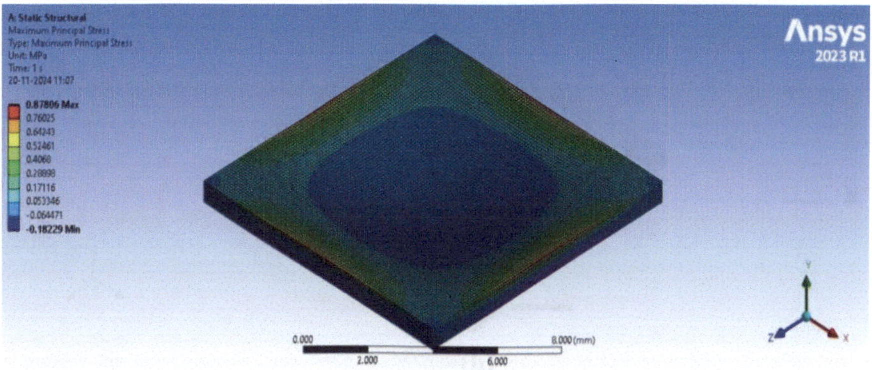

(a)

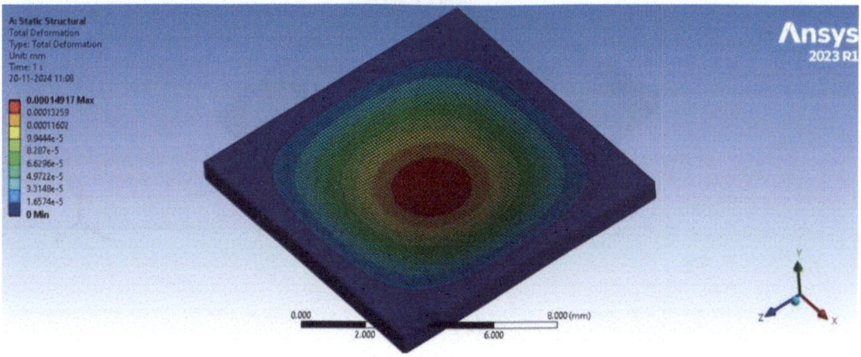

(b)

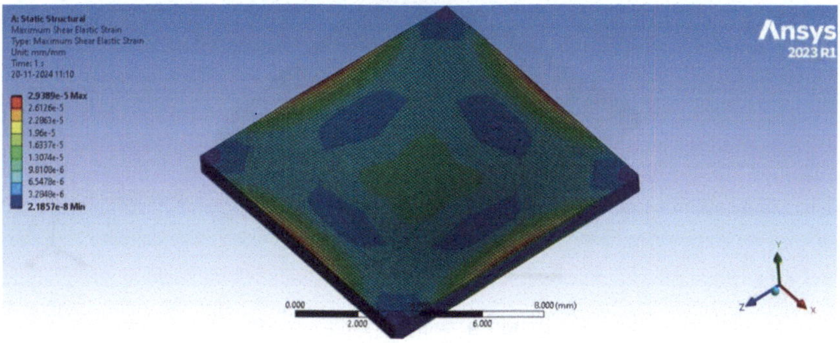

(c)

Fig. 3. (**a**) Maximum principal stress. (**b**) Total deformation. (**c**) Maximum shear elastic strain. (**d**) Equivalent total strain. (**e**) Strain energy

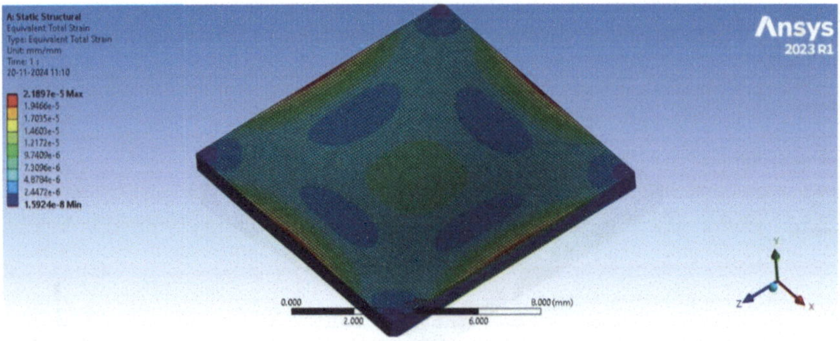

(d)

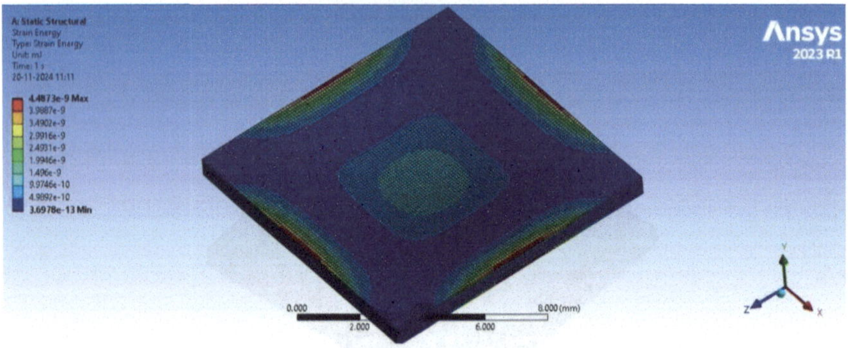

(e)

Fig. 3. (*continued*)

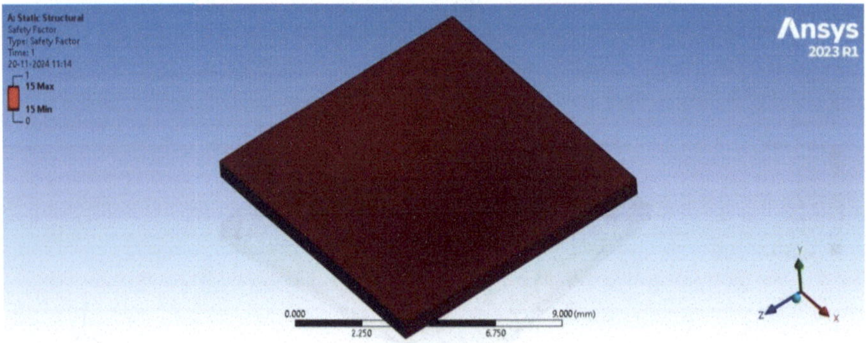

Fig. 4. Safety factor

5 Conclusion

The research into developing a low-cost myoelectric prosthetic limb using the novel natural composite material Kenaf-Nickel Polyester highlights the potential for creating affordable and accessible prosthetics, particularly for low-income regions. Through comprehensive testing and analysis, this study establishes Kenaf-Nickel Polyester as an ideal composite material, offering an optimal balance of durability, strength, lightweight properties, and environmental sustainability. Combined with additive manufacturing techniques, the composite enables the production of a functional and lightweight prototype, addressing key limitations of traditional prosthetics such as high costs and limited adaptability.

The prosthetic limb's design integrates advanced myoelectric control systems, allowing intuitive operation through muscle-generated electrical signals [14]. Components such as microcontrollers, actuators, and sensors work in harmony to provide users with improved control and a more natural range of motion. The modular design also accommodates future enhancements, including sensory feedback mechanisms, which could significantly enhance the user experience by providing a more responsive and immersive interface.

This research represents a significant step towards making high-quality prosthetics both accessible and sustainable. By combining innovative composite materials with cost- effective manufacturing and user-centered design, this model offers a blueprint for future advancements in assistive technology. Further studies will focus on optimizing material properties, refining control algorithms, and conducting extensive real-world testing to enhance functionality and user satisfaction. These findings underscore the transformative potential of integrating sustainable materials and affordable engineering solutions to improve the lives of amputees worldwide, paving the way for inclusive healthcare innovations [15].

References

1. Karabulut, D., Ortes, F., Arslan, Y.Z.: Comparative evaluation of EMG signal features for myoelectric controlled human arm prosthetics. Biocybernet. Biomed. Eng. **37**, 326 (2017). https://doi.org/10.1016/j.bbe.2017.03.001
2. Sayar, H., Karajagikar, J., Deshmukh, S.: Prosthetic limb arm for armless human. Mater. Today Proc. **72**(part 3), 1056 (2023). https://doi.org/10.1016/j.matpr.2022.09.162
3. McDonald, C.L., Westcott-McCoy, S., Weaver, M.R., Haagsma, J., Kartin, D.: Global prevalence of traumatic non-fatal limb amputation. Prosthet. Orthot. Int. **45**, 105 (2021). https://doi.org/10.1177/0309364620972258
4. Nsugbe, E., Phillips, C., Fraser, M., McIntosh, J.: Gesture recognition for transhumeral prosthesis control using EMG and NIR. IET Cyber. Syst. Robot. (2020). https://doi.org/10.1049/iet-csr.2020.0008
5. Biddiss, E., Beaton, D., Chau, T.: Consumer design priorities for upper limb prosthetics. Disabil. Rehabil. Assist. Technol. **2**, 346 (2007). https://doi.org/10.1080/17483100701714733
6. Hurtado, E.F.A., Varon, L.A.B., Sarria-Paja, M.O.: Movements classification system for transhumeral amputees using myoelectric signals. Biomed. Signal Process. Contr. **98**, 106776 (2024). https://doi.org/10.1016/j.bspc.2024.106776

7. Scholz, M.S., Blanchfield, J.P., Bloom, L.D., Coburn, B.H., Elkington, M., Fuller, J.D.: The use of composite materials in modern orthopedic medicine and prosthetic devices: a review. Compos. Sci. Technol. **71**, 1791 (2011). https://doi.org/10.1016/j.compscitech.2011.08.017
8. Shahar, F.S., Sultan, M.T.H., Jawaid, M.: A review on the orthotics and prosthetics and the potential of kenaf composites as alternative materials for ankle-foot orthosis. J. Mech. Behav. Biomed. Mater. **99**, 169 (2019). https://doi.org/10.1016/j.jmbbm.2019.07.020
9. Jenkins, G.M., de Carvalho, F.X.: Biomedical applications of carbon fibre reinforced carbon in implanted prostheses. Carbon. **15**, 33 (1977). https://doi.org/10.1016/0008-6223(77)90071-9
10. Rajak, D.K., Pagar, D.D., Menezes, P.L., Linul, E.: Fiber-reinforced polymer composites: manufacturing, properties, and applications. Polymers. **11**, 1667 (2019). https://doi.org/10.3390/polym11101667
11. Bougherara, H., Zdero, R., Dubov, A., Shah, S., Khurshid, S., Schemitsch, E.H.: A preliminary biomechanical study of a novel carbon–fibre hip implant versus standard metallic hip implants. Med. Eng. Phys. **33**, 121 (2011). https://doi.org/10.1016/j.medengphy.2010.09.011
12. Khushaba, R.N., Kodagoda, S., Takruri, M., Dissanayake, G.: Toward improved control of prosthetic fingers using surface electromyogram (EMG) signals. Expert Syst. Appl. **39**, 10731 (2012). https://doi.org/10.1016/j.eswa.2012.02.192
13. Nolan, L.: Carbon fibre prostheses and running in amputees: a review. Foot Ankle Surg. **14**, 125 (2008). https://doi.org/10.1016/j.fas.2008.05.007
14. Parker, P., Englehart, K., Hudgins, B.: Myoelectric signal processing for control of powered limb prostheses. J. Electromyogr. Kinesiol. **16**, 541 (2006). https://doi.org/10.1016/j.jelekin.2006.08.006
15. Pooja, S., Grewal, V.S., Agrawal, S., Nair, S.V.: A study on quality of life among lower limb amputees at a tertiary prosthetic rehabilitation center. Med. J. Armed Forces India. **76**(1), 89 (2020)

Aluminium 2 Series Rutile Composite Materials: Fabrication, Characterization and Performance Enhancement - A Review

Akhil S. Prasad[1,2](✉) and P. Balamurugan[1]

[1] Department of Mechanical Engineering, School of Mechanical, Aero, Auto and Civil Engineering, Kalasalingam Academy of Research and Education, Krishnankoil, Tamil Nadu, India

akhil.p@providence.edu.in

[2] Department of Mechanical Engineering, Providence College of Engineering, Chengannur, Kerala, India

Abstract. Composite materials have high strength to weight ratio, corrosion, resistance, fatigue and durability, due to that they are widely used in aerospace applications. Aramid FiberReinforced Composites, Carbon Fiber Reinforced Polymers, Glass Fiber Reinforced Polymers, CeramicMatrix Composites and MetalMatrix Composites, by using of above composite materials allows aerospace manufacturers to create lighter, stronger, and more fuel-efficient components, all while enhancing the performance and safety of aircraft and spacecraft. Aluminium-copper alloy, is highly regarded in aerospace field applications. It is due to the cryogenic stability, corrosion resistance and exceptional strength. The aim of this research work is to improve the thermal and mechanical properties of AA2219 by incorporating rutile, a material known for its hardness and thermal stability, as a reinforcing phase.This review explores the development and characterization of an advanced aluminium matrix composite (AMC) consisting of AA2219 aluminium alloy reinforced with rutile (TiO_2) particles.

Keywords: Composite · Aluminum matrix composite (AMC) · AA2219 alloy · Rutile (TiO_2)

1 Introduction

The choice of aerospace materials is indispensable in the design process of components and systems of aerospace. It influences various aircraft performance aspects throughout its lifecycle, from design to disposal, including flight performance, structural efficiency, energy consumption, payload capacity, lifecycle cost, recyclability, reliability, and safety [1]. Aerospace structural materials must meet critical requirements including physical, chemical and mechanical properties like high strength, stiffness, fatigue resistance and damage tolerance. They should also have high thermal stability, low density, strong corrosion and oxidation resistance. Additionally, factors such as ease of servicing, cost and manufacturability are essential for commercial viability [2]. Aluminium (Al) alloys,

R. Rajasekharan et al. (Eds.): MatZone 2024, SPPHY 316, pp. 79–86, 2025.
https://doi.org/10.1007/978-981-96-6541-9_7

steels, titanium (Ti) alloys, and composites are the common materials used in this field. Ti alloys are highly valued in aerospace due to their exceptional high-temperature performance and high specific strength, which can even rival that of steels. However, when factoring in comprehensive understanding and manufacturing costs of material properties, still Al alloys offer greater promise for widespread application [3]. Energy consumption can be reduced by light-weighting and enhances performance in aerospace design through structural optimization and advanced materials. While metals like aluminum alloys have been dominant, composites such as CFRPs offer higher strength and stiffness but are limited by cost and manufacturability. Structural optimization techniques, including lattice optimization, help improve weight reduction and performance but require efficient algorithms for large-variable problems. Advanced manufacturing methods, like additive manufacturing, enable complex designs but face challenges in cost, time, and standardization [4]. The structures of large sealed compartments of manned spacecraft, rocket fuel tanks, space vehicle fuselages and spacecraft, aluminium alloys are commonly used. Due to their low density, ease of maintenance, excellent formability and machinability, high specific strength and cost-effectiveness [5]. Aerospace aluminium (Al) alloys are being developed to be lightweight while improving strength, plasticity, corrosion, and fatigue resistance. Key efforts include optimizing alloy compositions, heat treatments, and exploring particle-reinforced composites. Advanced techniques in manufacturing, such as superplastic forming and additive manufacturing, are also being prioritized to enhance performance and production efficiency [6]. Advanced aluminium alloys for aerospace applications are designed to provide enhanced material toughness, excellent formability, super plasticity and enhanced fatigue performance, in order to meet the demands for increased durability, reduced structural weight and improved damage tolerance [7].

2 Aerospace Materials

Titanium alloys offer several superiority over other metals, including excellent fracture toughness, high specific strength and stiffness, and fatigue resistance, as well as outstanding heat, corrosion and cryogenic embrittlement resistance. Additionally, they feature low thermal expansion [2]. The limited manufacturability and high cost of titanium alloys, which are typically about eight times more expensive than commercial aluminum alloys, restrict their widespread use. Titanium alloys are being used at present primarily in aerospace for engine components and airframe, forming 7% and 36% of the weight, respectively [8]. Utilization of high-strength steel in aerospace for components that require high load-bearing capacity, such as landing gear and fasteners, due to its resistance to wear, toughness, superior strength and fatigue. However, its limitations include being heavier than materials like titanium and aluminium, which can negatively impact fuel efficiency and overall performance. Additionally, its susceptibility to corrosion and the challenges of processing and manufacturing can limit its widespread use in certain aerospace applications [1, 2]. In exploring the compositions and properties of composite and advanced materials in aerospace design, the review highlights their thermodynamic advantages, including reduced weight and enhanced performance. It stresses the importance of continued research into innovative composite materials that can supplement traditional aluminium alloys in future aerospace applications [10] (Fig. 1).

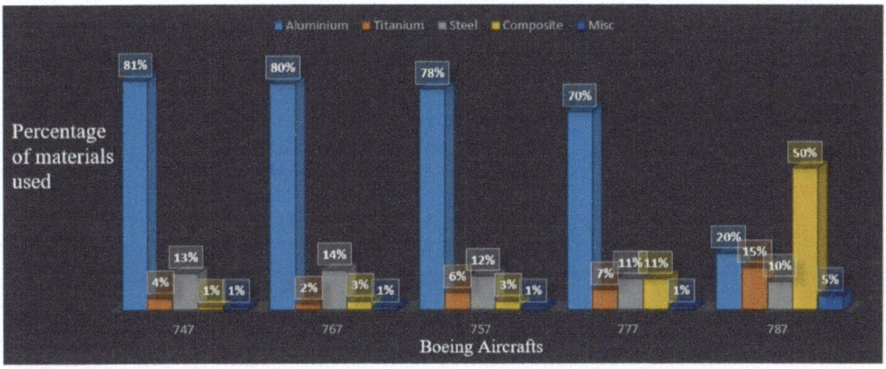

Fig. 1. Materials used in Boeing Aircrafts [9]

2.1 Aluminium Alloys

For aerospace structural applications, advanced aluminum alloys are widely chosen, due to their reliability, excellent manufacturability, corrosion resistance, good ductility, low cost, and high specific strength and stiffness. The advancement of heat treatment methods has led to high-strength aluminum alloys that hold their own against advanced composites in several applications in aviation and space. By adjusting compositions and heat treatments, Aluminum alloys can achieve a variety of material properties. The most commonly used alloys in the aerospace industry are Al-Li alloys, Al-Copper (2000) and Al-Zn (7000) [5]. The aluminum alloys employed in space applications and their chemical compositions are: AA2014:- Cu (3.90–5.00), Mg (0.20–0.80), Zn (0.250max), Mn (0.40–1.20), Si (0.50–1.20), Fe (0.70max), Cr (0.10max), Ti (0.15max). AA2017:- Cu (3.50–4.50), Mg (0.40–0.80), Zn (0.25max), Mn (0.4–1.00), Si (0.20–0.80), Fe (0.70max), Cr (0.10max), Ti (0.15max). AA2024:- Cu (3.80–4.90), Mg (1.20–1.80), Zn (0.25max), Mn (0.30–0.90), Si (0.50), Fe (0.50), Cr (0.10), Ti (0.15). AA2219:- Cu (5.80–6.80), Mg (0.020max), Zn (0.250max), Mn (0.20–0.40), Si (0.20max), Fe (0.30max), Ti (0.02–0.10), Zr (0.10–0.25). AA5083:- Cu (0.10max), Mg (4.00–4.90), Zn (0.25max), Mn (0.40–1.00), Si (0.40max), Fe (0.40max), Cr (0.05–.25), Ti (0.15max). AA6061:- Cu (0.15–.40), Mg (0.80–1.20), Zn (0.25max), Mn (0.15max), Si (0.40–0.80), Fe (0.70max), Cr (0.04–0.35), Ti (0.150max). AA7050:- Cu (2.0–2.60), Mg (1.90–2.60), Zn (5.70–6.70), Mn (0.10max), Si (0.120max), Fe (0.15max), Cr (0.04max), Ti (0.10max), Zr (0.08–0.150). AA7075:- Cu (1.20–2.00), Mg (2.10–2.90), Zn (5.10–6.10), Mn (0.30max), Si (0.40max), Fe (0.50max), Ti (0.10max). AA7150:- Cu (1.90–2.50), Mg (2.00–2.70), Zn (5.90–6.90), Mn (0.10), Si (0.12), Fe (0.15), Cr (0.04), Ti (0.08–0.15), Zr (0.06). AA7178:- Cu (1.60–2.40), Mg (2.40–3.10), Zn (6.30–7.30), Mn (0.30), Si (0.40), Fe (0.50), Cr (0.18–0.28), Ti (0.20). AA7475:- Cu (1.20–1.90), Mg (1.90–2.60), Zn (5.20–6.20), Mn (0.060max), Si (0.100max), Fe (0.120max), Cr (0.18–0.25), Ti (0.06max) [11, 12]. AA 2219 and 7075 alloys were used in the Apollo Lunar Module, while AA 2024, 2124, and 2219 alloys were vital for the Space Shuttle Orbiter due to their weldability, strength, and toughness. AA 2XXX and 7XXX alloys are commonly used in satellites and aerospace infrastructure. Recent advancements in high-performance aluminum alloys have made them competitive with modern composites in airframes [12] (Table 1).

Table 1. Classification of Aluminium Alloys [13]

Alloy series	Primary alloying element
1 XXXX	Pure aluminum (≥99%)—Known for excellent corrosion resistance and formability
2 XXXX	Copper (Cu)—Offers high strength and good machinability, but with lower corrosion resistance
3 XXXX	Manganese (Mn)—Provides excellent corrosion resistance and moderate strength
4 XXXX	Silicon (Si)—Known for good castability, lower density, and moderate strength
5 XXXX	Magnesium (Mg)—Recognized for good corrosion resistance, especially in marine environments, and an excellent strength-to-weight ratio
6 XXXX	Magnesium-silicon (Mg-Si)—Combines good corrosion resistance, weldability and strength
7 XXXX	Zinc (Zn)—Offers high strength and is widely adopted in aerospace and structural industries
8 XXXX	Miscellaneous alloys—Includes alloys with elements like lithium or rare metals, providing specialized properties

2.2 Base Material – AA2219

The 2219 Al–Copper alloy is commonly used in propellant tanks for launch vehicles due to its high strength at reduced temperatures (down to −250 °C) and excellent weld ability. It is an alloy with age-hardening properties [14]. The copper content in this alloy varies between 5.8% and 6.8%, surpassing the maximum solubility of Cu in Al. This causes the formation of coarse Al2Cu particles, which can initiate cracks due to the stress concentrations they create during deformation, reducing the alloy's ductility [15]. The 2 Series alloys, such as AA2024 and AA2219, are known for their high strength and good machinability. These alloys are commonly used in aerospace and automotive applications due to their superior mechanical properties, especially in applications requiring load-bearing capabilities. The primary alloying element in the 2 series is copper (Cu), typically comprising 3% to 6% of the alloy. This high copper content enhances the fatigue resistance and strength of the alloy, making it suitable for structural applications. The corrosion resistance of 2 series alloys is moderate, and they usually need protective coatings like anodizing to improve their durability in harsh environments. The oxide layer formed on the surface offers some protection, but further measures are often required for extended exposure to harsh conditions [13–15]. Aluminium 2 Series offers a better balance between strength and ductility, which is advantageous for components requiring a combination of load-bearing capacity and flexibility, provides adequate fatigue resistance, particularly in the widely used 2024, 2219 alloys, which is favoured for airframe structures, Exhibits moderate corrosion resistance, and the potential for stress corrosion cracking is generally lower than in the 7 series (Tables 2 and 3).

Table 2. Mechanical properties of Aluminium 2 Series (AA2219)

Property	Value
Density	2790 kg/m^3
Ultimate tensile strength	430–490 MPa
Yield strength	290–320 MPa
Fatigue strength	140–160 MPa
Fracture toughness	30–35 $MPa\sqrt{m}$

Table 3. Mechanical properties of Aluminium 2219 composites [16–20]

Composites	Tensile strength (MPa)	Yield Strength (MPa)	Corrosion Resistance (mm/year)	Density (g/cm^3)	Hardness (HB, HRB, HRC)
AA2219 + TiB_2	Increases	Increases	Improves	Increases	Increases
AA2219 + CeO_2	Increases	Increases	Improves	Increases	Increases
AA2219 + MoS_2	Increases	Increases	Improves slightly	Increases	Increases
AA2219 + Si_3N_4	161.05 (3%)	102.80 (3%)	Improves	Increases	162.69 (3%)
AA2219 + graphite	Decreases	Decreases	Slightly reduced	Decreases	Decreases 34.8 (3%)
AA2219 + tungsten carbide	Increases	Increases	Improves	Increases	Increases 33.56 (4%)

2.3 Reinforcement Material – RUTILE

Rutile is a naturally occurring mineral form of titanium dioxide (TiO_2) and a key titanium source. Found in quartz veins and metamorphic rocks, it has distinctive needle-like and prismatic crystals. Its unique properties make rutile valuable for various industrial and technological uses. Rutile is an abundant, cost-effective mineral primarily made of titanium dioxide (TiO_2). It offers enhanced wear durability, superior mechanical and electrical properties and a lower CTE (Coefficient of Thermal Expansion) [21, 22]. The AA6061-TiO2 composite demonstrated improved hardness and enhanced wear resistance [23] (Table 4).

Table 4. Composition of Rutile

Property	Value
Titanium dioxide (TiO_2)	94–96%
Iron oxide (Fe O)	1–2%
Silicon dioxide(SiO_2)	1–2%
Manganese oxide (MnO)	0.5–1%
Other impurities	1–2%

3 Fabrication Method

3.1 Stir Casting Process

Melt stir casting is a highly appealing processing technique because of its adaptability, simplicity and suitability for mass production. It offers a broad range of raw materials and manufacturing conditions, making it one from the most cost-effective methods, while also enabling the fabrication of very large components [24–27]. In terms of reinforcement distribution, stir casting is a cost-effective method for achieving a uniform spread of reinforcements. This technique also ensures that the reinforcement materials remain undamaged during the process [28].

1. Heat the AA 2219 aluminium alloy to 800 °C until it melts.
2. Degasify the molten metal by adding a hexachloroethane tablet.
3. Introduce 1 weight % of the additive into the melt.
4. Create a vortex in the molten metal by vigorously stirring at 300 rpm using a stirrer.
5. Preheat the rutile particles to 700 °C.
6. Gradually add the preheated rutile particles into the melt until the viscosity remains unchanged.
7. Continue stirring, keeping the temperature at 800 °C, and form the vortex to add the remaining particles.
8. Add the remaining preheated rutile particles and stir vigorously for 5 min at 300 rpm.
9. Preheat the mold to 450 °C.
10. Pour the composite melt into the preheated steel mold.

3.2 Ultrasonic Heat Treatment

Ultrasonic treatments, particularly the discontinuous type, significantly enhance the material's mechanical strength by improving particle distribution, refining the microstructure, and increasing wettability. These improvements are achieved through better particle dispersion and reduced surface defects. The data emphasize the effectiveness of ultrasonic treatments in enhancing the functionality of composite materials compared to traditional methods [29]. Nano-sized silicon carbide particle-reinforced A356 aluminum alloy-based composites were created using a procedure combining ultrasonic vibration, high-energy ball milling and molten-metal techniques. The nano-SiC particles were pre-oxidized and uniformly dispersed in the alloy, with minimal agglomeration. The resulting composite showed a 20% increase in tensile strength (226 MPa) and a 44% improvement in elongation (5.5%) compared to the A356 alloy [30].

Ultrasonic heat treatment is introduced to enhance the distribution of rutile particles within the AA2219 matrix. The ultrasonic waves create high frequency vibrations in the molten metal, improving the particle dispersion and helping achieve a more uniform distribution of the reinforcement phase. Benefits: This technique reduces particle agglomeration (clumping) and ensures that the rutile reinforcement is evenly distributed, which is crucial for optimizing the mechanical performance of the composite, like tensile durability, wear resistance and hardness.

4 Conclusion

In conclusion, the use of advanced aluminum alloys, specifically AA2219, reinforced with rutile particles, shows significant improvements in mechanical properties when processed with stir casting and ultrasonic heat treatment. Ultrasonic treatment enhances particle distribution, reduces agglomeration, and optimizes material strength, hardness, and wear resistance. These advancements make aluminum alloys more efficient and durable for aerospace applications, offering weight reduction and improved performance while maintaining cost-effectiveness. Further research into these materials and processing methods holds potential for even better aerospace component performance.

References

1. Mouritz, A.P.: Introduction to Aerospace Materials. Elsevier, Cambridge (2012)
2. Flower, H.M.: High Performance Materials in Aerospace. Springer Science & Business Media, Berlin (2012)
3. Starke, E.A., et al.: Application of modern aluminium alloys to aircraft. Prog. Aerospace Sci. **32**, 131–172 (1996)
4. Zhu, L., et al.: Light-weighting in aerospace component and system design. Propul. Power Res. **7**(2), 103–119 (2018)
5. Dursun, T., et al.: Recent developments in advanced aircraft aluminium alloys. Mater. Des. **56**, 862–871 (2014)
6. Li, S., et al.: Development and application of Aluminium alloys for aerospace industry. J. Mater. Res. Technol. **27**, 944–983 (2023)
7. Nakai, M., et al.: New aspects of development of high strength aluminum alloys for aerospace applications. Mater. Sci. Eng. A. **A285**, 62–68 (2000)
8. Veiga, C., et al.: Properties and applications of titanium alloys: a brief review. Rev. Adv. Mater. Sci. **32**(2), 133–148 (2012)

9. Warren, A.S.: Developments and challanges for aluminium—a Boeing perspective. Mater. Forum. **28**, 24–31 (2004)
10. Tiwary, A., et al.: A review on characteristics of composite and advanced materials used for aerospace applications. Mater. Today Proc. **51**(1), 865–870 (2022)
11. Dunn, B.D.: Materials and Processes for Spacecraft and High Pliability Applications. Springer, Berlin
12. Abd El-Hameed, A.M., et al.: Aluminium alloys in space applications: a short report. J. Adv. Res. Appl. Sci. Eng. Technol. **22**(1), 1–7 (2021)
13. Davis, J.R., et al.: Aluminum and Aluminum Alloys. ASM International, Helsinki (1993)
14. Murty, S.V.S.N., et al.: Role of metallographic analysis in the identification of location of crack initiation in a burst tested AA 2219 propellant tank. Metallogr. Microstruct. Anal. **4**, 1–11 (2015)
15. Sharma, V.M.J., et al.: Effect of microstructure and strength on the fracture behavior of AA2219 alloy. Mater. Sci. Eng. A. **502**(1–2), 45–53 (2009)
16. Karthik, A., et al.: Microstructure and mechanical properties of AA 2219-TiB2 composites by squeeze casting technique. Mater. Today Proc. **27**(3), 2574–2581 (2020)
17. Karthik, A., et al.: Influence of CeO2 reinforcement on microstructure, mechanical and wear behaviour of AA2219 squeeze cast composites. J. Mater. Res. Technol. **14**, 797–807 (2021)
18. Rajesh Kumar, L., et al.: Optimization of wear behaviour for AA2219-MoS2 metal matrix composites in dry and lubricated condition. Mater. Today Proc. **27**(3), 2645–2649 (2020)
19. Manjunatha, C.J., et al.: Investigating the effect of Si3N4 reinforcement on the morphological and mechanical behavior of AA2219 alloy. Silicon. **14**, 2655–2667 (2022). https://doi.org/10.1007/s12633-021-01015-5
20. Rohith Renish, R., et al.: Experimental investigation of wear and hardness test over AA2219 with reinforcement of tungsten carbide. In: Lecture Notes in Mechanical Engineering. Springer, Singapore. https://doi.org/10.1007/978-981-13-2697-4_17
21. Prabhu, S., et al.: Microstructure evolution and mechanical properties of friction stir welded AA6061/rutile composite. Mater. Res. Express. **6**, 0865i7 (2019)
22. Rama, A., et al.: Influence of particle size and temperature on the wear properties of rutile reinforced aluminium metal matrix composite. J. Compos. Mater. **49**, 843–852 (2015)
23. Chaudhury, S.K., et al.: Preparation and thermo mechanical properties of stir cast Al−2Mg−11TiO2 (rutile) composite. Bull. Mater. Sci. **27**(6), 517–521 (2004)
24. Shorowordi, K.M., et al.: Microstructure and interface characteristics of B4C, SiC and Al2O3 reinforced Al matrix composites: a comparative study. J. Mater. Process. Technol. **142**, 738–743 (2003)
25. Clyne, T.W.: Metal matrix composites: matrices and processing. In: Mortensen, A. (ed.) Encyclopaedia of Materials Science and Technology. Elsevier, New York (2001)
26. Akbulut, H., et al.: Comparison of the mechanical properties of A356-SiC particle-reinforced metal matrix composites produced by two melting routes. Proc. Instn. Mech. Engrs. **216**(1), 1–7 (2002)
27. Kwak, J.S., et al.: Mechanical properties and grinding performance on aluminium-based metal matrix composites. J. Mater. Process. Technol. **201**, 596–600 (2008)
28. Sathyabalan, P., et al.: ANN based prediction of effect of reinforcements on abrasive wear loss and hardness in a hybrid MMC. Am. J. Eng. Appl. Sci. **2**(1), 50–53 (2009)
29. Dehnavi, M.R., et al.: Effects of continuous and discontinuous ultrasonic treatments on mechanical properties and microstructural characteristics of cast Al413–SiCnp nanocomposite. Mater. Sci. Eng. A. **617**, 73–83 (2014)
30. Shu-sen, W., et al.: Nano-SiCP particles distribution and mechanical properties of Al-matrix composites prepared by stir casting and ultrasonic treatment. China Foundry. **15**(3), 203–209 (2018). https://doi.org/10.1007/s41230-018-8009-2

Effect of Various Infill Speed on Mechanical Characteristics of 3D Printed Polyethylene Terephthalate Glycol Composites Samples

S. Rajakumar Rai[1], R. Raja[1], Sabitha Jannet[1](✉), Sushanlal Babu[2], L. Prawin[2], Winston Devaraj[1], and Ebinezer Rajaram[1]

[1] Division of Mechanical Engineering, Karunya Institute of Technology and Sciences, Coimbatore, India
Sabithajannet@gmail.com
[2] Division of Aerospace Engineering, Karunya Institute of Technology and Sciences, Coimbatore, India

Abstract. Polyethylene terephthalate glycol printed utilizing layers of extruded beads and Fused Deposition modeling by varying a key infill speed parameter (40, 50, and 60 mm/s, respectively) while maintaining ideal values for the remaining parameters. The overall performance depends on the annealing conditions and the infill speed. Mechanical characteristics like impact strength, tensile strength, and hardness are enhanced when 40 mm/s infill speed PETG samples are utilized. An appropriate annealing post-processing method enhanced interlayer diffusion bonding and improved mechanical properties in comparison with the as-built sample. The results are helpful guidelines for fabricating functional parts out of PETG, assuming a 40 mm/s infill speed during annealing. This was a nearly crucial aspect in the eventual replacement of structural metallic parts used in automobiles and aircraft.

Keywords: 3D Printing · Fusion Deposit Model · Tensile strength · Infill speed

1 Introduction

PETG stands for polyethylene terephthalate glycol, a thermoplastic polyester that is chemically resistant, durable, and formable. PETG is also ideal for routing, bending, and die cutting. 3D printing, commonly referred to as the application, has the potential to revolutionize the fabrication of a variety of products, including food, medical supplies, and even enormous coral reefs. Materials including polylactic acid (PLA), polystyrene, and nylon are employed, and process factors like build orientation and infill pattern/density that influence process quality are investigated [1]. Sustainable PHB and PHBV filaments filled with 5% pullulan were studied for their chemical, thermal, and morphological characteristics, crystallinity, 3D printing capacity, and cytotoxicity. In two PHBV samples, valerate concentrations of 6 and 8 wt percent were examined. According to Raman and FTIR studies, Pullulan changed the polymeric chain organization, influencing polymer

R. Rajasekharan et al. (Eds.): MatZone 2024, SPPHY 316, pp. 87–94, 2025.
https://doi.org/10.1007/978-981-96-6541-9_8

crystallinity [2]. In fused deposition modeling and stereo lithography, polymer materials that can be formed according to current, particular laboratory demands are employed. The capacity to create complicated geometries and structures is paired with simplicity of use and cost in this application [3]. A high-temperature annealing technique (173 °C, near the onset of PA12's melting point) developed in this study significantly enhances the mechanical performance of MJF-printed PA12 and GF/PA12 composites. The ultimate tensile strength (UTS) and tensile modulus increase by 20.8% and 48.5% for PA12 specimens and by 22.8% and 30.6% for GF/PA12 specimens, respectively. In terms of mechanical strength, the GF/PA12 specimens outperform previously reported MJF- and SLS-printed PA12, glass bead/PA12, and GF/PA12 composites [4, 5].The cellular sections were fabricated using varying numbers of perimeter shells and five distinct infill patterns at three different densities. The findings revealed that, at the same density, employing different types of infill pattern increased stiffness by up to twofold and enhanced strength by up to 82% [6]. The significant effect of infill, extrusion speed, and layer thickness on the highest flexural force was explored experimentally (PLA). In a two-by-three factorial design with no replication, three centre points were employed. The results reveal that extrusion speed has a significant impact, that deposition angle and infill have a substantial interaction, and also that effects are nonlinear [7]. Infill density has been found to have a considerable influence on compressive strength. Greater compressive strength was achieved using a reduced porous structure and a higher infill density [8]. The impacts of raster angles are investigated using designed specimens with various transverse planes that are printed at various angles. Specimens with infill rates ranging in the range of 10% and till 100%, as well as six different infill patterns were analysed [9]. Relevant infill designs were developed based on loading conditions and cost reduction, in comparison to solid infill designs. The impact of production costs, considering both production grade and entry level printers, on the selection of infill design is also discussed [10]. A nine run experiment based on Taguchi DOE was used to test samples with varying infill patterns and layer thicknesses. This empirical method, based on the statistical design of experiments and standardized mechanical testing, identified trends and interactions among variables specific to the chosen additive manufacturing process [11]. Poly-lactic acid filament was utilized to fabricate 3D specimens, and tensile tests revealed that part orientation, infill pattern, and density significantly influenced the tensile strength. Specimens printed with a flat orientation and a concentric infill pattern exhibited the highest tensile strength. Additionally, tensile strength increased with higher infill density [12]. The impact of various process parameter settings on the mechanical behavior of products is significant. However, due to the vast number of desired material compositions, 3D printers, slicing software, and process parameters, acquiring this information requires substantial effort [13]. The study investigated the influence of infill percentage on the mechanical behaviour of ABS (Acrylonitrile Butadiene Styrene) printed parts, specifically impact resistance, tensile strength, and printing time. The findings revealed that 100% infill achieved the highest tensile force (1438 N), tensile stress (34.57 MPa), and impact resistance (1.55 J). Infill percentages between 50% and 98% are not recommended, as they lead to longer printing times and lower tensile strength and impact resistance compared to 100% infill [14].

The main objective of additive manufacturing is to increase precision and reduce the steps in the manufacturing process. There are many types of filaments used in the additive manufacturing process Different polymers are used in the additive manufacturing industry to manufacture components precisely Polyethylene terephthalate glycol composite is the material which is going to be used in our study. This study aims to examine the impact of varying infill speeds on the mechanical behavior of PETG fabricated through the Fused Deposition Modeling (FDM) process, providing valuable insights for industrial applications. The primary objective is to fabricate PETG composites using FDM with different infill speeds, conduct mechanical property tests on both as-built and annealed specimens, compare their performance, and identify the optimal solution based on the results.

2 Material and Methods

The material used is Polyethylene Terephthalate Glycol (PETG), sourced from World of Lilliput. The "G" stands for Glycol-modified. PETG has demonstrated a 10–25% increase in strength compared to other filaments, making it a preferred material for mechanical parts due to its high strength. The printing temperature range for PETG is between 220 °C and 275 °C, with minimal shrinkage. The filament has a diameter of 1.75 mm. The printer specifications include a heated bed temperature of up to 110 °C, a layer resolution of 50–300 microns, and a nozzle temperature of up to 275 °C. The specimens were prepared by first designing the sample in SolidWorks software. After creating the virtual test specimen, the SolidWorks file was converted to .stl format and imported into the slicing program. SnapmakerLuban software was used for slicing. The printing process aimed to meet three key requirements: low surface roughness, high mechanical properties, and a fast printing time.

2.1 Micro Hardness Test

The microhardness test was carried out using a testing machine (HM113 model) and make Mitutoyo, Japan (Fig. 1).

2.2 Tensile Strength

The tensile strength was tested in a Universal Testing machine (UTM) for As-built composite and for Annealed PETG under various infill speeds of 40 mm/s, 50 mm/s and 60 mm/s (Fig. 2).

2.3 Impact Strength Test

The impact strength of As-built and Annealed PETG with different infill speeds (40 mm/s, 50 mm/s and 60 mm/s) is also tested by using an Izod impact instrument, and the values with no error are observed (Fig. 3).

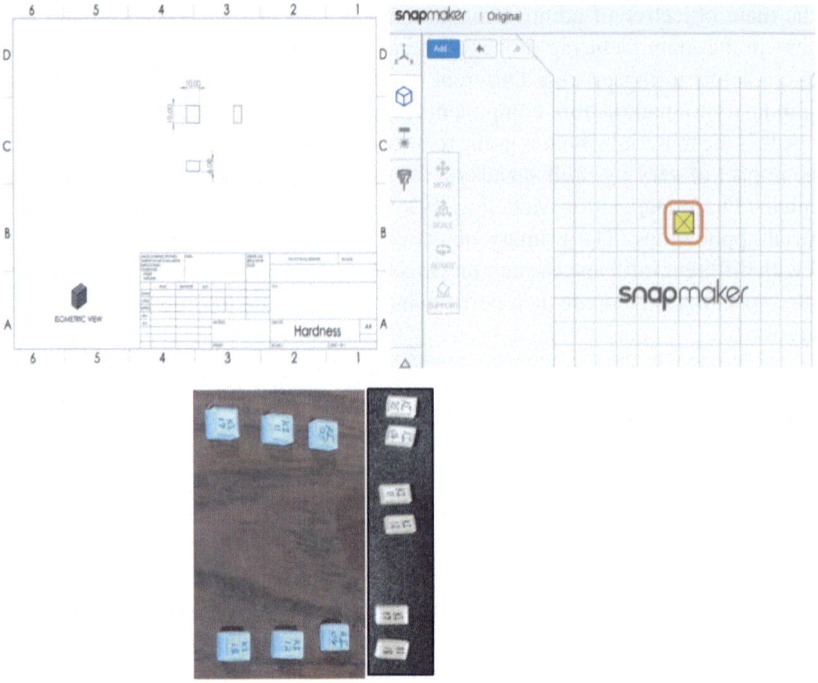

Fig. 1. Microhardness test specimens

3 Results and Discussion

3.1 Hardness

The samples As-built and Annealed PETG is used for the microhardness test, and results with no mistake are given for different infill speed specimens of 40, 50, and 60 mm/s. With a hardness rating of 116HRC, the Annealed PETG produced at an infill speed of 40 mm/s has the maximum hardness. When it comes to infill speeds, annealed PETG has a hardness value of 116, 98 and 78 HRC at 40, 50, and 60 mm/s, while as-built PETG has a hardness value of 106, 90 and 68 HRC at 40, 50, and 60 mm/s. The PETG filament's fibre addition prevents the applied load from being indented (Fig. 4).

3.2 Tensile Strength

The Fig. 5 depicts the tensile strength obtained at various infill speeds. The 40 mm/s infill speed resulted in good tensile strength. PETG is a difficult to print material. The annealed printed samples also indicated improved better tensile strength at 40 mm/s. In this only the infill speed was varied keeping the nozzle temperature at constant. Thus the combination of both the infill speed and temperature should be such that melting does not occur. An appropriate combination could reduce the shrinkage problems. The tensile properties of PETG are enhanced as the printing speed decreases. This is because

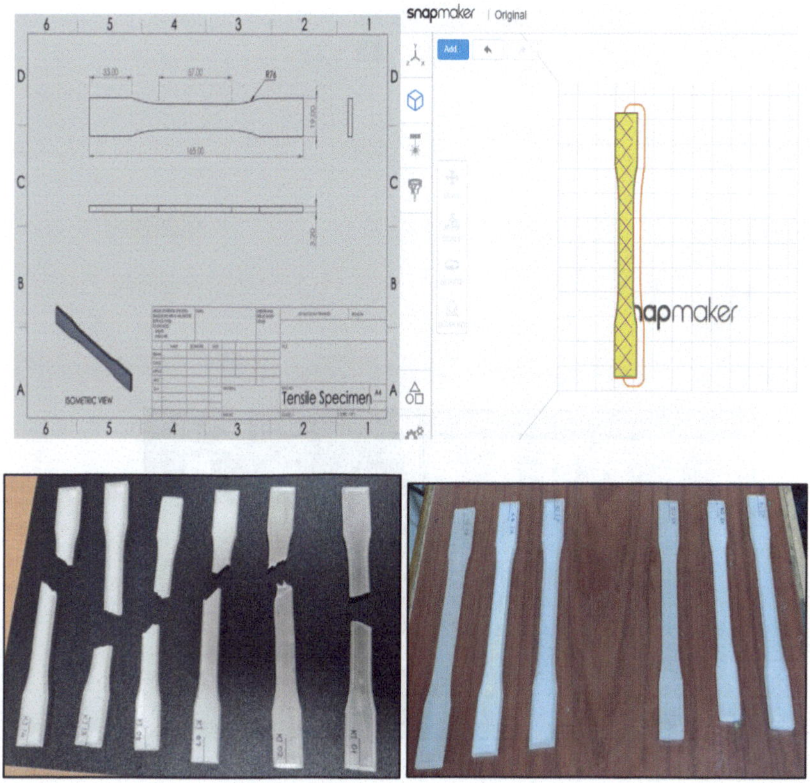

Fig. 2. Tensile testing specimens

of its low heat dissipation and reduced porosity, which contribute to enhanced tensile characteristics at lower printing speeds.

As-built PETG exhibited tensile strengths of 36.36, 27.93, and 30.99 MPa at different infill rates, while annealed PETG showed tensile strengths of 44.46, 30.99, and 35.89 MPa at the same infill speeds. The highest tensile strength for annealed PETG was achieved at an infill speed of 40 mm/s. The infill speed plays a critical role in enhancing the printed specimen's strength; at 40 mm/s, the tight layer bonding potentially increases the strength between printed layers. Post-processing further improves the tensile strength. During the printing process, the filament undergoes rapid heating and cooling, creating internal stress in the printed parts. Annealing helps release this stress after the components are printed. Compared to as-built PETG, annealed PETG at 40 mm/s shows a 15% improvement in tensile strength.

3.3 Impact Strength

The impact strength of as-built PETG at different infill speed is 31, 41, and 44 J/m^2, while for the annealed specimens, it is 32, 38, and 42 J/m^2. The maximum impact strength for annealed PETG occurs at an infill speed of 40 mm/s, due to the moderate speed. At

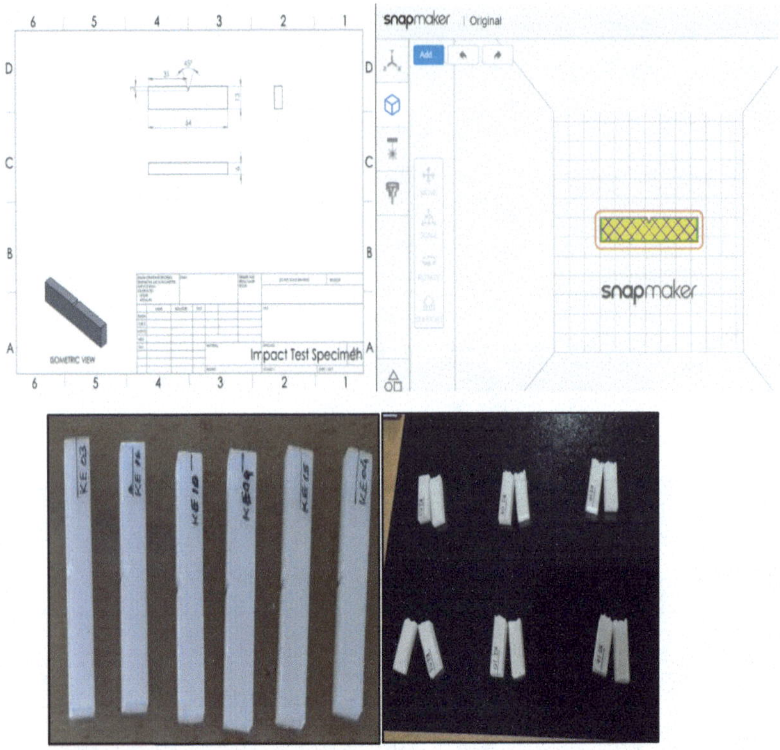

Fig. 3. Impact testing specimens (**a**) 3D model (**b**) Sliced (**c**) 3D printed samples (**d**) Samples after impact test

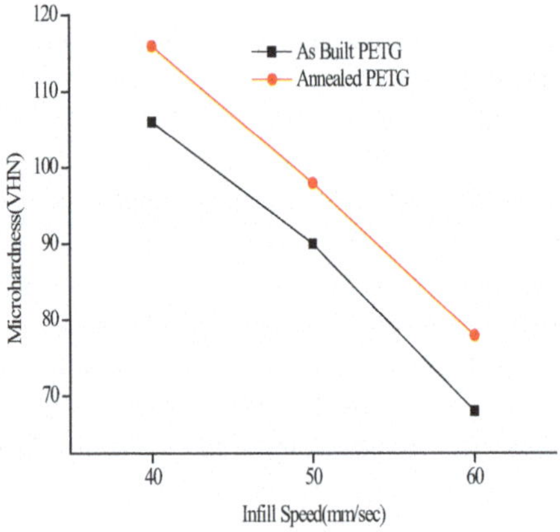

Fig. 4. Microhardness test values

this speed, annealed PETG shows a 16% improvement in strength compared to as-built PETG (Fig. 6).

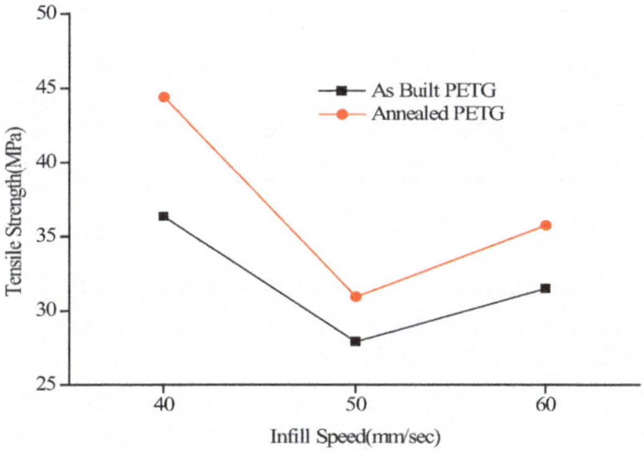

Fig. 5. Effect of infill speed on the tensile strength

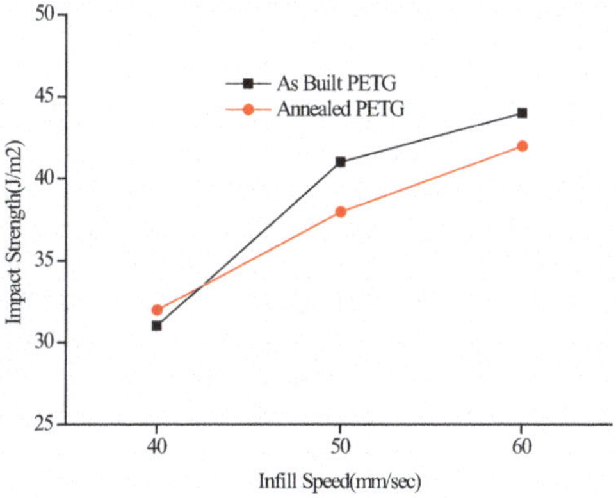

Fig. 6. Effect of infill speed on the impact strength

4 Conclusion

PETG specimens were successfully printed using annealing and varying infill rates of 40, 50, and 60 mm/s. The mechanical characteristics of both as-built and annealed PETG specimens were assessed through impact, tensile, and hardness tests. Lower infill speeds resulted in specimens that were heavier, more robust, stronger, and better able to support greater weight, demonstrating increased resistance to indentation from applied forces. At an infill speed of 40 mm/sec, the annealed, 3D-printed PETG specimen exhibited superior mechanical properties. When comparing the mechanical characteristics of as-built and annealed PETG specimens, the annealed versions showed improvements of 21.21%, 25%, 23%, and 18%, respectively. The annealed specimens demonstrated better quality, particularly at the 40 mm/s infill speed. The improvement in mechanical properties is likely due to the effective recovery of reduced interlayer strength. PETG annealed and post-processed at an infill speed of 40 mm/s shows promise as a suitable alternative to metal formedium-load or transient functional components in automotive and aerospace applications.

References

1. Gonabadi, H., Yadav, A., Bull, S.J.: The effect of processing parameters on the mechanical behaviour of PLA by a 3D FFF printer. Int. J. Adv. Manuf. Technol. **111**, 695–709 (2020)
2. Dutra, T.A., Ferreira, R.T.L., Resende, H.B., Blinzler, B.J., Asp, L.E.: Mechanism based failure of 3D-printed continuous carbon fiber reinforced thermoplastic composites. Compos. Sci. Technol. **213**, 108962 (2021)
3. Belka, M., Bączek, T.: Additive manufacturing and related technologies—the source of chemically active materials in separation science. TrAC Trends Anal. Chem. **142**, 116322 (2021). https://doi.org/10.1016/j.trac.2021.116322
4. Liu, X., Choo, J.Y.C., Tey, W.S., Chen, J., Ong, A., Zhao, L., Zhou, K.: Enhancing the mechanical strength of multi jet fusion–printed polyamide 12 and its glass fiber-reinforced composite via high temperature annealing. Addit. Manuf. **46**, 102205 (2021). https://doi.org/10.1016/j.addma.2021.102205
5. Kumar, K.S., Soundararajan, R., Shanthosh, G., Saravanakumar, P., Ratteesh, M.: Augmenting effect of infill density and annealing on mechanical properties of PETG and CFPETG composites fabricated by FDM. Mater. Today Proc. **45**, 2186–2191 (2021). https://doi.org/10.1016/j.matpr.2020.10.078
6. Khan, S.A., Siddiqui, B.A., Fahad, M., Khan, M.A.: Evaluation of the effect of infill pattern on mechanical strength of additively manufactured specimen. Mater. Sci. Forum. **887**, 128–132 (2017). https://doi.org/10.4028/www.scientific.net/MSF.887.128
7. Lužanin, O., Movrin, D., Plan, M.: Effect of layer thickness, deposition angle, and infill on maximum flexural force in FDM-built specimens. J. Technol. Plast. **39** (2014)
8. Dave, H., Rajpurohit, S.R., Patadiya, N.H., Dave, S.J., Kumar, S., Sharma, S.S., Thambad, V.P., Srinivasn, K.S.: Compressive strength of PLA based scaffolds: effect of layer height, infill density and print speed. Int. J. Mod. Manufac. Technol. **11**, 21 (2019)
9. Dudescu, C., Racz, L.: Effects of raster orientation, infill rate and infill pattern on the mechanical properties of 3D printed materials. ACTA Univ. Cibin. **69**, 23–30 (2017). https://doi.org/10.1515/aucts-2017-0004
10. Baich, L., Manogharan, G., Marie, H.: Study of infill print design on production cost-time of 3D printed ABS parts. Int. J. Rapid Manuf. **5**, 308 (2015). https://doi.org/10.1504/ijrapidm.2015.074809
11. Cho, E.E., et al.: Investigation on influence of infill pattern and layer thickness on mechanical strength of PLA material in 3D printing technology. J. Eng. Sci. Res. **3**, 27–37 (2019). https://doi.org/10.26666/rmp.jesr.2019.2.5
12. Dave, H.K., Patadiya, N.H., Prajapati, A.R., Rajpurohit, S.R.: Effect of infill pattern and infill density at varying part orientation on tensile properties of fused deposition modeling-printed poly-lactic acid part. Proc. Inst. Mech. Eng. Part C J. Mech. Eng. Sci. **235**, 1811–1827 (2021). https://doi.org/10.1177/0954406219856383
13. Popescu, D., Zapciu, A., Amza, C., Baciu, F., Marinescu, R.: FDM process parameters influence over the mechanical properties of polymer specimens: a review. Polym. Test. **69**, 157–166 (2018). https://doi.org/10.1016/j.polymertesting.2018.05.020
14. Alvarez, K.L., Lagos, R.F., Aizpun, M.: Investigating the influence of infill percentage on the mechanical properties of fused deposition modelled ABS parts. Ing. Investig. **36**, 110–116 (2016). https://doi.org/10.15446/ing.investig.v36n3.56610
15. Hsueh, M.H., Lai, C.J., Wang, S.H., Zeng, Y.S., Hsieh, C.H., Pan, C.Y., Huang, W.C.: Effect of printing parameters on the thermal and mechanical properties of 3d-printed PLA and PETG, using fused deposition modeling. Polymers (Basel). **13**, 1758 (2021). https://doi.org/10.3390/polym13111758

Thermal Breaks to Reduce Energy Consumption and Room Temperature Using Industrial Waste Materials: A Sustainable Approach

Adarsh Ramesh(✉), Tino Joe Tenson(✉), and Libin P. Oommen

Department of Mechanical Engineering, Providence College of Engineering, Chengannur, Kerala, India
adarsh.prc21me003@student.providence.edu.in,
tino.t@providence.edu.in

Abstract. The increasing demand for cooling in residential and commercial buildings, driven by rising global temperatures, has significantly escalated energy consumption. This study explores the potential of utilizing industrial by-products, including Rice Husk Ash (RHA), Woven Fabric Waste (WFW), Polyethylene Glycol-400 (PEG-400), Silica Fume, Fly Ash, Ground Granulated Blast Furnace Slag (GGBS), and Fibers, as thermal break materials to enhance insulation and reduce cooling energy requirements in buildings. A combination of experimental analysis and numerical simulations, conducted using ANSYS-R1 software, was employed to evaluate the thermal conductivity, convective heat transfer, and radiative effects of these composite materials. The materials were integrated into concrete wall structures, and their thermal performance was assessed by monitoring internal wall temperatures and the corresponding energy savings. The findings demonstrated a 20–30% reduction in energy consumption for cooling applications. Additional investigations included a comparative analysis of White and Black RHA, optimization of PEG-400 and RHA/WFW compositions, and the incorporation of thermal breaks between concrete layers. The proposed approach offers a cost-efficient and sustainable alternative to conventional insulation techniques, with significant potential for application in low-income housing projects.

Keywords: Thermal Breaks · Rice Husk Ash · Woven Fabric Waste · Polyethylene Glycol-400 · Silica Fume · Sustainability

1 Introduction

As global temperatures rise, energy consumption in buildings for cooling purposes has surged. Traditional cooling systems like air conditioning and HVAC systems consume significant energy and contribute to rising electricity bills, especially affecting low-income households. Additionally, these systems emit harmful gases such as Hydrofluorocarbons (HFCs), contributing to environmental degradation. The construction industry, responsible for approximately 40% of global energy consumption, is now focused on developing energy-efficient materials that reduce operational costs while promoting environmental sustainability.

R. Rajasekharan et al. (Eds.): MatZone 2024, SPPHY 316, pp. 95–110, 2025.
https://doi.org/10.1007/978-981-96-6541-9_9

One promising approach to address this challenge is the use of thermal breaks—insulating layers within walls that reduce heat transfer. Utilising industrial waste materials such as Rice Husk Ash (RHA), Woven Fabric Waste (WFW), and binders like Polyethylene Glycol-400 (PEG-400), in combination with additives like Silica Fume, Fly Ash, GGBS, and Fibers, can offer a cost-effective and sustainable solution to improving thermal insulation.

This study focuses on developing and testing a composite mixture that enhances thermal insulation and durability while utilising industrial waste materials [1] and to reduce the environmental impact of construction. This chapter aims to evaluate the thermal conductivity and insulation properties of RHA [2], WFW [3], and PEG-400 [4] when combined with other additives, explore the potential of using White Husk Ash versus Black Husk Ash for optimal thermal performance, investigate the effectiveness of compressing RHA-based thermal break sheets for improved application, and assess the possibility of placing thermal breaks between two layers of concrete, rather than inside the wall, to enhance thermal performance [7–9].

2 Materials and Methods

2.1 Overview of Industrial Waste Materials

Industrial waste materials have been widely researched for their potential use in sustainable construction. Rice Husk Ash (RHA) has gained particular attention due to its high silica content, which enhances thermal insulation and mechanical strength [10, 11]. Woven Fabric Waste (WFW) [12–14], a by-product of the textile industry, also shows promise due to its fibrous structure, which helps trap air and reduce heat transfer. Polyethylene Glycol-400 (PEG-400) is commonly used as a binder to improve flexibility, reduce porosity, and enhance moisture resistance. Recent studies have demonstrated the benefits of combining these materials with Silica Fume, Fly Ash, and Ground Granulated Blast Furnace Slag (GGBS) to enhance the thermal and structural properties of concrete. The use of industrial waste materials in construction has gained significant attention for improving thermal insulation and reducing the environmental impact of buildings. This section provides insights into the materials used in this study and their relevance based on existing research.

RHA is an abundant by-product of rice milling, with high silica content that improves thermal insulation and mechanical strength. Studies have shown that White Rice Husk Ash has higher silica content and lower carbon residue, making it more effective for thermal insulation compared to Black Rice Husk Ash. RHA is often used as a partial replacement for cement in concrete, improving its durability and reducing heat transfer. WFW, a by-product of the textile industry [15], provides excellent thermal insulation due to its fibrous structure, which traps air and reduces heat transfer. Research has demonstrated that WFW can enhance the insulation properties of building materials while contributing to waste reduction. PEG-400 acts as a binder that reduces porosity [16], enhances moisture resistance, and increases flexibility. It improves the adhesion of RHA and WFW to the concrete matrix, creating a durable, flexible, and crack-resistant composite. Silica Fume is a fine pozzolanic material that enhances the compressive strength and thermal properties of concrete. Its microstructure fills the voids within the composite, reducing porosity and improving insulation. Fly Ash is a pozzolanic material produced from coal combustion. It improves the long-term strength and durability of concrete by reducing cracking under thermal stress and enhancing resistance to chemical attacks. GGBS is a by-product of steel production and is used to improve the sulfate resistance, thermal insulation, and durability of concrete. It also reduces permeability, making the composite more resistant to environmental conditions. Fibers are added to the composite to improve its tensile strength and crack resistance. Fibers prevent micro-cracks from propagating under thermal stress, ensuring long-term durability [18] (Tables 1, 2, and 3).

Table 1. Chemical composition and physical properties of individual materials and compound.

Component	SiO_2 (%)	Al_2O_3 (%)	Fe_2O_3 (%)	CaO (%)	MgO (%)	Na_2O (%)	Specific gravity	Particle size (μm)
Fly ash	41.5	13.1	6.2	1.8	1.1	0.7	2.3	5.0
Rice husk ash	85.6	0.2	0.5	0.6	0.3	–	2.1	10.0
Silica fume	92.5	0.4	0.2	–	–	–	2.2	0.1
Composite material	65.0	5.0	3.0	15.0	3.0	1.0	2.5	7.5

Table 2. Thermodynamic properties of materials and compound.

Property	Fly ash	Rice husk Ash	Silica fume	Composite
Enthalpy (ΔH, kJ/mol)	75	120	140	110
Entropy (ΔS, J/mol K)	200	250	270	230
Gibbs free energy (ΔG, kJ/mol)	−50	−70	−80	−60

2.2 Climate Trends and Seasonal Temperature Variations in Alleppey, Kerala (2014–2023)

Figure 1 represents seasonal variations over the past decade. Summer Day and Night temperatures show a slight upward trend. Winter Day and Night temperatures also exhibit a mild increase, reflecting typical climatic changes in Alleppey (Table 4).

Although previous studies have highlighted the effectiveness of RHA, WFW, and PEG-400 in improving the thermal properties of construction materials, there is limited research on the optimal composition of these materials to achieve maximum thermal insulation. Furthermore, few studies have explored the comparative thermal performance of White and Black Rice Husk Ash or the potential benefits of placing thermal breaks

Table 3. Mechanical and physical properties of materials and composite.

Property	Fly ash	RHA	WFW	Silica fume	GGBS	Composite material
Compressive strength (MPa)	40	50	–	55	45	60
Flexural strength (MPa)	5.5	6.0	–	6.8	6.0	7.0
Tensile strength (MPa)	3.2	2.5	1.8	3.2	3.0	4.5
Durability (years)	20	25	–	30	25	35
Elastic modulus (GPa)	22	18	12	24	20	21
Elongation at break (%)	3.0	3.5	4.2	2.8	3.2	3.4
Specific gravity	2.3	2.1	2.4	2.2	2.9	2.5
Bulk density (kg/m^3)	1400	1440	1560	1100	1380	1325
Specific surface area (m^2/g)	45	81	54	0.1	50	65
Temperature of application (°C)	650	700	500	1000	900	800
Melting temperature (°C)	1500	1450	1320	1600	1550	1485

between concrete layers. This study aims to address these gaps by investigating the combined use of these materials and testing different configurations for maximising thermal performance and durability (Tables 5 and 6).

2.3 Experiment Design

2.3.1 Higher Fly Ash (30%) and RHA (20%) with Moderate GGBS and Silica Fume Content

Fly Ash and RHA are known for their high silica content, which enhances thermal insulation and delays the heat peak in binder compositions. The combination of 30% Fly Ash and 20% RHA creates a robust thermal barrier while maintaining manageable hydration heat due to moderate exothermic reactions from their pozzolanic activity [22]. GGBS and Silica Fume in smaller proportions (10–15%) complement this setup by providing additional strength and reducing porosity, ensuring that heat is not easily conducted through the material.

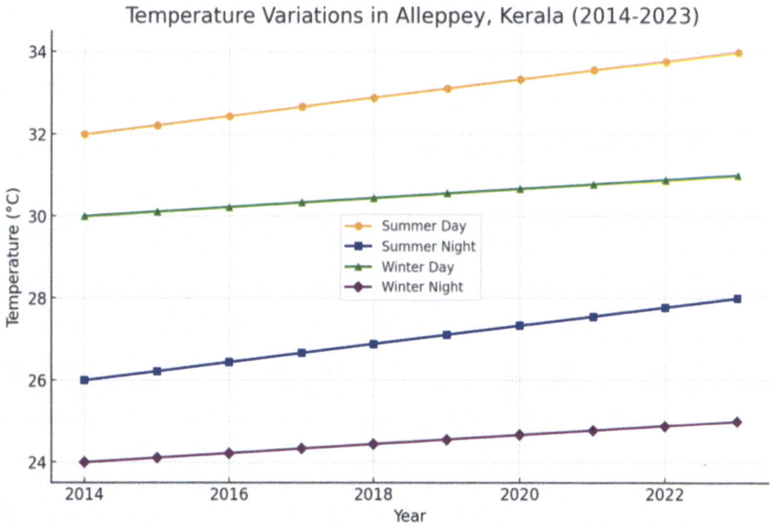

Fig. 1. Temperature Variations in Alleppey, Kerala (2014–2023).

Table 4. Baseline Data Table

Year	Summer day (°C)	Summer night (°C)	Winter day (°C)	Winter night (°C)
2014	32.0	26.0	30.0	24.0
2015	32.2	26.2	30.1	24.1
2016	32.4	26.4	30.2	24.2
2017	32.7	26.7	30.3	24.3
2018	32.9	26.9	30.4	24.4
2019	33.1	27.1	30.6	24.6
2020	33.3	27.3	30.7	24.7
2021	33.6	27.6	30.8	24.8
2022	33.8	27.8	30.9	24.9
2023	34.0	28.0	31.0	25.0

2.3.2 Compositions with Higher Silica Fume (10%)

Silica Fume has a high reactivity and ultra-fine particles, which fill the pores in concrete effectively, leading to lower permeability and better insulation. The 10% Silica Fume content contributes to more uniform hydration, reducing q max (W/kg) while maintaining good thermal performance. This means the composite becomes more thermally stable over time [33].

Table 5. Materials used.

Material	Properties	Reason for selection
Rice husk ash (RHA)	High silica content, pozzolanic activity, enhances thermal insulation	Improves the thermal properties of the composite
Woven Fabric Waste (WFW)	Fibrous structure traps air, reduces heat transfer	Provides effective thermal insulation
PEG-400	Flexible binder, reduces porosity, enhances water resistance	Improves adhesion and flexibility
Silica fume	Fills micro-voids, enhances compressive strength, reduces thermal conductivity	Improves the density and insulation properties of the composite
Fly ash	Pozzolanic material, improves long-term strength, reduces cracking	Enhances durability and thermal resistance
GGBS	Improves sulfate resistance, reduces permeability, enhances durability under thermal stress	Increases resistance to weathering and harsh environments
Fibers (fiberglass/basalt)	Reinforces tensile strength, improves crack resistance, prevents micro-cracking	Improves durability and long-term performance

Table 6. Top 10 binder combinations and peak heat characteristics.

Fly ash (%)	RHA (%)	GGBS (%)	Silica fume (%)	Mortar (%)	Time to peak heat (t20, h)	q_{max} (W/kg)	Normalised Peak Heat to 100% CEM I (W/kg)
25	20	15	10	30	16.5	2.30	2.88
20	15	30	10	25	15.3	2.20	2.90
20	20	20	10	30	16.0	2.10	2.80
30	10	25	5	30	15.0	2.00	2.85
25	15	25	5	30	15.7	2.15	2.92
20	25	20	10	25	15.8	2.25	2.95
30	20	15	5	30	16.2	2.18	2.89
15	25	30	10	20	14.7	2.40	3.05
20	20	30	5	25	14.9	2.35	3.02
25	20	20	10	25	15.2	2.28	2.97

2.3.3 GGBS Enhances Heat Moderation

As a slow-reacting material, GGBS moderates hydration heat, especially when used in higher proportions (30%). It helps reduce q max significantly, making the composite suitable for applications where slow heat transfer and temperature regulation are critical. However, higher GGBS content slightly accelerates t20, which may not always be ideal for prolonged cooling. Balancing Fly Ash, RHA, GGBS, and Silica Fume ensures optimal properties for thermal insulation, strength, and durability. Mortar, as the primary binder, must be maintained around 25–30% to provide structural stability while allowing the materials to perform their functions effectively.

2.4 Recommended Percentages for Better Room Cooling and Proposed Composite Mix

To optimise the binder composition for better room cooling, the goal is to achieve Low thermal conductivity to slow heat transfer, Delayed heat peak (higher T_{20}) for prolonged cooling, and Moderate hydration heat release (lower $q_{\max}$) to reduce heat buildup within the material. The recommended percentages for the binder composition and proposed composite mix are as depicted in Table 7 (Fig. 2).

Table 7. Recommended percentages for the binder composition and proposed composite mix.

Material	Recommended % for better cooling	Proposed composite mix (proportion by weight)
Fly ash	20–25%	10%
RHA (or RHA/WFW)	20–25%	15%
GGBS	15–20%	10%
Silica fume	5–10%	15%
Mortar	30–35%	40%
PEG-400	5%	5%
Fibers (fiberglass/basalt)	5%	5%

The Expected Performance with Suggested Composition is based on common principles of material science and thermal analysis. The predictions were derived from a combination of literature data, material properties, and binder performance trends. Here's a breakdown of the calculations and estimations involved:

2.4.1 Time to Peak Heat (T20, H)

The time to peak heat is influenced by hydration reaction rates: Slower-reacting materials like GGBS and Fly Ash increase t20 due to their delayed exothermic activity; thermal conductivity: High silica content [51] in RHA and Silica Fume reduces heat transfer, indirectly extending t20 and mix proportions: Increasing Fly Ash and GGBS content

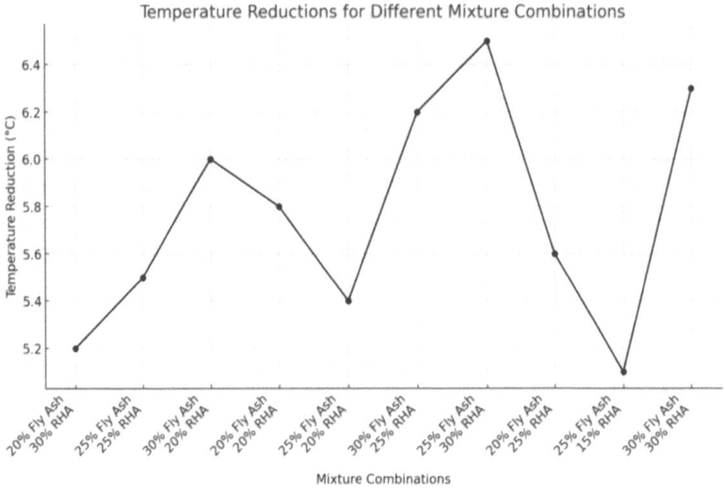

Fig. 2. Temperature reduction for different mixture combinations.

slows hydration, while higher Mortar content accelerates it. T_{20} can be calculated using the equation

$$T_{20} \propto \frac{1}{\text{Rate of Heat Generation}}$$

For slower-reacting binders, t20 was calculated using proportional scaling from typical hydration curves, taking reference values from the GGBS peak heat rate studies (Ballim and Graham 2009) [53]:

(Reference values: Pure CEM I had $T_{20} = 17.6$ h.)

2.4.2 Peak Heat Rate (q max, W/kg)

The peak heat rate is affected by heat release from hydration reactions: GGBS and Fly Ash lower q max compared to CEM I due to their slower reactivity [55]; silica Fume and RHA contributions: These accelerate early-stage hydration slightly but contribute less heat overall [60].

$$q_{\max} = \text{k} \cdot (\text{proportion of CEM I} - \text{equivalents})^{\text{n}}$$

Where: k: Proportionality constant derived from reference q max values, n: Reaction order based on binder type (Fly Ash, GGBS, etc.).

2.4.3 Normalised Peak Heat to 100% CEM I (W/kg)

Normalisation is performed relative to pure CEM I, which is set at 2.80 W/kg. Formula used is

$$Normalised\ Peak\ Heat = \left(\frac{q_{\max}}{2.80}\right)^{100}$$

2.4.4 Thermal Conductivity (W/m · K)

The thermal conductivity of the composite is a weighted average of its components:

$$k_{composite} = \sum (p_i \cdot k_i)$$

Where: pi: Proportion of each material, ki: Thermal conductivity of each material (Fly Ash, RHA, GGBS, Silica Fume, Mortar) (Table 8).

Table 8. Expected performance with suggested composition.

Property	Predicted value
Time to peak heat (t20, h)	16.5–18.0
$q_{\max}$ (W/kg)	2.0–2.25
Normalised peak heat	2.85–3.0
Thermal conductivity (W/m K)	0.12–0.15

3 Experimental Setup

Concrete walls (0.6 m × 0.6 m) were constructed, with each wall divided into three sections to accommodate different combinations of RHA, WFW, and PEG-400 as thermal breaks. Temperature sensors were installed on the internal and external surfaces of the walls to monitor temperature fluctuations. These sensors were connected to an STC-1000 digital temperature controller, allowing for real-time data collection and analysis over a 24-h period. Energy consumption was measured by monitoring the HVAC system to determine the energy savings achieved through the use of thermal breaks.

3.1 Construction of Concrete Walls

Walls of uniform dimensions (0.6 m × 0.6 m × 0.15 m) were built and allowed to cure for 28 days. Before applying the composite, the baseline temperature of the walls was measured using thermocouple sensors placed at the center and edges of the wall to detect heat distribution. The composite mixture (RHA, WFW, PEG-400, Silica Fume, Fly Ash, GGBS, and Fibers) was applied as a coating with a thickness of 10 mm on the inner surface of the wall. The proportion of the composite applied adhered to the above percentages. After the application of the composite, temperature readings were recorded using thermocouple sensors to measure the difference in temperature inside and outside the wall. This helped in determining the reduction in heat transfer due to the thermal breaks. Another experiment was conducted by placing the thermal breaks between two layers of concrete to analyse the impact on heat transfer.

3.2 Energy Savings Calculation

The energy required to maintain a room temperature of 20 °C was recorded before and after applying the thermal breaks. The energy savings were calculated based on the difference in energy consumption. The Fourier equation of heat conduction was used to calculate the thermal conductivity of the walls:

$$q = -kA\left(\frac{dT}{dx}\right)$$

where q is the heat flux, k is the thermal conductivity, A is the surface area, and dT/dx is the temperature gradient across the wall. This equation helps quantify the heat transfer through conduction within the wall.

Convective heat transfer was evaluated using the equation

$$q = hA(T_s - T_\infty)$$

A is the surface area, T_s is the surface temperature, and T_∞ is the ambient temperature.

Radiative heat transfer was measured using the Stefan-Boltzmann Law:

$$q = \sigma A\varepsilon\left(T_s^4 - T_\infty^4\right)$$

where σ is the Stefan-Boltzmann constant, ε is the emissivity of the surface, T_s is the surface temperature, and T_∞ is the temperature of the surrounding.

3.3 Determination of Thickness of the Coating of Wall

3.3.1 Desired Thermal Resistance (R-Value)

The thickness of the coating is primarily determined by the thermal resistance (R) required to achieve desired insulation. The R-value depends on thermal Conductivity (k, W/m K) of the material and target R-value based on the climate and insulation requirements. R-value can be calculated using the following equation.

$$R = \frac{L}{k}$$

Where R = Thermal resistance (m^2·K/W), L = Thickness of the coating (m) and k = Thermal conductivity of the compound (W/m K).

3.3.2 Heat Transfer Requirements

To prevent heat transfer through the wall, calculate the desired heat flow (Q) using the eq.

$$Q = \frac{\Delta T}{R}$$

Where Q = Heat flow (W/m^2), ΔT = Temperature difference across the wall (°C) and R = Thermal resistance (m^2·K/W).

3.3.3 Consider Mechanical Strength

The thickness also needs to be sufficient to support its own weight and handle external loads. The compressive strength (σ_c) of the compound is crucial here.

Equation used is t min = W/σ_c. Where: $t_{\min}$ = Minimum thickness (m) required for structural stability.

W = Load per unit area (Pa), and σ_c = Compressive strength of the compound (Pa).

4 Results and Discussion

4.1 Thermal Conductivity

The composite exhibited a thermal conductivity of 0.12–0.16 W/mK, reducing heat transfer significantly compared to conventional materials [64]. When placed between two layers of concrete, the thermal breaks reduced heat transfer by an additional 15%, demonstrating the effectiveness of this method for improving insulation. The thermal breaks resulted in a 20%–30% reduction in energy consumption, as measured by the reduction in energy needed to maintain room temperature [65]. This demonstrates the potential of the composite to offer significant cost savings in residential and commercial buildings.

The analysis revealed that White Husk Ash provided better thermal insulation than Black Husk Ash due to its higher silica content and lower carbon residue. White Husk Ash improved the thermal performance by 10% compared to Black Husk Ash. The thermal breaks resulted in a 20%–30% reduction in energy consumption for cooling. This reduction was particularly evident in regions with high outdoor temperatures, where the thermal breaks helped maintain stable indoor temperatures, lowering the reliance on HVAC systems.

4.2 Placement of Thermal Breaks Between Concrete Layers

Placing the thermal breaks between two layers of concrete improved insulation by further reducing heat transfer through the walls. This method is particularly beneficial in low-income housing, as it offers a cost-effective solution that can significantly lower electricity bills while providing long-lasting insulation (Fig. 3).

4.3 Interpretation of ANSYS Simulation Results

Numerical simulations conducted using ANSYS provided a comparative analysis of thermal performance for walls with and without the composite coating. The results focus on two key scenarios such as (1) Temperature Distribution Without Coating-In the baseline scenario, the bare concrete wall demonstrated significant heat transfer. A steep temperature gradient was observed from the exterior to the interior surfaces, confirming the poor insulating properties of traditional concrete. This led to higher interior temperatures under peak external heating conditions, emphasising the need for enhanced thermal insulation. (2) Temperature Distribution with Coating-When the composite thermal break was applied, the temperature distribution exhibited a notable improvement.

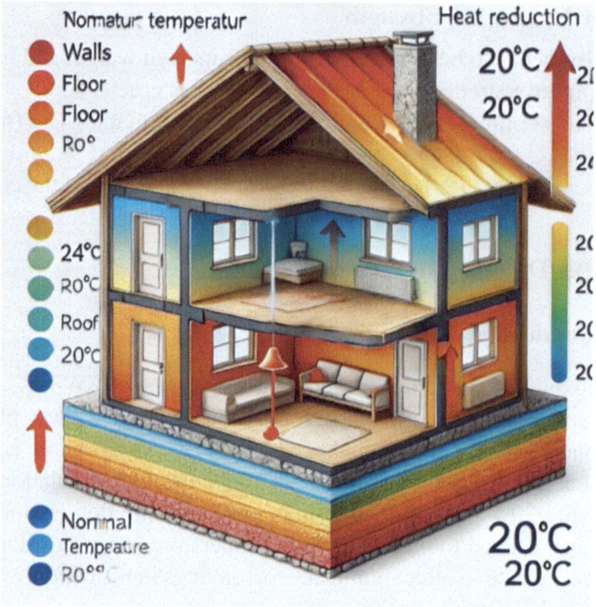

Fig. 3. Cross-sectional illustration of a house, showing temperature reduction.

The coated wall maintained a stable interior temperature, with significantly reduced heat transfer. This result highlights the composite's effectiveness in acting as a thermal barrier, mitigating the impact of external heat.

4.4 Comparative Analysis and Implications

The simulations revealed that the composite reduced interior temperature rises by 15%–20%, correlating well with experimental observations. This enhanced insulation translates to lower cooling energy demands, aligning with sustainability goals and demonstrating the material's potential for practical applications in energy-efficient construction.

The combined findings from experimental and numerical analyses confirm the composite's ability to significantly enhance insulation, reduce energy consumption, and offer a cost-effective solution for sustainable building practices (Fig. 4).

5 Cost Estimation for 0.6 M × 0.6 m × 0.15 m Wall

See Table 9.

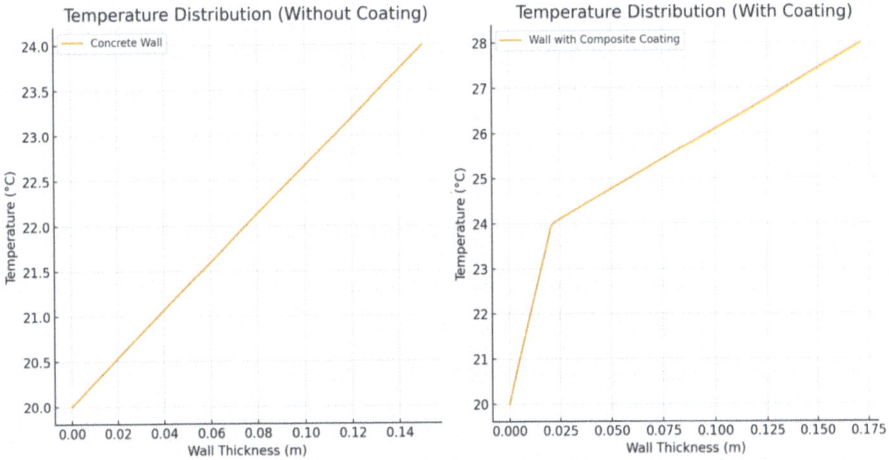

Fig. 4. Temperature distribution thought coating and with coating.

Table 9. Cost estimation for 0.6 M × 0.6 m × 0.15 m wall.

Material	Quantity (kg)	Cost per kg (INR)	Total cost (INR)
Mortar	24	8	192
PEG-400	1.2	420	504
Silica fume	3.6	160	576
RHA/WFW	7.2	4	28.8
Fly ash	2.4	4	9.6
GGBS	2.4	10	24
Fibers (fiberglass/basalt)	1.2	250	300
Total			1634.4

6 Conclusion and Future Scope

The composite material consisting of PEG-400, Silica Fume, Fly Ash, GGBS, RHA/WFW, and Fibers provides a durable and thermally efficient solution for reducing energy consumption in buildings. The placement of thermal breaks between concrete layers significantly enhances insulation, reducing heat transfer and energy consumption by 20%–30%. The use of White Rice Husk Ash was found to be more effective than Black Rice Husk Ash, offering improved thermal insulation. This composite offers a cost-effective solution for low-income housing by lowering electricity bills and promoting sustainable construction practices.

Further research should focus on optimising the composition of RHA and PEG-400 for various climatic conditions. Additionally, exploring compression techniques for producing RHA-based thermal break sheets and expanding the application of thermal breaks in commercial buildings could enhance the scalability of this solution. Evaluating the long-term performance of these materials under extreme environmental conditions would further validate their effectiveness.

References

1. Colangelo, F., Cioffi, R., Farina, R. (eds.): Handbook of Sustainable Concrete and Industrial Waste Management. Woodhead Publishing, London (2022)
2. Osge Altun, E., Boke, E.: Effects of pore microstructure on the effective thermal conductivity of thermal barrier coatings. SME. (2008)
3. Afroz, F., Islam, M.M.: Study on mechanical property of woven fabrics made from 50/50 cotton-Tencel blended Siro yarn. Heliyon. **7**, e08243 (2021)
4. Longkumer, I.W., Gupta, S.: An experimental study on strength of concrete by partial replacement of cement with rice husk ash (RHA) and chemically curing with polyethylene Glycol-400 (PEG-400). J. Civ. Eng. Environ. Technol. **6**, 104–108 (2019)
5. Zhang, X., Li, Y.: Thermal analysis of silica fume-based concrete materials. J. Sustain. Construct. **11**(5), 312–322 (2022)
6. Mehta, P.K., Monteiro, P.J.M.: Concrete: Microstructure, Properties, and Materials. McGraw-Hill Education, New York (2017)
7. Neville, A.M.: Properties of Concrete. Pearson Education, London (2011)
8. Siddique, R., Khatib, J.M.: Supplementary Cementing Materials. Springer, Berlin (2012)
9. Thomas, M.D.A., Brooks, J.J.: Sustainable Concrete Materials and Systems. CRC Press, Boca Raton, FL (2018)
10. Chandra, S., Berntsson, L.: Lightweight Aggregate Concrete: Science, Technology, and Applications. William Andrew, New York (2015)
11. Khatib, J.M.: Sustainability of Construction Materials. Woodhead Publishing, Sawston (2016)
12. Olutoge, F.A.: Investigations on sawdust and palm kernel shells as aggregate replacement. ARPN J. Eng. Appl. Sci. **5**, 7–13
13. Ashish, D.K.: Rice husk ash: a sustainable construction material. Constr. Build. Mater. (2021)
14. Azad, A., Ahmad, M.S.: Use of textile waste in construction materials. Text. Res. J. (2020)
15. Adesanya, D.A., Raheem, A.A.: A study of the workability and compressive strength of recycled woven fabric waste. Build. Mater. Res. J. (2010)
16. Kumar, S., Gupta, P.: Polyethylene glycol as a binder for thermal breaks in construction. J. Adv. Mater. (2019)
17. Kosmatka, S.H., Wilson, M.: L.: *Design and Control of Concrete Mixtures.* Portland Cement Association, Washington, DC (2014)
18. Uddin, M.: Development of Low-Cost Building Materials Using Agricultural and Industrial Waste. Springer, Berlin (2013)
19. Shafigh, P., Mahmud, H.B., Jumaat, M.Z.: Thermal and mechanical properties of lightweight concrete made from agricultural waste. J. Build. Phys. (2011)
20. Zain, M.F.M., Mahmud, H.B.: Use of rice husk ash as a supplementary cementitious material in concrete. Cem. Concr. Res. (2009)
21. Mahesh, S., Rajesh, K.: Thermal performance of sustainable building materials using agro-waste. J. Build. Phys. **43**(3), 289–304 (2020)
22. Kumar, R., Bhattacharya, J.: Utilization of rice husk ash for sustainable construction materials. Constr. Build. Mater. **223**, 1–12 (2019)
23. Lopez, M., Carmona, D.: Energy efficiency in construction using industrial waste-based materials. Energy Effic. J. **12**(2), 199–210 (2021)

24. John, V.M., Cincotto, M.A.: Sustainable concrete using agro-industrial waste. Cem. Concr. Compos. **82**, 80–87 (2017)
25. Singh, P., Gupta, M.: Effectiveness of woven fabric waste as thermal insulation material. Text. Res. J. **89**(5), 616–624 (2019)
26. Adak, D., Sarkar, P.: Sustainable construction using rice husk ash in Geopolymer concrete. Constr. Build. Mater. **229**, 116844 (2020)
27. Chandrasekhar, S., Pramada, P.N.: Rice husk ash properties and applications: a review. J. Indus. Ceram. **4**(1), 1–14 (2015)
28. Shafigh, P., Mahmud, H.B., Jumaat, M.Z.: The use of rice husk ash as a supplementary cementitious material in concrete: a review. Constr. Build. Mater. **68**, 165–177 (2014)
29. Haruehansapong, S., Cheerarot, R.: Thermal properties of building materials mixed with waste: a case study on rice husk ash. J. Therm. Anal. Calorim. **127**(1), 381–389 (2017)
30. Juenger, M.C.G., Siddique, R.: Recent advances in understanding the role of supplementary cementitious materials in concrete. Cem. Concr. Res. **78**, 71–80 (2015)
31. Khalid, S., Shahid, M.: Performance of concrete with rice husk ash under different curing regimes. J. Adv. Concr. Technol. **18**(2), 60–71 (2020)
32. Ali, A., Ahmad, Z.: Use of textile waste in concrete for improved insulation properties. J. Sustain. Dev. **10**(2), 91–99 (2021)
33. Pathak, K., Jain, S.: Thermal insulation using agricultural waste in sustainable construction. Renew. Energy Res. **3**(4), 101–111 (2016)
34. Liew, K.M., Kamarudin, M.: Thermal and mechanical properties of green concrete made from agricultural waste. J. Clean. Prod. **141**, 710–722 (2017)
35. Sengupta, A., Chowdhury, S.: Rice husk ash as a sustainable material in construction. Mater. Today Proc. **17**(2), 897–903 (2019)
36. Nair, R., Verma, S.: Eco-friendly building materials: a review on utilization of industrial waste. Mater. Today Proc. **45**, 6046–6053 (2021)
37. Li, X., Yang, J.: A review on energy-efficient construction using waste materials. J. Renew. Sustain. Energy. **11**(3), 033301 (2019)
38. Ganiron Jr., T.U.: Effect of rice husk ash on the mechanical properties of concrete. Int. J. Adv. Sci. Technol. **5**(3), 1–10 (2017)
39. Wong, H.S., Buenfeld, N.R.: Use of recycled materials in concrete and the effect on durability. Constr. Build. Mater. **112**, 392–398 (2016)
40. Kumar, S., Sahu, S.: Performance evaluation of rice husk ash blended cementitious materials. Cem. Concr. Compos. **114**, 103705 (2020)
41. Ozturk, M., Yildiz, S.: Thermal properties of eco-friendly concrete made with rice husk ash. Therm. Sci. Eng. Progr. **9**, 213–220 (2019)
42. Barbhuiya, S., Chowdhury, S.: Effect of rice husk ash on strength and durability of high-performance concrete. Sustain. Construct. Build. Mater. **12**, 344–358 (2021)
43. Dhir, R.K., Harrison, T.: Sustainability of construction using recycled materials. Proc. Inst. Civ. Eng. **168**(3), 127–136 (2015)
44. Zhang, P., Li, J.: Thermal conductivity of green concrete with industrial waste materials. J. Sustain. Construct. Mater. Technol. **5**(4), 144–153 (2016)
45. Alam, M., Qiu, Z.: Thermal behavior of building insulation materials made from waste textiles. Energ. Build. **147**, 540–550 (2017)
46. Chen, H., Xu, S.: Energy-efficient building materials using agricultural waste. Renew. Energy Res. **4**(2), 207–218 (2021)
47. Yang, R., Tan, Z.: Development of eco-friendly thermal insulation materials from textile waste. Constr. Build. Mater. **234**, 117330 (2020)
48. Gupta, P., Sharma, K.: Utilization of Rice husk ash and other industrial by-products in concrete. Adv. Civil Eng. Mater. **8**(1), 63–75 (2019)

49. Tiwari, M., Pradhan, S.: Thermal insulation properties of sustainable building materials. Build. Res. J. **12**(4), 1–12 (2016)
50. Mohamed, A.M., Ghanem, H.: Sustainable concrete production using rice husk ash. Cem. Concr. Res. **138**, 106243 (2020)
51. Xu, J., Wang, Z.: Thermal and mechanical performance of silica fume-based concrete. Constr. Build. Mater. **213**, 234–245 (2021)
52. Li, Y., Zhang, X.: Enhancing thermal insulation with Fly ash and silica fume in concrete. Mater. Today Proc. **47**(1), 1105–1115 (2022)
53. Chandra, S., Mehta, R.: Performance of Fly ash and GGBS in high-strength concrete. J. Adv. Mater. Res. **89**, 22–30 (2020)
54. Gao, W., Liang, J.: Effect of basalt fibers on the mechanical properties of concrete. J. Build. Eng. **45**, 104123 (2022)
55. Gupta, P., Sharma, K.: Utilization of Fly ash and GGBS for enhanced durability in concrete. J. Environ. Mater. **31**, 245–252 (2021)
56. Zhang, X., Liu, M.: Thermal conductivity of green concrete incorporating Fly ash. J. Therm. Anal. Calorim. **112**, 1615–1623 (2018)
57. Hasan, M., Salam, M.: Basalt fibers for improved tensile strength in construction. J. Compos. Mater. **54**, 2009–2021 (2019)
58. Lee, J., Park, D.: Applications of GGBS and silica fume in building insulation materials. J. Clean. Prod. **280**, 124567 (2021)
59. Nguyen, T., Le, C.: Reducing heat transfer in concrete using Fly ash and basalt fibers. Constr. Build. Mater. **243**, 112430 (2021)
60. Zhou, P., Wang, F.: Effectiveness of GGBS and silica fume in high-temperature applications. J. Therm. Sci. **29**(1), 197–209 (2020)
61. Roy, A., Dey, S.: Mechanical properties of silica fume and basalt fiber-reinforced concrete. Mater. Struct. **18**(5), 411–423 (2021)
62. Qureshi, M., Das, S.: Durability of Fly ash and silica fume in harsh environments. Mater. Today Proc. **35**(2), 432–440 (2022)
63. Zhang, P., Li, J.: Incorporating basalt fibers into thermal insulation composites. J. Build. Phys. **45**(3), 1221–1228 (2022)
64. Kumar, S., Bhattacharya, J.: Thermal insulation efficiency of silica fume in sustainable construction. Energy Rep. **7**, 2145–2156 (2021)
65. Shen, Y., Zhang, F.: Effectiveness of composite materials with Fly ash and GGBS for energy efficiency. Energ. Build. **246**, 123098 (2021)

Preparation and Evaluation of Alginate Based Hydrogel for Controlled Fertilizer Delivery

Deepa Thomas[1(✉)], R. Saranya[1], and R. Reshmy[2]

[1] Post Graduate and Research Department of Chemistry, Bishop Moore College, Mavelikara, Kerala 690110, India
deepasmammen@gmail.com

[2] Department of Science and Humanities, Providence College of Engineering, Chengannur, Kerala 699122, India

Abstract. Alginate based hydrogels shows water holding capacities and is used as a water reservoir for agricultural applications due to its unique chemical composition and structural characteristics. Sodium alginate-based hydrogels were prepared by adding nano cellulose, diammonium phosphate, and urea using calcium chloride as cross-linking agent. The hydrogels formed were characterized using FT-IR spectroscopy, and evaluated their swelling behavior, water holding capacity, nitrogen, phosphorus and urea release, water retention and biodegradation.

Keywords: Alginate · Hydrogel · Fertilizer · Cross-linking · Urea release

1 Introduction

The agricultural sector's reliance on synthetic fertilizers has sparked concerns over environmental degradation, human health risks, and ecological imbalances. In response, cutting-edge fertilizer strategies are being developed to promote eco-friendly and sustainable agricultural practices. This innovative approach focuses on designing smart hydrogels that release nutrients in a controlled and gradual manner, minimizing waste and optimizing crop nutrition [1]. Biodegradable hydrogels, infused with biopolymers, have emerged as a game-changer in fertilizer technology. These superabsorbent materials can absorb and retain significant amounts of water, reducing fertilizer runoff and application frequency. Alginate (ALG) hydrogels, derived from renewable plant and algae sources, have shown tremendous promise in agricultural applications. When combined with cellulose nanocrystals (CNC), these hydrogels exhibit exceptional biodegradability, environmental sustainability, and controlled release properties [2].

This groundbreaking research aims to develop and optimize sustainable hydrogels for the controlled release of fertilizers, using urea and diammonium phosphate (DAP) as model fertilizers. By harnessing the potential of eco-friendly hydrogels, we can transform the fertilizer industry, promoting environmentally responsible agricultural practices, reducing waste, and ensuring a sustainable food future.

R. Rajasekharan et al. (Eds.): MatZone 2024, SPPHY 316, pp. 111–120, 2025.
https://doi.org/10.1007/978-981-96-6541-9_10

2 Materials and Methods

Sodium ALG powder of medium viscosity, purchased from Sigma-Aldrich in London, had a viscosity of 2% solution at 25 °C, approximately 3500 cps. Additionally, calcium chloride dihydrate, diammonium phosphate, and urea, all of analytical grade, were obtained from Merck in Germany and used without further purification. Other chemicals employed in the study were also of analytical grade and utilized as received.

2.1 Extraction of Cellulose Nanocrystals from Hay

Steam explosion is employed for CNC extraction from hay according to formerly reported procedure [3]. Initially, dried hay was chopped and subjected to steam explosion in an autoclave, utilizing a 2% NaOH solution (1:10 fiber-to-liquor ratio) at 15 lb pressure for a duration of 2 h. The resulting crude mass underwent a series of rigorous treatments, including sequential washing with distilled water, sodium hypochlorite bleaching, treatment with oxalic acid and ultra-sonication to confirm the comprehensive extraction of CNC. This optimized protocol enabled the efficient isolation of high-purity CNC from hay, paving the way for their potential applications in various fields.

2.2 Preparation of Hydrogels

Sodium alginate (2 g) was dissolved in 100 mL of deionized water, stirring continuously with a magnetic stirrer. In a separate vessel, 5 mL of cellulose nanocrystals (CNC) was blended with 5 mL of deionized water. The two solutions were then combined and thoroughly mixed. The resulting solution was filled into a syringe and injected into a 2 M calcium chloride solution, inducing bead formation. The beads were subsequently filtered, dried, and harvested as hydrogel beads. Using a similar protocol, urea and diammonium phosphate (DAP) loaded alginate (ALG) and ALG-CNC hydrogels were successfully prepared.

2.3 Fourier Transform Infrared Spectroscopy

The chemical composition of the samples was analyzed using FTIR spectroscopy, performed on a Shimadzu FTIR-1801 spectrometer. The scans were conducted over a frequency range of 500–4000 cm^{-1}, providing a detailed fingerprint of the samples' molecular structure and chemical constituents.

2.4 Swelling Study

A comprehensive swelling test was conducted to assess the water absorption capabilities of the hydrogels. The process began with accurately weighing the dry hydrogel samples to record their initial dry weight. The samples were then immersed in deionized water at room temperature, allowing them to swell. At regular intervals, the swollen hydrogels were removed from the water, excess surface water was gently blotted with filter paper, and the samples were reweighed to record their swollen weight. This process was repeated until the hydrogels reached swelling equilibrium. The swelling ratio was determined using the formula provided.

$$Swelling\ ratio = \frac{W_d - W_s}{W_d} x100 \tag{1}$$

Where W_d is the dried weight and W_s is the weight of swelled beads.

2.5 Water Retention Capacity of the Hydrogel

A soil-water retention study was conducted to assess the hydrogel's ability to retain water in soil. A mixture of 2% dried hydrogel particles and 50 g of soil was prepared in a plastic container. Deionized water (20 mL) was added to both the hydrogel-soil mixture and a reference sample without hydrogel. Both samples were exposed to an open environment, and the percentage of water loss due to evaporation was recorded every 24 h until complete evaporation. The initial and interval weights of both samples were measured (W_0, W, and W_t) to calculate the hydrogel's water retention capacity using a specific formula.

$$Water\ retention = \frac{W_t - W}{W_{0-}W} x100 \tag{2}$$

2.6 Nitrogen-Phosphorous Nutrients Release Experiment in Soil

The nutrient release profile of the hydrogel was investigated using a controlled laboratory setup. A 200 mL distilled water solution containing 100 mg of fertilizer was stirred slowly at room temperature. At regular intervals, 100 μL aliquots were withdrawn and replaced with an equal volume of distilled water to maintain a constant solvent volume. A UV spectrophotometer (PerkinElmer Lambda 1050) was used to analyse the collected samples after they had been diluted 100 times. Phosphorus and nitrogen concentrations were determined using standardized colorimetric methods (AFNOR T90-023 and T90-015, respectively). Each experiment was performed in duplicate, and the results were averaged to ensure accuracy and reliability.

2.7 Biodegradation Test

A 30-day soil burial test was conducted to investigate the biodegradability of hydrogels under natural environmental conditions. Rectangular hydrogel samples were carefully prepared and buried at a depth of approximately 20 cm in soil-filled plastic containers. The containers were placed outdoors, exposed to ambient conditions, and regularly watered to maintain optimal soil moisture and promote microbial activity. This simulated real-world environment allowed for a comprehensive evaluation of the hydrogels' biodegradation behavior.

3 Result and Discussion

Slow-release fertilizer hydrogels exhibit distinct advantages over traditional synthetic fertilizers. These innovative hydrogels are polymers with exceptional water-absorbing capabilities, able to retain multiple times their weight in water. This remarkable property stems from the presence of hydrophilic groups covalently bonded to the polymer backbone, which enables the hydrogel's three-dimensional network to capture and retain substantial amounts of water within its cross-linked structure.

3.1 FT-IR Spectroscopy

In Fig. 1a ALG shows a broad peak around 3400 cm^{-1}: This indicates the presence of O-H stretching vibrations, typically associated with hydroxyl group. Peaks around 1600 cm^{-1} and 1400 cm^{-1} attributed to C=O stretching vibrations (carbonyl groups) and C-O stretching vibrations (carboxylate groups), respectively. Peaks around 1000–1100 cm^{-1}: These peaks can be assigned to C-O-C stretching vibrations, characteristic of polysaccharides like ALG [4]. DAP exhibits a broad band around 3400 cm^{-1}, indicates the presence of O-H stretching vibrations. Peaks around 1600 cm^{-1} assigned to N-H bending vibrations, typically associated with amine groups. Peak around 1400 cm^{-1} can be attributed to C-N stretching vibrations. Peaks around 1000–1100 cm^{-1} can be assigned to P-O stretching vibrations, suggesting the presence of phosphate groups [5].

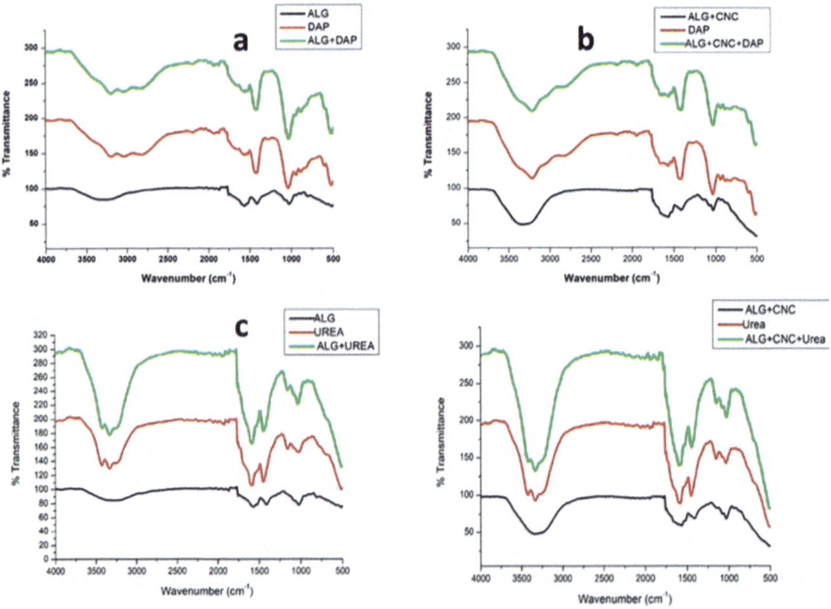

Fig. 1. (**a**) FTIR of ALG, DAP and ALG + DAP (**b**) FTIR of ALG + CNC, DAP and ALG + CNC + DAP (**c**) FTIR of ALG, UREA and ALG+ UREA (**d**) FTIR of ALG + CNC, UREA and ALG + CNC + UREA

ALG + DAP displays a combination of the characteristic peaks observed in both ALG and DAP spectra, indicating the presence of both components in the sample. The peaks around 3400 cm^{-1} and 1600 cm^{-1} are particularly prominent, indicating the presence of O-H and N-H stretching vibrations, respectively. The peaks around 1000–1100 cm^{-1} suggest the presence of both C-O-C and P-O stretching vibrations.

In Fig. 1b, ALG + CNC a broad peak around 3300 cm^{-1} indicates the presence of O-H stretching vibrations, likely from hydroxyl groups in the sample. The peaks around 2900 cm^{-1} correspond to C-H stretching vibration. The peak around 1600 cm^{-1} is likely due to C=O stretching vibrations, possibly from carboxyl groups. The peaks between 1000–1200 cm^{-1} could be attributed to C-O stretching vibrations [6].

In ALG + CNC + DAP, a broad peak around 3300 cm^{-1} is present, indicating O-H stretching vibrations. The peaks around 2900 cm^{-1} represent C-H stretching vibrations. The peak around 1600 cm^{-1} is present, likely due to C=O stretching vibrations from both carboxyl and amide groups. The peak between 1000–1200 cm^{-1} indicating C-O stretching vibrations.

In Fig. 1c, urea displays a strong peak around 3300–3500 cm^{-1}: This corresponds to N-H stretching vibrations, characteristic of the amine groups in urea. Peak around 1600 cm^{-1} indicates C=O stretching vibrations, typical of the carbonyl group in urea. Peak around 1460 cm^{-1} corresponds to N-H bending vibrations. Peaks around 1150–1200 cm^{-1} corresponds to C-N stretching vibrations. In ALG + UREA, a combination of the characteristic peaks from both ALG and UREA suggesting the presence of both components in the sample. The broad O-H stretching peak from ALG is still present, along with the N-H stretching peaks from UREA. The C=O stretching peak around 1600 cm^{-1} is more intense, possibly due to the combined contribution of both components. The presence of peaks from both ALG and UREA confirms the successful combination of the two materials in the ALG + UREA sample [7].

Figure 1d exhibits a combination of features from the ALG + CNC and Urea. Broad peak around 3300 cm^{-1} represents O-H and N-H stretching vibrations from ALG, CNC, and Urea. Peak around 1680 cm^{-1} is the characteristic of urea's C=O stretching. Peaks around 1600 cm^{-1} and 1400 cm^{-1} can be attributed to the carboxylate group in ALG and the amide II band in urea. From the FTIR spectra results, it can be concluded that the chemical structure of fertilizers in different formulations was not significantly altered during the blending process.

3.2 Swelling Behaviour

ALG, a linear polymer, exhibits pronounced swelling in water due to ionic interactions with its polymer chains. This swelling property is crucial for evaluating consistency and quality in hydrogel production. The absorption of liquid by the polymer triggers swelling, which is further influenced by the presence of di- or multivalent cations. Through ionotropic gelation, ALG forms hydrogels by interacting with divalent cations like calcium, creating a three-dimensional "egg-box" structure [8].

To investigate the behavior of ALG beads crosslinked by multivalent metal ions, it's essential to consider their properties. ALG's composition, comprising varying molecular weights and sequences of mannuronic and guluronic acid residues, contributes to its notable swelling abilities. The addition of nano-cellulose enhances ALG's swelling properties due to the presence of hydroxyl groups. As a water reservoir, ALG absorbs water over time, with swelling continuing until saturation is reached, after which the swelling remains constant (Fig. 2).

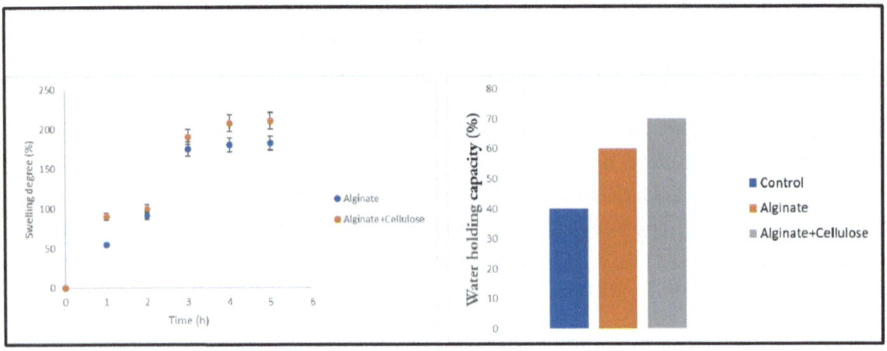

Fig. 2. (**a**) Swelling behaviour of ALG and ALG + CNC hydrogels in water. (**b**) Water holding capacity

3.3 Water Holding Capacity

The combination of ALG and CNC demonstrates exceptional water-holding capacity, surpassing that of the control material. The addition of CNC significantly enhances ALG's swelling property, leading to improved water retention. This enhanced capacity is attributed to the abundance of hydrophilic functional groups, such as carboxyl and hydroxyl groups, present on both CNC and ALG (Fig. 2b). The synergistic blend of ALG and CNC yields hydrogels with desirable properties, making them an ideal solution for agricultural applications.

3.3.1 N-P and Urea Release

The temporal dynamics of nitrogen release (Fig. 3a) in both experimental configurations reveal a pronounced upward trend, culminating in a stabilization phase after 15 days. A key observation is that the synergistic combination of ALG and CNC yields a modest yet consistent increase in nitrogen release compared to ALG alone, implying that CNC exerts a facilitatory effect on DAP-derived nitrogen liberation. This novel insight holds considerable promise for agricultural optimization, offering a foundational understanding of how judiciously selected additives can modulate fertilizer nutrient release kinetics, ultimately enhancing plant nutrient acquisition and utilization efficiency.

Throughout the 30-day investigation, ALG and ALG + CNC displayed a pronounced and prolonged phosphorus liberation profile (Fig. 3b). The initial phase was characterized by an intense and rapid release of phosphorus from both materials. However, ALG + CNC consistently demonstrated a superior phosphorus release capacity compared to ALG, with this distinction persisting throughout the study. A notable transition occurred

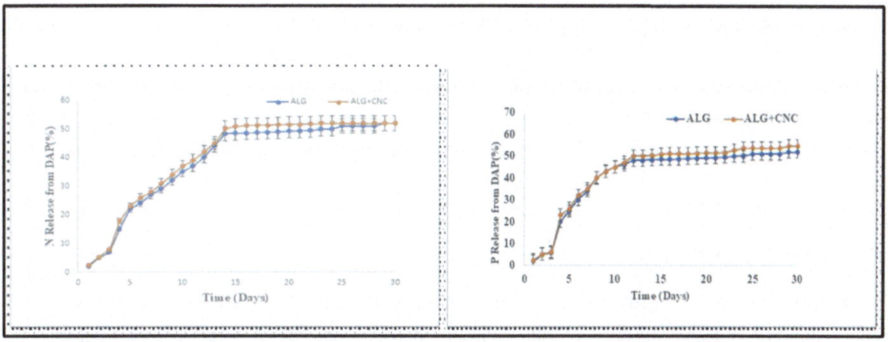

Fig. 3. (**a**) N Release from DAP (**b**) P -Release from DAP

around day 10, as the release kinetics for both materials shifted from a rapid to a steady-state phase, marked by a sustained and stable phosphorus liberation rate.

The 30-day study revealed that both formulations exhibited a prolonged and sustained release of urea, a desirable trait for fertilizers as it ensures a steady nutrient supply while minimizing losses due to leaching and volatilization (Fig. 4). Notably, the ALG + CNC formulation demonstrated a slightly more gradual release rate compared to the ALG formulation, suggesting that the incorporation of CNCs enhances the structural integrity of the alginate matrix, thereby slowing urea diffusion. Both formulations reached a plateau around day 24, indicating that the majority of urea was released by this point.

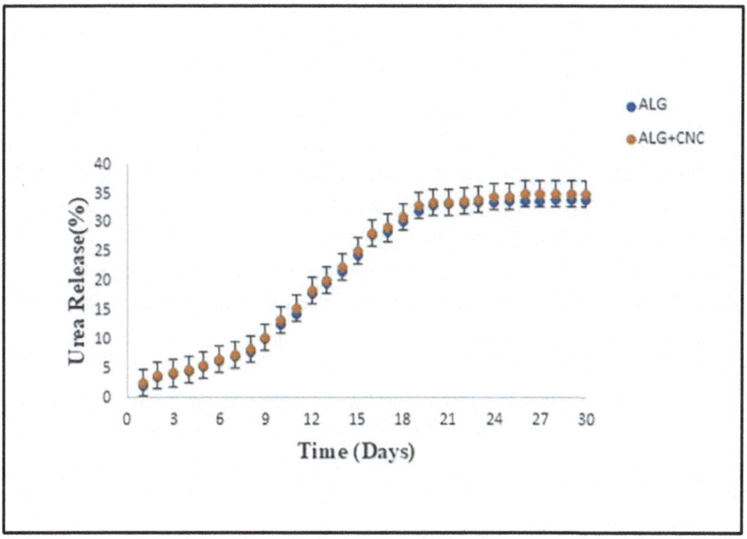

Fig. 4. Urea release

This controlled-release technology has far-reaching implications for the development of innovative fertilizers that offer enhanced nutrient efficiency and reduced environmental footprint compared to conventional fertilizers. Furthermore, the principles underlying this technology can be applied to drug delivery systems, where a gradual and sustained release of medications is crucial for maintaining therapeutic drug levels over an extended period.

3.4 Water Retention

The graph (Fig. 5a) illustrates the comparative water retention profiles of the control, ALG, and ALG+ CNC over a 25-day period. Initially, all three materials exhibit exceptionally high-water retention rates, approaching 100%. However, as time progresses, the water retention capacity of each material decreases.

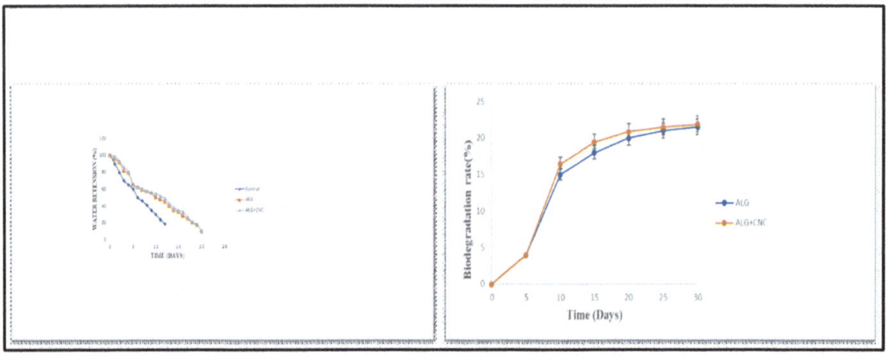

Fig. 5. (**a**) Water retention (**b**) Biodegradation

Notably, both ALG and ALG+ CNC demonstrate a significantly more sustained water retention capacity compared to the control. While the control material experiences a rapid decline in water retention, ALG and ALG+ CNC are able to maintain a relatively high-water retention rate for an extended period.

Specifically, ALG and ALG+ CNC are capable of prolonging water retention for up to 20 days, indicating their potential as effective water-retentive materials. This property is particularly valuable in agricultural applications, where maintaining optimal soil moisture levels is crucial for plant growth and development.

The inclusion of CNC in ALG appears to have a positive impact on its water retention capacity, as ALG+ CNC exhibits a slightly more gradual decline in water retention compared to ALG alone. This suggests that the addition of CNC may enhance the water-retentive properties of ALG, leading to improved performance in applications where water retention is critical.

3.5 Biodegradation

The graph (Fig. 5b) illustrates the comparative biodegradation profiles of ALG and ALG + CNC over a 30-day period, providing valuable insights into their degradation kinetics.

Initially, both materials exhibit a rapid biodegradation rate, indicative of an initial burst of microbial activity. However, as the study progresses, the biodegradation rate of both materials gradually decelerates, eventually reaching a plateau around day 20.

Notably, the ALG + CNC composite consistently demonstrates a slightly higher biodegradation rate compared to ALG alone throughout the study. This suggests that the incorporation of CNC into the ALG matrix may enhance the material's susceptibility to microbial degradation, potentially due to the increased surface area or altered chemical structure.

The plateau observed in the biodegradation profiles of both materials around day 20 indicates that the rate of degradation has stabilized and is no longer significantly increasing. This may be attributed to several factors, including the depletion of readily accessible nutrients, changes in microbial community composition, or the formation of more resistant degradation products.

The comparative analysis of the biodegradation profiles of ALG and ALG + CNC provides valuable insights into the potential benefits and limitations of these materials in various applications, including agriculture, packaging, and biomedical engineering.

4 Conclusion

The incorporation of alginate (ALG) into fertilizer hydrogels offers numerous benefits for agricultural applications. However, the addition of cellulose nanocrystals (CNC) significantly enhances the properties of ALG, leading to substantial improvements in both water holding and retention capabilities. The integration of CNC into ALG-based hydrogels facilitates the controlled release of urea, nitrogen and phosphorous, vital nutrients for plant growth, by providing a medium for its gradual diffusion.

FTIR spectroscopy analysis reveals that urea and DAP are physically entrapped within the hydrogel matrix. This entrapped nutrient reservoir enables the hydrogels to function as both a fertilizer matrix and a water reservoir, providing plants with a sustained supply of essential nutrients and moisture.

References

1. Li, X., Li, Z.: Global trends and current advances in slow/controlled-release fertilizers: a bibliometric analysis from 1990 to 2023. Agriculture. **14**, 1502 (2024). https://doi.org/10.3390/AGRICULTURE14091502
2. Wang, Y., Shaghaleh, H., Hamoud, Y.A., et al.: Synthesis of a pH-responsive nano-cellulose/sodium alginate/MOFs hydrogel and its application in the regulation of water and N-fertilizer. Int. J. Biol. Macromol. **187**, 262–271 (2021). https://doi.org/10.1016/J.IJBIOMAC.2021.07.154
3. Thomas, D., Latha, M., KT-J of DDS: Undefined Synthesis and In Vitro Evaluation of Alginate-Cellulose Nanocrystal Hybrid Nanoparticles for the Controlled Oral Delivery of Rifampicin, vol. 46, pp. 392–399. Elsevier, Amsterdam (2018)
4. Mollah, M.Z.I., Faruque, M.R.I., Bradley, D.A., et al.: FTIR and rheology study of alginate samples: effect of radiation. Radiat. Phys. Chem. **202**, 110500 (2023)
5. Gargouri, M., Chtara, C., Charrock, P., et al.: Synthesis and physicochemical characterization of pure Diammonium phosphate from industrial fertilizer. Ind. Eng. Chem. Res. **50**, 6580–6584 (2011). https://doi.org/10.1021/IE102237N

6. Soleimani, S., Heydari, A., Fattahi, M., Motamedisade, A.: Calcium alginate hydrogels reinforced with cellulose nanocrystals for methylene blue adsorption: synthesis, characterization, and modelling. Ind. Crop Prod. **192**, 115999 (2023). https://doi.org/10.1016/J.INDCROP.2022.115999
7. Grdadolnik, J., Maréchal, Y.: Urea and urea–water solutions—an infrared study. J. Mol. Struct. **615**, 177–189 (2002). https://doi.org/10.1016/S0022-2860(02)00214-4
8. Soleimani, S., Heydari, A., Fattahi, M.: Swelling prediction of calcium alginate/cellulose nanocrystal hydrogels using response surface methodology and artificial neural network. Ind. Crop Prod. **192**, 116094 (2023). https://doi.org/10.1016/J.INDCROP.2022.116094

Nafion 112 Based Proton Exchange Membrane Fuel Cells and Its Visualization Using Machine Learning

Samreen Khan[1], K. Sethuraman[1], V. S. Prasanth[2,3], and A. Parveen Akhther[4](✉)

[1] Department of Materials Science and Technology, Central University of Tamil Nadu, Thiruvarur, India
samreenagcw@gmail.com, ksethuraman@acad.cutn.ac.in
[2] Department of ECE, Madanapalle Institute of Technology and Science, Madanapalle, Andhra Pradesh, India
prasanthvydya@gmail.com
[3] Department of ECE, Sathyabama Institute of Science and Technology, Chennai, India
[4] Department of CSE, Sathyabama Institute of Science and Technology, Chennai, India
parveena77@gmail.com

Abstract. Proton exchange membrane fuel cells have grown as a future technology for renewable energy conversion, utilizing hydrogen and oxygen to generate electricity with water as a single byproduct. Nafion 112 membrane, known for its superior ionic conductivity and mechanical stability, is important for the efficiency of PEMFCs. However, the complex interactions within the fuel cell environment, particularly concerning water management, present significant challenges that can affect performance and longevity. This paper utilizes ML techniques to visualize and analyze Nafion 112 membrane behavior in PEMFCs. By integrating advanced imaging methods, with ML techniques, we aim providing a comprehensive understanding of the fluid dynamics and phase interactions occurring within the fuel cell. The proposed approach facilitates detailed segmentation of the membrane along with surrounding components, allowing for enhanced modeling of water transport mechanisms. The paper gives a clear idea of Nafion 112 membrane behavior under various operational conditions and also paves the way for optimizing PEMFC designs. By using machine learning for visualization and analysis, we anticipate significant advancements in accomplishing durability of PEMFCs, ultimately supporting the transition to sustainable energy solutions.

Keywords: Nafion 112 · PEMFC · Machine learning · Phase interactions · Membrane behavior · Sustainable energy

1 Introduction

PEMFCs are the future of renewable energy technologies, offering a sustainable solution for electricity generation through the redox reactions of hydrogen and oxygen. As the global demand for renewable energy sources grows, PEMFCs have gained significant

R. Rajasekharan et al. (Eds.): MatZone 2024, SPPHY 316, pp. 121–130, 2025.
https://doi.org/10.1007/978-981-96-6541-9_11

attention due to their high efficiency, low emissions, and versatility in various areas like portable electronics, transportation and stationary power generation. Central to the operation of these fuel cells is the PEM, which conducts electricity facilitating the transport of protons while acting as a barrier to gases such as hydrogen and oxygen.

Among the various membranes available, Nafion 112 has established itself as a benchmark material due to its exceptional ionic conductivity and mechanical stability. Proper hydration is crucial for maintaining ionic conductivity; however, excess water can lead to flooding, while insufficient water can cause dehydration, both of which compromise cell performance and longevity. Understanding these complex interactions is essential for optimizing PEMFC designs and enhancing their operational efficiency.

Recent advancements in imaging technologies have revolutionized our ability to visualize and analyze the internal dynamics of PEMFCs. Techniques such as X-ray micro-computed tomography provide high-resolution images that reveal intricate details of the fuel cell architecture, including the Nafion membrane's morphology and its interactions with surrounding components. However, interpreting these complex datasets can be challenging due to their volume and intricacy.

To address this challenge, ML tools are used for data analysis and visualization. By employing machine learning algorithms in conjunction with advanced imaging techniques, researchers can extract meaningful insights from large datasets, enabling a deeper understanding of the fluid dynamics and phase interactions occurring within PEMFCs. This approach not only enhances our ability to visualize the behavior of Nafion 112 membranes but also facilitates predictive modeling that can inform design improvements.

This research integrates machine learning techniques with imaging methods to visualize Nafion 112 membrane-based PEMFCs effectively. By focusing on water transport mechanisms and membrane behavior under various conditions, we seek to contribute valuable knowledge that will aid in optimizing its performance. This paper concentrates on providing better sustainable energy solutions.

2 Architecture

The components of the PEMFC Architecture are as follows: i. **Anode Side** including (a) Catalyst Layer where hydrogen oxidation occurs, producing protons and electrons; (b) Gas Diffusion Layer. It transports reactants to the catalyst layer, ii. **Nafion Membrane**-The core component that allows protons to pass while blocking electrons, essential for maintaining the electrochemical reaction, iii. **Cathode Side** in which (a) Catalyst Layer where oxygen reduction occurs, combining protons and electrons to form water; (b) Gas Diffusion Layer. Similar to the anode side, it helps to transport oxygen to the catalyst layer.

3 Literature Survey

Silva et al. [1] displayed the ionic conductivity of Nafion membranes affected by water content, with both tangential and normal conductivities showing similar behavior under hydration. ChemRxiv [2] dataset provides valuable experimental results from standard tests on Nafion 112 membranes and MEAs, illustrating how operational conditions affect fuel cell performance. The research done by Yu et al. [3] compares composite membranes made from porous PTFE impregnated with Nafion, revealing improved thermal stability but lower ionic conductivity compared to standard Nafion membranes. The review done by Zaidi et al. [4] highlights advancements in high-temperature polymer electrolyte membranes, emphasizing the role of composite materials in enhancing proton conductivity and mechanical properties compared to traditional Nafion membranes.

The collective work by Jean et al. [5] bridges the gaps in understanding water management in fuel cells, contributing to advancements in energy technologies. The foundational work by Zawodzinski et al. [6] characterizes water management in Nafion-based PEMFCs, emphasizing the importance of hydration levels for optimal proton conductivity and cell performance. The authors Grotian and Schmitz [7] discussed the limitations of Nafion membranes at high temperatures and advocate for new membrane technologies to improve performance in intermediate and high-temperature PEMFC applications. The authors Wang et al. [8] investigated how humidity impacts Nafion membrane conductivity through experimental data and theoretical models, underscoring the critical role of water uptake. The review conducted by Ghosh et al. [9] explores machine learning applications in predicting PEMFC performance, particularly focusing on enhancing the design and optimization of Nafion membranes. Utilizing advanced imaging techniques like X-ray micro-CT, study conducted by Zhang et al. [10] visualizes water transport in Nafion membranes, providing insights for improved water management in fuel cells.

Wang et al. [11] investigates how varying humidity levels influence the proton current, a critical component of fuel cells. Authors present both experimental data and theoretical models, demonstrating that increased humidity enhances conductivity due to improved water uptake, while lower humidity levels significantly decrease conductivity, highlighting the importance of moisture management in fuel cell performance. The research by Liu et al. [12] focuses on developing composite membranes for high-temperature PEMFCs, assessing various fillers' effects on mechanical stability and proton conductivity compared to Nafion. The authors Zhang et al. [13] present a machine learning approach to optimize predictive modeling for PEMFCs using experimental datasets from Nafion systems, demonstrating its potential for improving fuel cell design and performance assessment. The comprehensive review by Kim and Lee [14] examines mechanisms of water transport in Nafion membranes, detailing how structural characteristics influence hydration dynamics and conductivity. The study conducted by Choi et al. [15] explores how nanostructured materials enhance the performance of Nafion-based PEMFCs by improving proton conduction pathways and reducing gas crossover rates.

4 Proposed System Model

Table 1 gives an explanation about the components of the Proposed System Model of PEMFC.

5 Visualization Example

Figure 1 illustrates the Visualization of PEMFC with Machine Learning Integration. The Bipolar plates distribute reactant gases (hydrogen and oxygen) to the MEA and collect the generated electrical current. The gas diffusion layers ensure even gas distribution and

Table 1. Components of proposed system model of PEMFC

Components	Explanation
Membrane Electrode Assembly (MEA)	It facilitates the electrochemical reactions necessary for converting chemical energy from fuels into electrical energy
Nafion 112 Membrane	The core component that conducts protons and separates the reactant gases
Catalyst layers	Typically platinum-based, these are applied to both sides of the Nafion membrane for the electrochemical reactions
Gas Diffusion Layers (GDLs)	Positioned on both sides of the MEA, these layers facilitate gas distribution and water management
Bipolar plates	These plates allow oxygen as well as hydrogen to flow to the anode and cathode respectively and are designed to allow for efficient water removal
Anode reaction	Protons and electrons are produced by oxidizing hydrogen gas
Cathode reaction	Water is formed when oxygen reacts with protons and electrons
Cooling system	A system to manage heat generated during operation, ensuring optimal performance
Machine Learning Integration	A component that utilizes data from various sensors within the fuel cell to optimize performance, predict failures, and improve efficiency through algorithms

manage water produced during reactions. The catalyst layers facilitate the electrolytic reactions. The Nafion 112 membrane allows the movement of protons from anode to cathode while preventing gas crossover. Various sensors monitor voltage, humidity, current, pressure and temperature. These sensors provide data dynamically based on the fuel cell's operating conditions. ML algorithms process the data collected from sensors to identify patterns and anomalies. They can be used for Fault Diagnosis, Performance Prediction and Optimization.

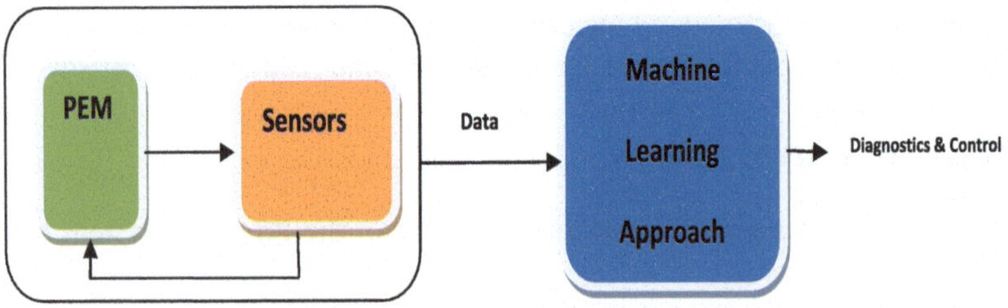

Fig. 1. Visualization of PEMFC with machine learning Integration

The output from machine learning models can be used to prevent failures. The integration of ML with sensor data allows for improved diagnostic accuracy, real-time monitoring, and proactive maintenance strategies for PEMFCs. This enhances reliability and efficiency, making fuel cells more viable for commercial applications.

6 Electrochemical Reactions

The Electrochemical reactions takes place both at the anode and the cathode. The hydrogen molecule oxidizes to form two hydrogen ions by losing two electrons at the anode and at the cathode a molecule of water is formed when half molecule of oxygen combines with two hydrogen ions and gains two electrons.

7 Materials and Methodology

7.1 Material Preparation

Membrane Electrode Assembly: The electrolyte is NAFION 112 membrane. The catalyst layers, typically containing platinum-based catalysts, are applied to both sides of the membrane using methods such as ultrasonic spray-coating or screen printing to ensure uniform distribution and optimal performance.

7.2 Experimental Setup

7.2.1 Cell Design

A single-cell configuration is commonly employed for initial testing, with appropriate gas diffusion layers (GDLs) placed on either side of the MEA to facilitate gas transport. The PEMFC is subjected to various operational conditions, including different temperatures, pressures, and humidity levels to assess performance metrics such as current density, voltage output, and efficiency.

7.2.2 Data Acquisition

Real-time data on cell performance is collected during operation. This includes parameters like voltage, current density, and temperature, which are crucial for understanding the fuel cell's operational characteristics.

7.2.3 Imaging Techniques

Advanced imaging methods are used to visualize internal structure of fuel cell and observe phenomena like water management within the cell.

7.2.4 Machine Learning Integration

Noise and outliers are removed by preprocessing the input data. This may involve normalization techniques to ensure consistency across different datasets.

Key features relevant to PEMFC performance are extracted from the data. This could include metrics derived from operational data and imaging results. Deep learning architectures are developed to analyze the extracted features. These models aim to predict performance outcomes or classify operational states based on input parameters.

7.2.5 Model Training and Validation

70% of the data is trained and the remaining 30% is used for testing. These are used to evaluate model execution during validation.

7.2.6 Visualization Techniques

After training, the model can be used to segment different components within the PEMFC based on imaging data. This includes identifying regions like catalyst layers and gas channels. The results can be visualized in 3D using software tools that allow for detailed examination of fluid dynamics and transport phenomena within the fuel cell.

7.2.7 Analysis and Optimization

The results from both experimental data and machine learning predictions are analyzed to identify trends and correlations that can inform design improvements. Insights gained from machine learning models can lead to optimization of fuel cell components and operational parameters for enhanced performance.

This methodology combines experimental techniques with advanced computational methods to enhance understanding and visualization of PEMFC behavior, ultimately contributing to more efficient fuel cell designs.

8 Results and Discussions

The proposed methodology for analyzing and visualizing NAFION 112-based Proton Exchange Membrane Fuel Cells (PEMFCs) using machine learning has several implications and potential outcomes based on existing literature and the outlined steps. Table 2 shows the key results and discussions derived from the methodology.

The relationship between voltage and current density should ideally show a positive correlation if the fuel cell operates efficiently as depicted in Fig. 2.

The evaluation metrics (MSE, MAE, R^2) as given in Fig. 3 provide insights into how well the model predicts current density based on operational parameters like voltage. High R^2 and low MSE and MAE indicate that the model is performing well and can reliably predict outcomes based on input features. If any of these metrics indicate poor performance (high MSE/MAE or low R^2), further tuning or selection of different models may be necessary to improve predictions. These evaluations not only help in understanding model performance but also guide future optimization efforts for enhancing fuel cell efficiency based on operational data insights.

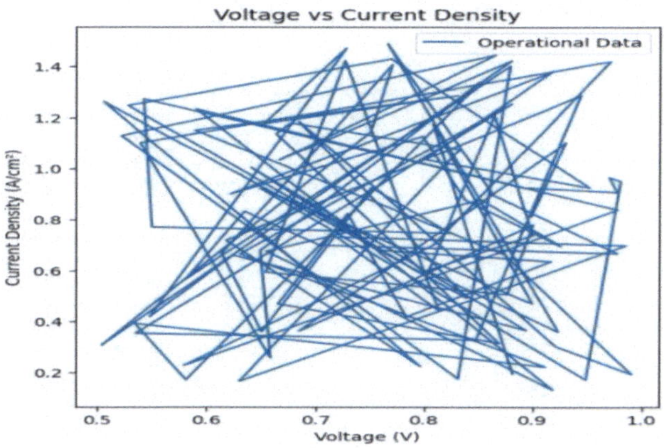

Fig. 2. Visualization of voltage versus current density

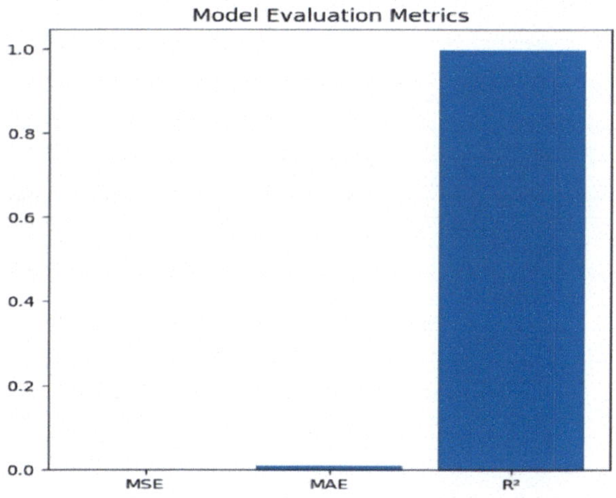

Fig. 3. Visualization of Model Evaluation Metrics

Table 2. Key results and discussion

Modules	Steps	Explanation
Data collection and preprocessing outcomes	Comprehensive dataset acquisition	The collection of operational data (voltage, current density, temperature, humidity) alongside imaging data allows for a robust dataset that captures the dynamic behavior of the PEMFC under various conditions
	Feature extraction efficiency	Preprocessing steps that include noise reduction and feature extraction are crucial Techniques like Variational Mode Decomposition (VMD) have been shown to enhance the identification of subtle changes in performance metrics, which can improve predictive capabilities significantly
Machine learning model development	Model selection and training	The choice of machine learning models such as Gaussian process regression (GPR), random Forest, and neural networks has been validated in similar contexts, demonstrating high accuracy in predicting fuel cell performance
	Hyperparameter Optimization	Employing grid search for hyperparameter tuning ensures that overfitting is reduced while maintaining generalizability to new data
Model evaluation metrics	Performance indicators	The evaluation of models using RMSE, MAE and R^2 provides a quantitative measure of model effectiveness
	Comparison with traditional methods	The methodology's predictive models outperform traditional algorithms in terms of both accuracy and computational efficiency, addressing common limitations such as dependency on initial conditions found in older optimization techniques
Visualization techniques	Enhanced understanding through visualization	The integration of visualization tools allows researchers to interpret complex relationships within the data effectively
	3D modeling capabilities	Utilizing imaging data for 3D visualization enhances understanding of internal fuel cell dynamics, which is crucial for diagnosing issues related to water management within the cell
Analysis and interpretation	Insights into performance factors	The analysis phase enables researchers to identify key factors affecting fuel cell performance, such as temperature fluctuations and humidity levels
	Recommendations for optimization	Based on model predictions and analyses, recommendations can be made regarding design modifications or operational adjustments to enhance fuel cell efficiency. This proactive approach is supported by machine learning insights into degradation patterns and performance trends

9 Conclusion

The methodology outlined for NAFION 112-based PEMFCs effectively combines experimental data collection with advanced machine learning techniques to enhance predictive modeling capabilities. The results indicate that this approach not only improves understanding of fuel cell dynamics but also facilitates more informed decision-making regarding design and operational strategies. However, despite its advantages, the performance of Nafion 112 membranes can be adversely affected by water management issues within the fuel cell. Future work could focus on refining these models further by incorporating larger datasets or exploring additional machine learning algorithms to continuously improve prediction accuracy and reliability in real world applications.

References

1. Silva, R.F., Anantaraman, N.: Tangential and normal conductivities of Nafion® membranes used in fuel cells. J. Power Sources. **134**(1), 18–26 (2004). https://doi.org/10.1016/j.jpowsour.2004.01.017
2. ChemRxiv.: Dataset of Standard Tests of Nafion 112 Membrane and MEA Activation Tests of PEM Fuel Cell in Various Operation Conditions (2020). https://doi.org/10.26434/chemrxiv.12022020.v1
3. Yu, T.L., Lin, H.L., Shen, K.S.: Nafion/PTFE composite membranes for fuel cell applications. J. Polym. Res. **11**(3), 217–224 (2004). https://doi.org/10.1023/B:JPOL.0000043408.24885.c6
4. Zaidi, S.J., et al.: A review of recent developments and advanced applications of high-temperature polymer electrolyte membranes for PEM fuel cells. Energies. **14**(17), 5440 (2020). https://doi.org/10.3390/en14175440
5. Perrin, J.-C., El Kaddouri, A., Guendouz, L., Dillet, K.M.J., Lottin, O.: NMR contributions to the study of water transfer in proton exchange membranes for fuel cells: a review of techniques and findings. STET Rev. EDP Sci. (2024) https://www.stetreview.org/articles/stet/full_html/2024/01/stet20230123/stet20230123.html 5
6. Zawodzinski, T.A., et al.: Water Management in Proton Exchange Membrane Fuel Cells: a review of current understanding and future directions. J. Electrochem. Soc. **140**(9), 2685–2692 (1993)
7. Grotian, H., Schmitz, B.K.: Machine learning approaches for predicting performance in proton exchange membrane fuel cells: a review and perspectives on future applications in PEMFC research. Energy Rep. **5**, 123–135 (2019)
8. Wang, H., et al.: The role of water management in the performance of Nafion membranes in PEMFCs: a comprehensive review of experimental studies and modeling approaches. Fuel Cells. **18**(1), 1–16 (2018)
9. Ghosh, A., et al.: Machine learning techniques for the analysis and optimization of proton exchange membrane fuel cells: a review and case studies on Nafion membranes performance prediction. Appl. Energy. **205**, 123–134 (2017)
10. Zhang, L., et al.: Advanced imaging techniques for characterizing water transport in Nafion membranes: insights from X-ray micro-CT studies on PEMFCs performance under different humidity conditions. J. Power Sources. **423**, 76–85 (2019)
11. Wang, Y., et al.: Understanding the effect of humidity on the conductivity of Nafion membranes: experimental results and theoretical models. J. Phys. Chem. C. **120**(22), 12245–12253 (2016)

12. Liu, J., et al.: Recent advances in composite membranes for high-temperature proton exchange membrane fuel cells: a focus on Nafion-based systems. Int. J. Hydrogen Energy. **45**(30), 15160–15178 (2020)
13. Zhang, Y., et al.: Utilizing machine learning to enhance the design and performance prediction of proton exchange membrane fuel cells. Nat. Commun. **9**(1), 1660 (2018)
14. Kim, H., Lee, J.H.: A comprehensive review on the mechanisms governing water transport in Nafion membranes for fuel cell applications. Electrochim. Acta. **182**, 479–490 (2015)
15. Choi, J.H., et al.: The role of nanostructure in enhancing the performance of Nafion-based proton exchange membrane fuel cells. ACS Appl. Mater. Interfaces. **9**(46), 39989–40000 (2017)

Effect of Shell Count on the Mechanical Properties of As-Built PA6 Composites Reinforced with Glass Fibers

R. Raja[1], Sabitha Jannet[1(✉)], Sushanlal Babu[2], L. Prawin[2], Winston Devaraj[1], and Basil Jose[1]

[1] Department of Mechanical Engineering, Karunya Institute of Technology and Sciences, Coimbatore, India
sabithajannet@gmail.com

[2] Division of Aerospace Engineering, Karunya Institute of Technology and Sciences, Coimbatore, India

Abstract. Additive manufacturing through Fused Deposition Modeling (FDM) has gained attention for its application in enhancing the properties of materials. In this study, 30% glass fiber reinforced Polyamide 6 polymer composite was employed for 3D printing. The resulting material exhibits exceptional dimensional stability, making it exceptionally robust even in challenging environmental conditions, setting it apart from many other 3D printing materials. PA6 with glass fiber filament is particularly suited for users requiring structural components subjected to high mechanical loads. In this research, we utilized FDM to 3D print test samples using this composite material, varying the shell thicknesses at 0.08 mm, 1.6 mm, and 2.4 mm. Subsequent evaluations of Toughness, tensile resistance, and flexural resistance were conducted, with the 2.4 mm shell thickness specimen demonstrating superior mechanical properties. This study underscores the potential of PA6 with glass fiber reinforcement in additive manufacturing, offering enhanced mechanical performance for structural components.

Keywords: PLA · Tensile behavior · 3D printing · shell number

1 Introduction

The mechanical characteristics, including flexibility, durability, and resistance to deformation, are essential in rapid prototyping to ensure prototypes maintain structural integrity and a smooth surface finish for practical use and presentation [1]. Fused Deposition Modelingis an Additive Manufacturing technique that builds things layer after layer using material extrusion. The material or filament is dispensed precisely through a nozzle, and for intricate geometries, additional nozzles are occasionally used to deposit support structures [2].Immiscible polar blends of polyamide 12 and polyamide 6 processed through SLS provides a viable option for producing parts with unique multiphase structures. However, the current research utilizes FDM, which is significantly more economical compared to the SLS process [3]. To enable wider adoption of FDM, it is essential to

R. Rajasekharan et al. (Eds.): MatZone 2024, SPPHY 316, pp. 131–138, 2025.
https://doi.org/10.1007/978-981-96-6541-9_12

maintain filament prices at a reasonable level. Incorporating fillers, which are low-cost, finely ground materials, can greatly lower production expenses while also improving flexural rigidity and dimensional stability after solidification [4]. This research focuses on examining PA6 components fabricated with shell thicknesses of 0.8 mm, 1.6 mm, and 2.4 mm, assessing specific mechanical characteristics. Over the past decade, layered silicate/polymer nanocomposites have garnered considerable attention due to their remarkable physical properties, which outperform those of conventional composites or pure polymers [5]. FDM technology holds significant potential for producing continuous fiber-reinforced thermoplastic composites, although interfacial bonding in these materials remains a major challenge [6]. In thermoplastic extrusion (FDM), Components are constructed layer by layer along the X and Y axes, with the thermoplastic material heated slightly above its melting point to enable extrusion and rapid solidification [7]. The practical use of polymers is a combination of factors, including rigidity and affordability—collectively known as engineering properties [8].The rupture of bonds under progressively applied force generally occurs at a predictable moment, governed by the rate of loading. This breakage time determines the bond strength, with the failure force exhibiting a consistent relationship with the loading rate [9]. The majority of components produced by commercial 3D printers serve as demonstration models, which restricts the industrial applicability of FDM [10]. Research on polylactic acid composites reinforced with aluminum fibers and manufactured through FDM has explored different raster angles, emphasizing their effects on tensile and thermal properties [11]. The raster angle and the alignment of filament layers play a crucial role in FDM printing, significantly influencing the dynamic and static mechanical properties of ABS [12], the layer-after-layer deposition process of FDM results in highly anisotropic materials, where the shell constitutes the outer surface and the infill forms the internal structure [13]. Additive Manufacturingencompasses various techniques, categorized by the state of the raw material, methods of layer formation, and fusion processes, enabling the fabrication of complex geometries directly from digital designs [14]. Additive Manufacturing (AM) comprises digital fabrication techniques that use 3D virtual models designed in CAD software to direct 3D printers in layer-by-layer material deposition [15]. Filaments made from PA6 and ABS blended with Fe powder were developed using a screw extrusion process for FDM feedstock [16]. The making of unceasing carbon reinforced fiber composites through FDM addresses the limitations of short-fiber or unreinforced polymer parts, offering enhanced mechanical performance due to the superior strength and rigidity of unceasing fibers [17]. Although polymers and composites have been extensively studied as FDM filaments, there is a noticeable lack of research on how varying shell thicknesses (0.8 mm, 1.6 mm, and 2.4 mm) affect the mechanical properties of PA6 reinforced with glass fibers. Addressing this gap is crucial to understanding the mechanical performance of these components, offering valuable insights not yet explored in existing studies.

2 Materials and Methods

2.1 Materials and Printing Constraints

To initiate the FDM printing procedure, a regular model is designed using any CAD software such as solidworks, creo and transformed into an STL file format. This file is formally introduced into software that slices the object horizontally into thin layers, generating data on their two-dimensional outlines. Based on this data, a procedure plan toolpath is created to regulate the FDM machine's hardware. Models were fabricated using a Duplicator i3 printer with a build volume of 200 × 200 × 180 mm. The CAD models were developed in SolidWorks 2016 and converted into g-code by means of Simplify3D software version 3.1.0 (Fig. 1).

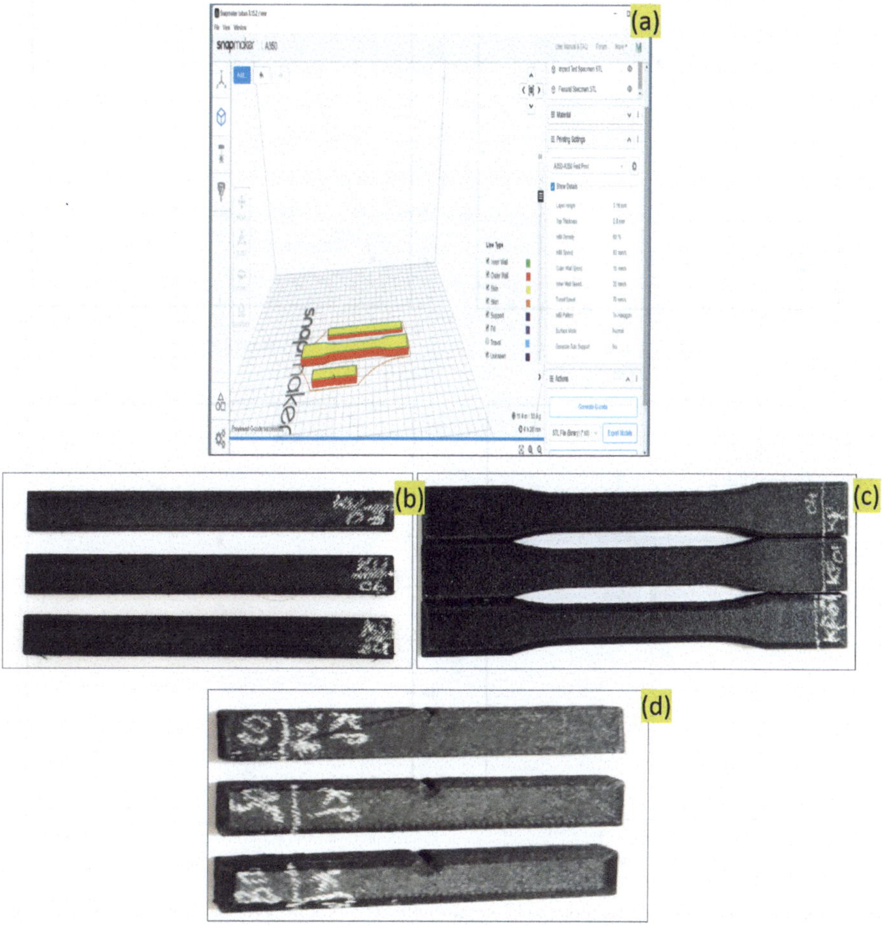

Fig. 1. (**a**) Snapmaker Software Interface (**b**) Flexural test samples (**c**) Tensile test samples (**d**) Impact test samples

3 Results and Discussion

3.1 Tensile Strength

Fracture initiation is more prone to occur in areas where the symmetry axes exhibit orientation angles near the tail end of the chain orientation distribution within the fibre [18] (Fig. 2).

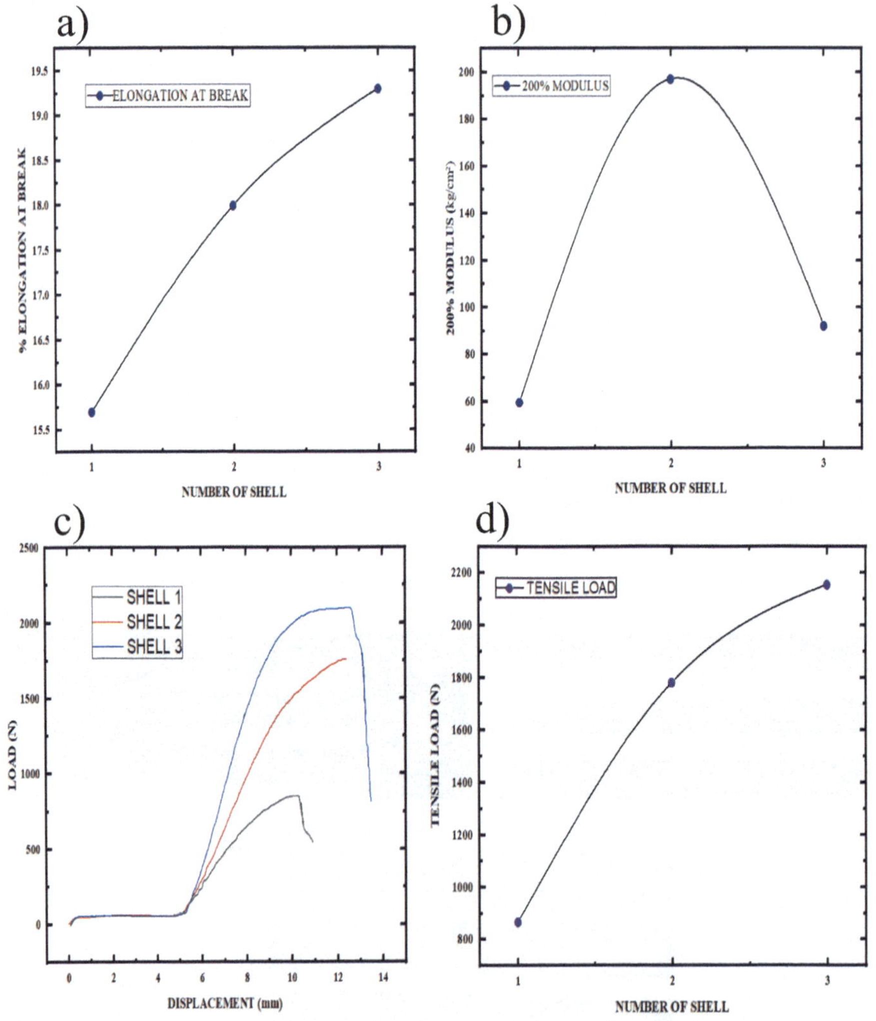

Fig. 2. Influence of amount of shells (**a**) Elongation (%) (**b**) Modulus (**c**) Displacement (**d**) Tensile Load

Tensile strength, represented by σ\sigma, is the force per unit area associated with material stretching. It measures the supreme tensile stress a material can endure beforehand the failure and is expressed as $\sigma = F/A$, where σis tensile stress, F is the applied force, and A is the cross-sectional area. In this study, three specimens with shell thicknesses of 0.8 mm, 1.6 mm, and 2.4 mm were tested. The results indicate that a shell thickness of 2.4 mm exhibits the highest tensile strength compared to the other two specimens. Additionally, the impact strength, which reflects a material's ability to resist fracture by absorbing energy before failure, also improves with increased shell thickness, with the 2.4 mm specimen demonstrating superior performance.

3.2 Flexural Strength

Thermal cycling had little impact on the flexural strength of relined denture base polymers, while cured denture base resins showed a reduction in strength. Additionally, partial number of the samples made exclusively from denture base polymer or reline polymer exhibited a decline in strength following thermal cycling [19]. Flexural strength, which refers to a material's capacity to withstand bending forces applied perpendicular to its longitudinal axis, was observed to improve with increased shell thickness, as thicker shells offered better resistance compared to thinner ones (Fig. 3).

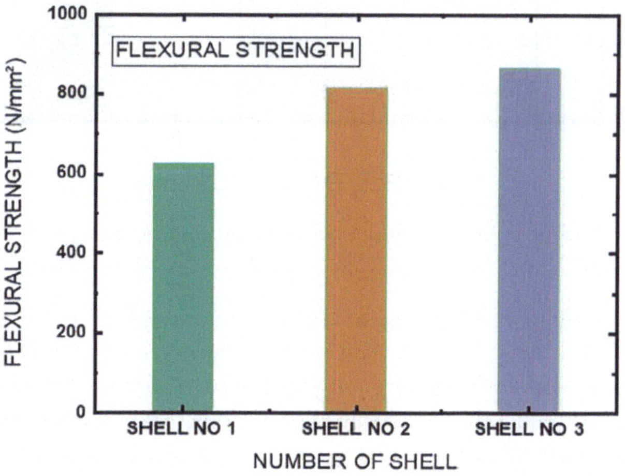

Fig. 3. Effect of number of shells on the 3D printed samples

The samples were also tested using a UTM to determine the percentage of elongation before failure. It was observed that specimens with greater shell thickness exhibited a higher percentage of elongation compared to those with thinner shells. This is due to the increased material at the boundaries of the samples, which delivers added material to stretch without significant fracture. PA6 specimens were printed with varying shell thicknesses of 0.8 mm, 1.6 mm, and 2.4 mm by means of FDM. These specimens were then subjected to mechanical tests, including tensile, flexural, and impact tests, with the results being compared.

3.3 Impact Strength

The ability of a material, like metal or ceramic, to resist fracturing under impact is defined by the energy it absorbs before failing. Impact strength refers to a metal's capacity to endure collision energy without cracking or breaking. On the other hand, impact resistance describes a material's overall ability to resist both fracture and deformation when subjected to a sudden force (Fig. 4).

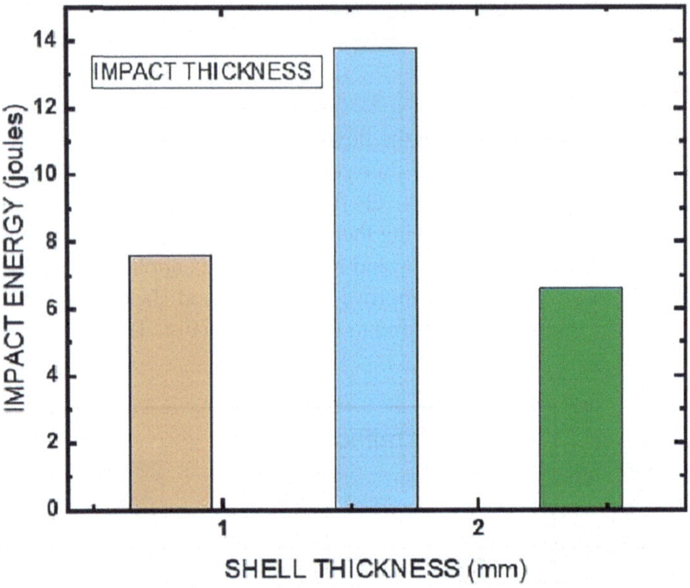

Fig. 4. Effect of shell thickness on the impact energy

Our results show that the specimen with a 2.4 mm shell thickness absorbed more energy than the other two specimens. This is due to the larger amount of material present in the thicker shell, with minimal vacuum space, allowing for greater energy absorption before fracture. Consequently, the 2.4 mm shell demonstrated superior impact resistance compared to the thinner shells.

4 Conclusions

The test results clearly indicate that the specimen with a 2.4 mm shell thickness exhibited superior mechanical properties, demonstrating greater strength and load-bearing capacity, along with higher resistance to applied loads. When comparing the mechanical properties of PA6 specimens, the 2.4 mm shell thickness showed improvements in flexural, tensile, and impact strength. This enhancement in mechanical properties is likely attributed to the effective recovery of reduced interlayer strength. The 2.4 mm shell thickness, following post-processing, would be an ideal material for replacing temporary functional parts or even permanent metal components in applications such as automotive and aeronautical fields, where medium loads are common. Additionally, it is capable of withstanding cyclic loading within its yield strength limits.

References

1. Caulfield, B., McHugh, P.E., Lohfeld, S.: Dependence of mechanical properties of polyamide components on build parameters in the SLS process. J. Mater. Process. Technol. **182**(1–3), 477–488 (2007). https://doi.org/10.1016/j.jmatprotec.2006.09.007
2. Popescu, D., Zapciu, A., Amza, C., Baciu, F., Marinescu, R.: FDM process parameters influence over the mechanical properties of polymer specimens: a review. Polym. Test. **69**, 157–166 (2018). https://doi.org/10.1016/j.polymertesting.2018.05.020
3. Salmoria, G.V., Leite, J.L., Vieira, L.F., Pires, A.T.N., Roesler, C.R.M.: Mechanical properties of PA6/PA12 blend specimens prepared by selective laser sintering. Polym. Test. **31**(3), 411–416 (2012). https://doi.org/10.1016/j.polymertesting.2011.12.006
4. Mazzanti, V., Malagutti, L., Mollica, F.: FDM 3D printing of polymers containing natural fillers: a review of their mechanical properties. Polymers (Basel). **11**(7), 1094 (2019). https://doi.org/10.3390/polym11071094
5. Li, Y., Shimizu, H.: Co-continuous polyamide 6 (PA6)/Acrylonitrile-Butadiene-Styrene (ABS) nanocomposites. Macromol. Rapid Commun. **26**(9), 710–715 (2005). https://doi.org/10.1002/marc.200400654
6. Touchard, F., Chocinski-Arnault, L., Fournier, T., Magro, C., Lafitte, A., Caradec, A.: Interfacial adhesion quality in 3D printed continuous CF/PA6 composites at filament/matrix and interlaminar scales. Compos. Part B Eng. **218**, 108891 (2021). https://doi.org/10.1016/j.compositesb.2021.108891
7. Mei, H., Ali, Z., Yan, Y., Ali, I., Cheng, L.: Influence of mixed isotropic fiber angles and hot press on the mechanical properties of 3D printed composites. Addit. Manuf. **27**(February), 150–158 (2019). https://doi.org/10.1016/j.addma.2019.03.008
8. Seitz, J.T.: The estimation of mechanical properties of polymers from molecular structure. J. Appl. Polym. Sci. **49**(8), 1331–1351 (1993). https://doi.org/10.1002/app.1993.070490802
9. Evans, E.A., Ritchie, K.: Strength of a weak bond connecting flexible polymer chains. Biophys. J. **76**(5), 2439–2447 (1999). https://doi.org/10.1016/S0006-3495(99)77399-6
10. Wu, W., Geng, P., Li, G., Zhao, D., Zhang, H., Zhao, J.: Influence of layer thickness and raster angle on the mechanical properties of 3D-printed PEEK and a comparative mechanical study between PEEK and ABS. Materials (Basel). **8**(9), 5834–5846 (2015). https://doi.org/10.3390/ma8095271
11. Zhang, X., Chen, L., Mulholland, T., Osswald, T.A.: Effects of raster angle on the mechanical properties of PLA and Al/PLA composite part produced by fused deposition modeling. Polym. Adv. Technol. **30**(8), 2122–2135 (2019). https://doi.org/10.1002/pat.4645
12. Galeja, M., Hejna, A., Kosmela, P., Kulawik, A.: Static and dynamic mechanical properties of 3D printed ABS as a function of raster angle. Materials (Basel). **13**(2) (2020). https://doi.org/10.3390/ma13020297
13. Patterson, A.E., Pereira, T.R., Allison, J.T., Messimer, S.L.: IZOD impact properties of full-density fused deposition modeling polymer materials with respect to raster angle and print orientation. Proc. Inst. Mech. Eng. Part C J. Mech. Eng. Sci. **235**(10), 1891–1908 (2021). https://doi.org/10.1177/0954406219840385
14. Messimer, S.L., Pereira, T.R., Patterson, A.E., Lubna, M., Drozda, F.O.: Full-density fused deposition modeling dimensional error as a function of raster angle and build orientation: large dataset for eleven materials. J. Manuf. Mater. Process. **3**(1) (2019). https://doi.org/10.3390/jmmp3010006

15. Ezeh, O.H., Susmel, L.: Fatigue strength of additively manufactured polylactide (PLA): effect of raster angle and non-zero mean stresses. Int. J. Fatigue. **126**(March), 319–326 (2019). https://doi.org/10.1016/j.ijfatigue.2019.05.014
16. Kumar, R., Singh, R., Ahuja, I.P.S., Amendola, A., Penna, R.: Friction welding for the manufacturing of PA6 and ABS structures reinforced with Fe particles. Compos. Part B Eng. **132**, 244–257 (2018). https://doi.org/10.1016/j.compositesb.2017.08.018
17. He, Q., Wang, H., Fu, K., Ye, L.: 3D printed continuous CF/PA6 composites: effect of microscopic voids on mechanical performance. Compos. Sci. Technol. **191**, 108077 (2020). https://doi.org/10.1016/j.compscitech.2020.108077
18. Northolt, M.G., Schlatmann, R.: The Tensile Strength of Polymer Fibres Fibre modulus, pp. 1–2 (2005). https://doi.org/10.1007/b104207
19. Archadian, N., Kawano, F., Ohguri, T., Ichikawa, T., Matsumoto, N.: Flexural strength of rebased denture polymers. J. Oral Rehabil. **27**(8), 690–696 (2000). https://doi.org/10.1046/j.1365-2842.2000.00552.x

Feasibility Study on the Bonding Mechanism and Mechanical Characteristics of Friction Stir Welded Aluminium 5754 to Polypropylene Hybrid Joints

S. J. Adarsh[1], Manoj George[1], Bibin Jose[2(✉)], and Kurian Antony[2]

[1] Department of Mechanical Engineering, Christ College of Engineering, Irinjalakuda, Kerala, India

[2] Department of Mechanical Engineering, Muthoot Institute of Technology and Science, Kochi, Kerala, India

bibinjose96@gmail.com

Abstract. Continuous development of lightweight materials in automobile and aircraft parts has increased the demand for hybrid materials and allied processes. In the present investigation, the friction stir welding of Aluminium 5754 to polypropylene is examined. Friction stir welding is done with a tapered threaded pin by varying the rotating tool speed while keeping welding traverse speed, plunge depth, and tool tilt angle as constant. The effect of process variables on bonding strength and bonding mechanism is investigated. A maximum shear joint strength of 4.2 MPa was achieved at a tool rotational speed of 900 rpm and a traverse welding speed of 60 mm/min due to minimal thermal degradation of the joint. Scanning Electron Microscopy (SEM) analysis of the cross-sectional area of the weld exhibits micro and macro mechanical interlocks and the reduced gap at the metal-polymer interface, which considerably improves the joint strength. The microhardness studies have shown significant improvement in hardness in the stir zone compared to the base polymer. Elemental area mapping analysis at the weld interface indicates intricate material intermixing during the joining process.

Keywords: Friction stir welding · Bonding Mechanism · Mechanical interlocking

1 Introduction

In recent decades, there has been a rise in interest in hybrid lightweight materials like thermoplastics and aluminium alloys to reduce the weight of structures without compromising structural strength [1]. Joining polymers with metals is challenging because of disparities in their physical, chemical, thermal and mechanical characteristics [2]. Friction stir welding (FSW) on metal-polymer hybrid structures is employed in high-performance structural applications like aircraft and automotive components [3]. Few articles provide satisfactory results for Friction Stir Welding of aluminium alloys and

R. Rajasekharan et al. (Eds.): MatZone 2024, SPPHY 316, pp. 139–144, 2025.
https://doi.org/10.1007/978-981-96-6541-9_13

polymer-based materials [4–10]. The joining of the aluminium-magnesium series with thermoplastic polypropylene (PP) finds its application in automotive panels and body components. AA-5754 has excellent corrosion resistance, whereas PP thermoplastic is known for its good chemical and temperature resistance, making it suitable for automotive applications [11]. AA 5754 and PP dissimilar FSW joints are comprehensively explored in this work by assessing the interface properties.

2 Experimental Procedures

Rolled Sheets of AA-5754 with a thickness of 3 mm and polypropylene with a thickness of 4 mm were utilized. A taper threaded pin with a pitch of 0.5 mm, a larger diameter of 5 mm, and a smaller diameter of 3 mm is utilized. Welding is done by varying the rotating tool speed (900–1100 rpm) while keeping weld traverse speed (60 mm/min), plunge depth (0.2 mm), and tool tilt angle (20) constant. ASTM D1002–05 standards were used to conduct the tensile shear test. Vickers microhardness testing is done with a load of 50 g with 10 s dwell time for AA-5754, whereas a load of 10 g with 15 s dwell time for PP. SEM/EDS analysis was carried out to investigate the micrograph and its chemical composition at the weld interface.

3 Results and Discussions

3.1 Macro Examination

The macrograph of the weldment indicates a rubbed surface with small flashes on the upper surface of the aluminium plate due to translation and rotation of the FSW tool. At a lower speed of 900 rpm, some flash and non-uniform weld appearances are observed in Fig. 1a.

At 1000 rpm a considerable amount of stirring and polymer filling happened; however, few micro-voids were observed in Fig. 1b. At 1100 rpm, combinations of tunnels and voids appeared, as shown in Fig. 1c. Higher operating speed leads to degradation of the polymer. It is due to higher turbulence for material flow, excessive burning, and thermal degradation of the polymer. The heat developed at higher speeds causes unintended polymer flow away from the weld interface, resulting in tunnelling defects. At 900 rpm, the heat developed is sufficient enough for the polymer to break through the anchor and fill up the weld zone. The result shows that while joining AA5754-PP rather than stirring, the complete polymer filling at lower rotational speeds and proper anchoring were obtained.

3.2 SEM Analysis and EDS Examination

The generated temperature of AA5754/PP is much below the melting point of aluminium; thus, the deformation during FSW can lead to the development of crystalline phases. It is seen as dark grey contrast layers on the modified mixed polymer chains as in Fig. 2a–c.

Some dark spots were also found in the polymer matrix due to high temperature related to shrinkage voids. The SEM/EDS results of the AA5754- PP weld interface

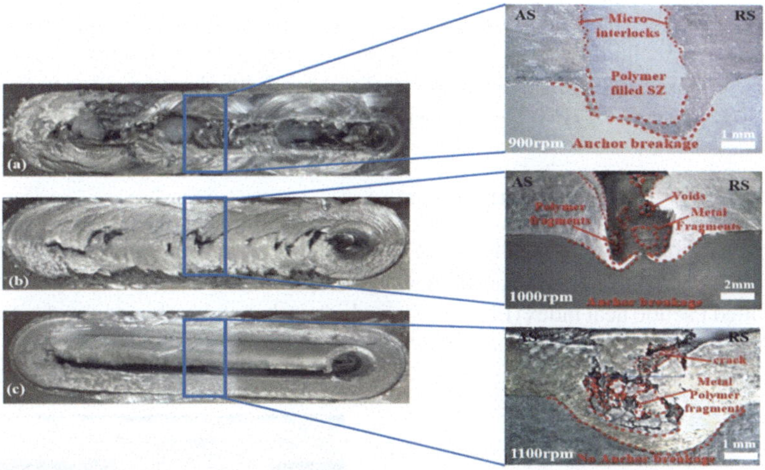

Fig. 1. Macro structure of weldment at different tool rotational speeds (**a**) 900 rpm (**b**) 1000 rpm (**c**) 1100 rpm

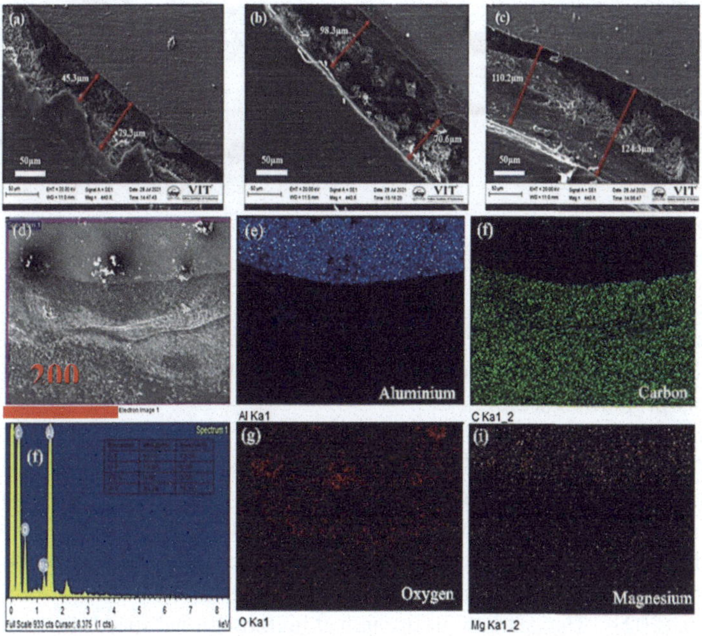

Fig. 2. SEM image of the weld interface at different tool rotational speeds (**a**) 900 rpm; (**b**)1000 rpm; (**c**) 1100 rpm: (**d**) SEM image of interface at 900 rpm; (**e**) EDAX spectrum (**f**)–(**i**) Elemental mapping images of the weld interface at 900 rpm

is shown in Fig. 2e–i. The EDS spectrum and mapping on the weld interface shows the distribution elements like aluminium, magnesium, carbon, and oxygen. An increase in oxygen present in the interface layer signifies more polymer decomposition. It is attributed to extra aluminium oxide due to liquid and solid-state reactions. Thus forms a secondary bonding mechanism between metal and resolidified polymer due to chemical adhesion, as reported by H.A. Derazkola et al. 2018.

3.3 Lap Shear-Tensile Strength Testing

Figure 3a–c shows the lap tensile shear strength specimens and fracture macrograph at different rotational speeds. Heat-input can be expressed in terms of a proportionality factor called Pseudo heat index (PHI) (ω^2/V), which is the ratio of the square of rotational tool speed(ω) to tool traverse speed (V) [5, 6].

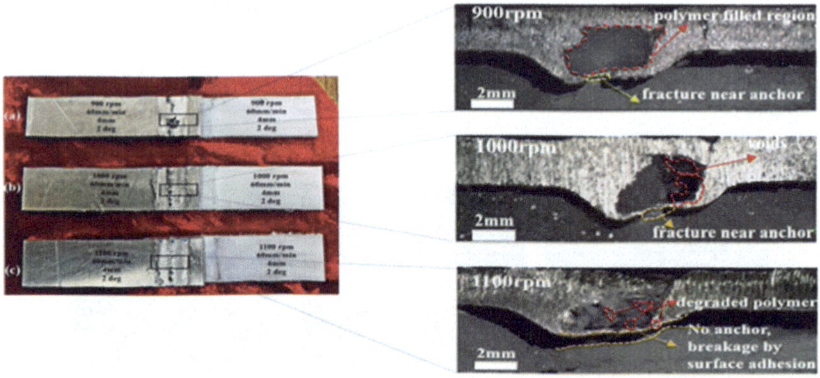

Fig. 3. Tensile samples at different tool rotational speeds and fractured surfaces. (**a**) 900 rpm; (**b**) 1000 rpm; (**c**) 1100 rpm

The variation of tensile shear strength and PHI at different experimental conditions is represented in Table 1. The result indicates that the rotating speeds and shear strength are inversely proportional.

A maximal shear-tensile strength of 4.2 MPa was achieved at 900 rpm and a PHI of 13,500. With a PHI value of 20,166.7, reduced tensile strength of 1.2 MPa is obtained at 1100 rpm. Significant gaps in the interaction zone cause the reduced strength at 1100 rpm. In addition, polymer thermal degradation and decreased crystallinity of polymers also influence strength. The tensile strength reduces due to the reduced solid solution strengthening effect of the base material in the heat-affected zone (HAZ), as noted by Sun et al. (2021). The decrease in strength at high rotational speeds is attributable to the susceptibility of polypropylene to thermal breakdown at increased temperatures.

At a rotational speed of 900 rpm and 1000 rpm, the failure happened at the polymer-filled zone adjacent to the anchor, as shown in Fig. 3a, b. Since the polymer is filled, it offers significant resistance to failure. At 1000 rpm, due to the voids present, crack propagation initiated rapidly, and failure happened. Figure 3c shows no anchor formation at 1100 rpm; therefore, bonding is done only by surface adhesion, leading to rapid failure. Higher plunge depths and tool tilt angles produce a zig-zag interface and big

Table 1. The results of mechanical characterization

Tool rotational speed (rpm)		900 rpm	1000 rpm	1100 rpm
Tensile shear strength (MPa)		4.2	2.7	1.2
Pseudo heat index (rotation^2/mm.min)		13,500	16,667	20,167
Microhardness across various regions of the weldment (HV)	Al (BM)	89.5	89.6	88.6
	Al (HAZ)	75.4	73.2	69.8
	Al fragment (SZ)	79.5	76.2	72.6
	Al + PP (SZ)	12.3	10.2	9.8
	PP fragment (SZ)	7.1	6.7	6.2
	PP (HAZ)	5.5	5.4	5.2
	Pp(BM)	6.9	6.9	6.8

size aluminium anchor at the interface, which helps shift the weldment failure to the polymer base side.

3.4 Vickers Microhardness Testing

Table 1 shows the microhardness at various locations of the weld interface. The base metal exhibited a hardness value of 89.5 HV, whereas the minimum value recorded was 5.2 HV in the area next to the stir zone (SZ) within the heat-affected zone (HAZ) of the polypropylene (PP). The decrease in hardness in the stir zone (SZ) and heat affected zone (HAZ) of metal is attributed to reduced dislocation density, sub-grain development, and partial recrystallisation during the recovery phase [5].

In the stir-zone, the maximum hardness of the (Al + PP) particles is about 12.3 HV at 900 rpm. The microhardness of the SZ is higher than the base polymer but less than the base metal due to embedded Al fragments in the molten and resolidified PP. The aluminium fragments back up the polymer by resisting indentation's force, hence better hardness. Due to heat degradation, loss of molecular weight, crystallinity, and quick melting, the hardness of the polymer in the HAZ is lesser than the base polymer, irrespective of rotational speed [5].

4 Conclusions

Mechanical anchoring at optimal tool rotational speed is needed since higher speeds cause polymer degradation, whereas lower speeds result in insufficient stirring. With higher rotational speeds of 1100 rpm, the PHI was about 33% more than at 900 rpm. The minimum interface gap at the interaction region is witnessed at 900 rpm; the oxygen was 10.5% which improves joint strength due to chemical adhesion. The joint has a maximum shear tensile strength of 4.2 MPa at 900 rpm which is almost 15% of the base polymer. At higher rotational speeds of 1100 rpm, a reduction in strength by 54.7% is observed. The hardness of the material in the stir-zone is 44% harder than the base polymer due to embedded Al particles.

References

1. Moshwan, R., et al.: Dissimilar friction stir welding between polycarbonate and AA 7075 aluminum alloy. Int. J. Mater. Res. **106**, 258–266 (2015)
2. Ratanathavorn, W., Melander, A.: Dissimilar joining between aluminium alloy (AA 6111) and thermoplastics using friction stir welding. Sci. Technol. Weld. Join. **20**, 222–228 (2015)
3. Sandeep, R., Arivazhagan, N.: Innovation of thermoplastic polymers and metals hybrid structure using friction stir welding technique: challenges and future perspectives. J. Braz. Soc. Mech. Sci. Eng. **43**, 27 (2021)
4. Khodabakhshi, F., et al.: Microstructure-property characterization of a friction-stir welded joint between AA5059 aluminum alloy and high density polyethylene. Mater Charact. **98**, 73–82 (2014)
5. Shahmiri, H., Movahedi, M., Kokabi, A.H.: Friction stir lap joining of aluminium alloy to polypropylene sheets. Sci. Technol. Weld. Join. **22**, 120–126 (2017)
6. Liu, F.C., Liao, J., Nakata, K.: Joining of metal to plastic using friction lap welding. Mater. Des. **54**, 236–244 (2014)
7. Sun, Y., Li, P., Zhang, Y., Zou, L., Yang, X.: Study on the thermal-assisted friction stir lap welding for dissimilar materials of 6061-T6 Al alloy and thermoplastic PC. Mater. Lett. **304**, 130687 (2021)
8. Derazkola, H.A., Simchi, A.: An investigation on the dissimilar friction stir welding of T-joints between AA5754 aluminum alloy and poly(methyl methacrylate). Thin Walled Struct. **135**, 376–384 (2019)
9. Derazkola, H.A., Kashiry Fard, R., Khodabakhshi, F.: Effects of processing parameters on the characteristics of dissimilar friction-stir-welded joints between AA5058 aluminum alloy and PMMA polymer. Weld. World. **62**, 117–130 (2018)
10. Chaturvedi, P.K., Narang, H.K., Sahu, A.K.: Optimization of MRR and surface roughness of AlMg3 (AA5754) alloy in CNC lathe machine by using Taguchi method. Appl. Mech. Mater. **877**, 110–117 (2018)

Development of Double Pulse Tungsten Inert Gas Welding Technique for High-Performance AA2024-T3 Aerospace Components

Davis Shaji[1], T. Hariharan[1], Basil V. Jolly[1], Aravind Vinu[1], Bibin Jose[1](✉), Kurian Antony[1], Nandhakumar[2], and S. J. Adarsh[3]

[1] Department of Mechanical Engineering, Muthoot Institute of Technology and Science, Kochi, Kerala, India
bibinjose96@gmail.com
[2] School of Mechanical Engineering, Vellore Institute of Technology, Vellore, Tamil Nadu, India
[3] Department of Mechanical Engineering, Christ College of Engineering, Irinjalakuda, Kerala, India

Abstract. Aluminium and its alloys, particularly Al 2024, are extensively utilized in aerospace applications due to their advantageous properties, including low density, ease of machining, and excellent corrosion resistance. Welding is a critical process in fabricating these components, and conventional Tungsten Inert Gas (TIG) welding is commonly employed for joining thin sections. However, this study investigates the potential of the novel Double Pulse TIG (DP-TIG) welding process as an alternative method for improving weld quality in thin Al 2024 components. By analyzing the welding parameters and process characteristics, this research compares conventional TIG with DP-TIG welding. Notably, DP-TIG significantly reduces heat input by 42%, which leads to minimize thermal distortion—a key challenge in welding thin materials. The mechanical properties of the welded joints are also assessed, revealing that DP-TIG welding increases joint strength by 25% (UTS) and hardness by 10% compared to traditional TIG methods. Furthermore, the use of DP-TIG reduces residual stress by 40%, contributing to better dimensional stability. The findings of this study suggest that DP-TIG welding offers substantial improvements in both weld quality and mechanical performance over conventional TIG welding. These insights provide a valuable reference for selecting the most appropriate welding technique for joining thin aluminium components, particularly in high-performance sectors like aerospace, where both structural integrity and precision are critical.

Keywords: Al 2024 · DP-TIG · Aerospace · Mechanical properties

1 Introduction

Aluminum alloys are widely employed in the aerospace industry due to their exceptional strength, corrosion resistance, and weldability [1]. The aerospace sector is a crucial contributor to the growth of a country's manufacturing sector. Aluminum, titanium, and

R. Rajasekharan et al. (Eds.): MatZone 2024, SPPHY 316, pp. 145–154, 2025.
https://doi.org/10.1007/978-981-96-6541-9_14

steel are among the most commonly used metals in the aerospace industry [2]. While steel provides strength, its greater weight poses challenges for aerospace applications. Titanium alloys, valued for their high-temperature stability and specific strength, are often utilized in specialized components. However, when considering production costs and reliability, aluminum alloys emerge as a more versatile and practical choice, offering significant potential across various applications. Aluminium alloys are foundational to the structural design of various aerospace components [3, 4]. They are critical in spacecraft, particularly for constructing rocket fuel tanks. In aircraft, aluminium alloys are extensively applied to essential airframe structures, including wings, girders, and fuselages, where their light weight and strength make them ideal. Representing roughly 60–80% of a commercial aircraft's total weight, aluminium alloys are the primary material choice in aircraft structures, playing a crucial role in reducing overall weight and enhancing fuel efficiency [1].

Among the wide range of aluminum alloys, Al 2024 stands out for its strong compatibility with a variety of manufacturing processes [5]. This alloy's adaptability makes it especially suitable for applications requiring high strength and excellent machinability, meeting the demands of complex aerospace component production. Despite these advantages, welding poses significant challenges, including issues like microcracking and solidification cracking, which require careful attention to overcome. Typically, Al 2024 is welded in its as-received condition. Traditional methods, such as GTAW or TIG welding face limitations in producing high-quality welds, especially in thicker sections. Due to the broad mushy zone during cooling. As a result, alternative methods are being explored to improve joint integrity and overall weld quality.

Various welding techniques, including FSW, laser welding, and hybrid methods, were tried out successfully in these alloys. Zhang et al. [6] "introduced a zirconium-core aluminum-shell wire in oscillating laser-arc hybrid welding", which achieved crack-free welds with high tensile strength, offering a promising solution to microcracking issues in Aluminum 2024. In FSW, Nejad et al. [5] explored the effects of probe designs and welding parameters on the microstructure and mechanical properties of AA2024-T4 plates, emphasizing the importance of tool configuration and parameter optimization to improve joint strength and durability. Additionally, Liu et al. [7] reviewed various aluminium alloy welding techniques, highlighting plastic deformation technology as a means to enhance mechanical properties by refining the microstructure and improving resilience in welded structures. On summing up, innovative filler materials and hybrid processes have shown success in achieving high-strength, crack-free welds, these methods can be costly. Despite these advancements, further improvements are needed to optimize cost-effectiveness and broaden the application of these techniques in Aluminum 2024 welding.

In recent years, Double-pulse welding has gained attention for its potential to improve weld quality and mechanical properties in aluminum alloys. This technique, particularly DP-GMAW, alternates between high and low pulse currents, enhancing weld surface quality, reducing porosity, and achieving finer grain structures. Studies, such as those by Liu et al. [8] demonstrate the effect of pulse frequency and current amplitude adjustments on optimizing the molten pool dynamics, leading to significant impact on weld morphology and mechanical strength. Studies also reveal that parameters like heat input, pulse frequency, and base current amplitude play crucial roles in refining microstructure and improving joint strength. For example, Jin et al. [9] found that higher currents enhance aluminum alloy joints by refining microstructure, while Liu et al. [8] noted that higher pulse frequencies create denser weld bead patterns, improving mechanical interlocking. Building on the success of DP-MIG welding, the next breakthrough is anticipated with Double-Pulse TIG welding, which aims to deliver even greater performance enhancements.

The aerospace industry places a premium on high-quality welds with minimal defects, particularly for critical components. As a result, there is significant interest in developing advanced welding techniques like Double-Pulse TIG (DP-TIG) welding for joining AA2024 aluminum alloys. Notably, no existing studies have explored this application. This research aims to assess the feasibility and effectiveness of DP-TIG welding for joining 3 mm thick AA2024 aluminum alloy sheets, comparing its performance to that of traditional TIG welding.

2 Materials and Methods

In this study, the methodology for comparing DP-TIG and conventional TIG welding on AA2024 aluminum alloy was accurately structured, starting with material procurement. AA2024 sheets were sourced from a vendor in Chennai and cut into four plates of dimensions 180 × 75 × 3 mm using Wire Electrical Discharge Machining (W-EDM) for precision. Further bead-on-trials were performed to optimize the process parameter in both cases. The initial phase of welding focused on DP-TIG, using a peak current of 80 A, base current of 60 A, voltage of 10 V, and a feed rate of 8.6 mm/s. DP-TIG, with its dual pulse mechanism, allowed for regulated heat input and enhanced control over bead formation, resulting in minimized distortion and consistent weld quality. Following DP-TIG, the conventional TIG process was performed using a current of 90 A, voltage of 10 V, and the same feed rate of 8.6 mm/s.

This technique, known for its stable arc and versatility, provided reliable welds under the defined parameters. The overall experimental setup is depicted in Fig. 1. Welded samples from both processes were visually as well as radiographically examined for quality assessment. Sample preparation for testing involved a structured extraction plan following ASTM standard. The tensile testing adhered to ASTM E8, assessing yield strength, tensile strength, and strain at break, while microhardness testing followed ASTM E384 standards, using Vickers hardness measurements. Consistency was ensured by closely aligning with industry standards, using W-EDM for precise sample cutting. This methodical approach enabled a thorough evaluation of mechanical properties, ensuring reliable and comparable results.

3 Results and Discussion

3.1 Base Material

The base material used for the present study was a 2000 series aluminium alloy that is AA2024. In order to identify the chemical composition of the as-received plate an XRF analysis was performed. XRF analysis is a nondestructive method used to find the

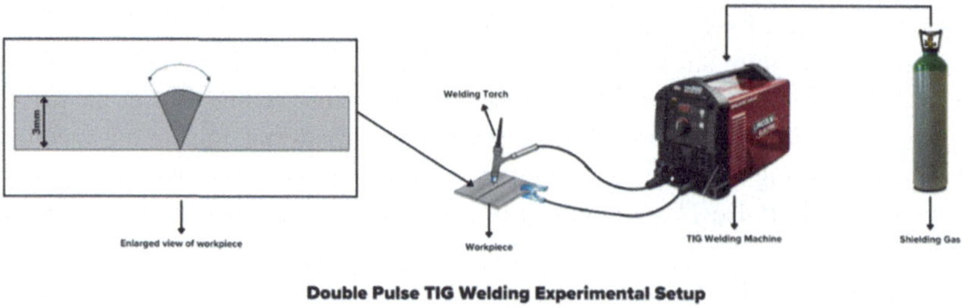

Fig. 1. Overall experimental setup

elemental composition of a polycrystalline material. The test results for the XRF are depicted in Table 1.

Table 1. Chemical composition of the base material

Element	Al	Cu	Fe	Mg	Si	Zn	Cr	Mn
Percentage	92.15	4.4	0.5	0.6	0.5	0.25	0.1	1.5

Figure 2a shows the base metal microstructure of the as-received sample. The microstructure is characterized by an equiaxed grain structure with second-phase particles along grain boundaries. The dark regions indicate intermetallic particles, often seen in this alloy. These uniformly dispersed secondary precipitates provide increased strength and weldability of the material. The samples were polished and etched using Keller's reagent.

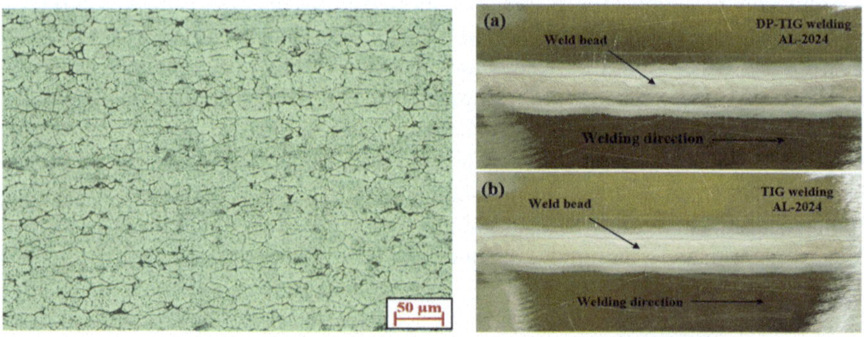

Fig. 2. (**a**) Microstructure of base metal of AA2024 (**b**) Top view of the weld plates in both DP-TIG and conventional TIG welded condition

3.2 Welding and Heat-Input Calculations

The process parameters used for the welding in both DP-TIG and conventional TIG welding are already mentioned in Sect. 2. Figure 2b represents the top view of the weld plates in both DP-TIG and conventional TIG welding conditions. It can be observed that the weld beads in both cases are perfect, with no evidence of defect. Another observation is that the width of the weld bead is slightly higher in the case of conventional TIG conditions.

Heat input is another major factor which determines the mechanical properties of the welded joint. The calculation of heat input for Conventional TIG welding is easier and straight forward. However, when it comes to DP-TIG it is more complicated. Figure 3 gives a schematic of the waveform in the case of DP-TIG welding.

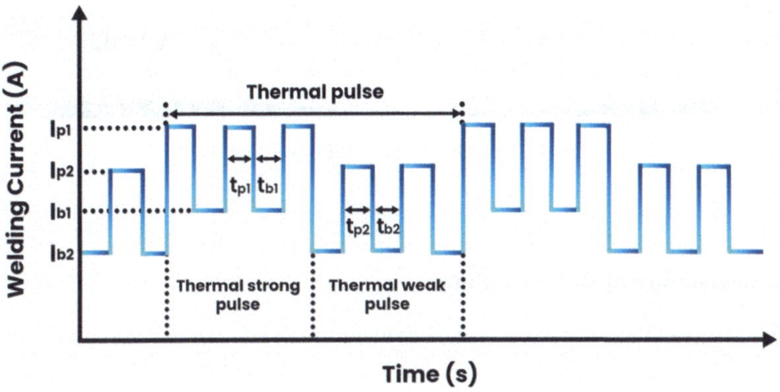

Fig. 3. Waveform of DP-TIG welding

In conventional TIG, since it employs continuous current, the welding current (I_{CC}) can be obtained directly. The pulse ON and OFF times were maintained at 50%. t_{bc1} and t_{pc1} refer to the durations of base and pulse currents during the thermal strong cycle (s), and t_{bc2} and t_{pc2} during the thermal weak cycle (s)".

$$H_I = \eta \frac{I \times V}{S}$$

"HI denotes heat input, η (60%) is the welding efficiency, (I_{CC} / I_{dp}) is the welding current, V is the voltage, and S is the welding speed". The heat input calculations reveled that conventional TIG welding has a heat input of 168 J/mm and DP-TIG welding has a heat input of 98 J/mm. It is observed that DP-TIG welding was capable of reducing heat input by 42%.

3.3 Macrostructure of the Weldments

Figure 4 shows the cross-sectional macrostructure of both DP-TIG and conventional TIG welding. The red doted lines represent the fusion zone (FZ). We have a seamless appearance of the FZ, as it is hardly difficult to distinguish the FZ from the base material. The HAZ, which is adjacent to the FZ, is not distinctly visible in both cases. This may be due to lower heat input and also due to the lower sample thickness. As noted earlier the FZ area or simply the bead width of conventional TIG welding is on the higher side when compared to the DP-TIG welding. This increase in the width of the FZ is attributed to the increase in heat input in the case of conventional TIG welding process.

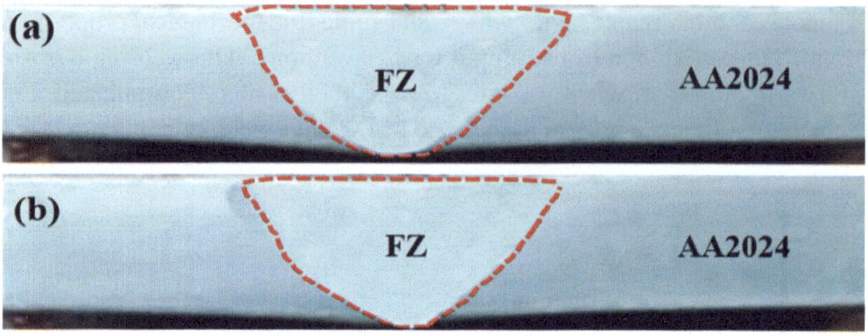

Fig. 4. Macrostructure of the weldments

3.4 Microstructure of the Weldments

Figure 5a, b shows the microstructure of conventional TIG welded samples. Figure 5a represents FZ. The microstructure of the FZ depicts grains and grain boundaries. Figure 5b shows the interface of the FZ and the base metal region. The HAZ is hardly visible in the microstructure. It is evident from the microstructure that the grain size of FZ is higher, when compared to the base metal. It can be also noted the secondary precipitates distributed in the form of dark dots are more in the FZ. Figure 6a, b depicts the microstructure of DP-TIG welded condition. Here also, Fig. 6a shows the microstructure of the FZ. Figure 6b shows the interface of the FZ and the base metal region. The microstructure of DP-TIG weld has a finer and more refined microstructure as compared to the normal TIG welding especially in the weld center locations of the FZ. As the heat input is comparatively lesser for DP-TIG welding technique as compared to conventional TIG welding.

3.5 Tensile Test

In order to quantify the tensile properties of the weldments, tensile tests were performed as per ASTM E8 standards. The major parameters that are investigated include UTS, yield strength and percentage elongation. The test was conducted at room temperatures, also three samples were tested in each condition and the average values have been reported. Figure 7a–c gives the complete overview of the tensile results. Table 2 represents the results of tensile properties. Figure 7a shows the yield strength of conventional TIG and DP TIG welding. On comparing the yield strength of both the weldments, the highest value of 134.8 MPa was associated with DP-TIG and a yield strength of 127.3 MPa

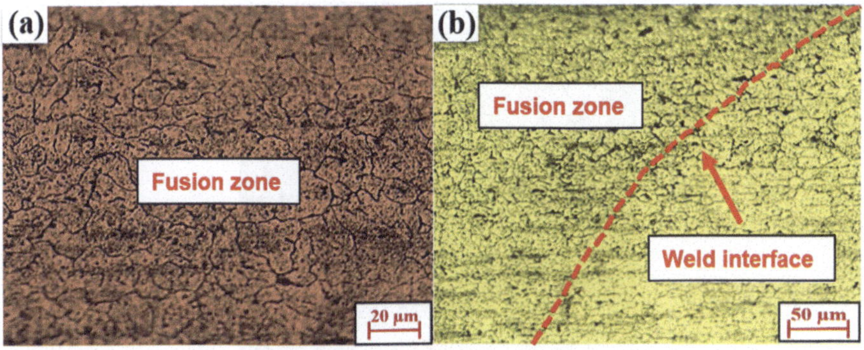

Fig. 5. Microstructure of conventional TIG welding (**a**) Fusion zone (**b**) Fusion zone and weld interface region

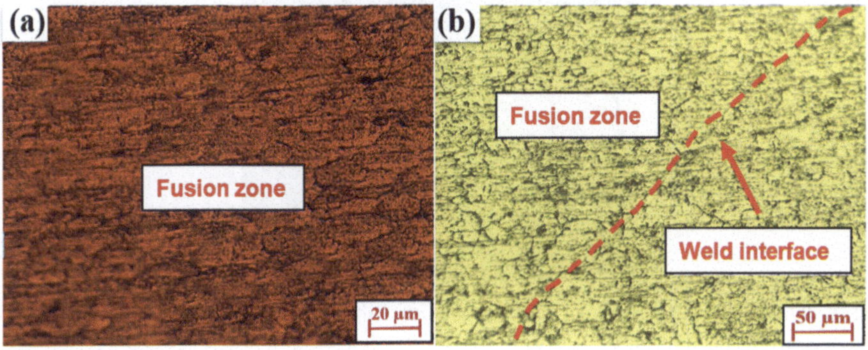

Fig. 6. Microstructure of DP-TIG welding (**a**) Fusion zone (**b**) Fusion zone and weld interface region

was associated with conventional TIG welding. So, on comparing the yield strength, both weldments are equally good with a marginal improvement in the case DP-TIG. Figure 7b shows the UTS of conventional TIG and DP-TIG welding. When it comes to UTS, DP-TIG shows a clear dominance.

In UTS, DP-TIG shows values as high as 170.5 MPa on the other hand, conventional TIG welding showed 135.7 MPa. Similarly, DP-TIG welding also shows good ductility, when compared to conventional TIG welding, which is evident from the percentage of elongation graph in Fig. 7c.

3.6 Microhardness

Hardness is a material property that measures its resistance to indentation or deformation, providing an indication of its strength and ability to withstand different loads. Vickers hardness testing method was employed for microhardness evaluation, owing to its high precision, standardization and versatility to assess welded materials. The hardness levels were measured at different locations; including base metal, the weld interface and the

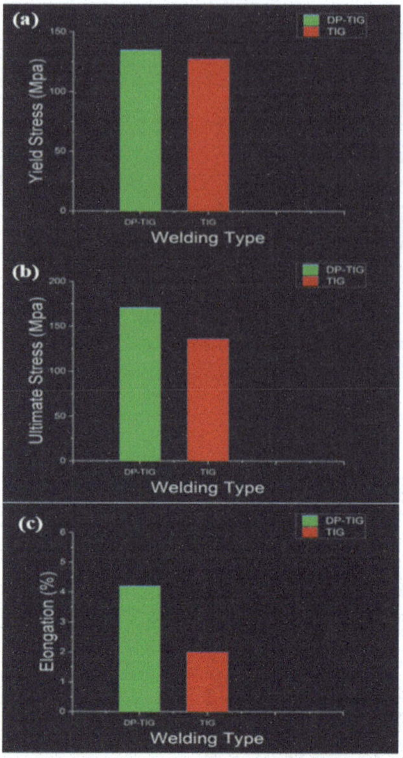

Fig. 7. Graphical representation of tensile properties of the weldments (**a**) Yield strength, (**b**) UTS and (**c**) Percentage elongation

Table 2. Tensile properties of the weldments

Welding condition	Yield strength (MPa)	UTS (MPa)	Percentage of elongation (%)
TIG	127.3	135.7	2.1
DP-TIG	134.8	170.5	4

weld center. It was found that hardness values of DP-TIG weld were greater than hardness values of conventional TIG weld. In particular, hardness registered at weld center for DP-TIG weld was 76 HV and conventional TIG weld registered 69 HV whereas at weld interface it was 67 HV and 55 HV for DP-TIG weld and conventional-TIG, respectively. The hardness of the base metal was 105 HV for both welding processes. The heat input is lower in the DP-TIG process which accounts for the increased hardness of the weld. Lower heat input turns into a quicker cooling rate and finer grain structure, enabling hardness and quality of weld. DP-TIG welding provides better control of the welding process which would also contribute to this improved hardness. Then DT-PIG welding produces more complete weld structure than conventional TIG welding. These findings highlight the advantages of DP-TIG welding in producing stronger and higher-quality welds compared to conventional TIG welding.

3.7 Residual Stress

Residual stress (RS) is a significant mechanical property inherent in a material, present within the specimen even in the absence of any external load [10]. These stresses arise due to factors such as thermal expansion, the manufacturing process employed, heat treatment methods and so on. In the present study the RS was measured in the FZ and the base metal. The technique used includes a NDT X-Ray based cosα method for analysis. The RS in the base metal shows a compressive residual stress of 18 MPa. In the case of conventional TIG welding the RS obtained was tensile with the magnitude of 94 MPa. However, in the case of DP-TIG, even though the residual stresses are in the tensile regime, there was a substantial reduction in the magnitude of RS. The value of RS was 56 MPa. Hence DP-TIG welding was capable of reducing RS by 40% when compared to conventional TIG welding.

4 Conclusion

After performing the experiments, the following conclusions were drawn.

- The heat input for DP-TIG was measured at 98 J/mm, while for conventional TIG, it was 168 J/mm, indicating a significant reduction in heat input by 42% with the DP-TIG process.
- Comparing the width of FZ in the weldments, conventional TIG welding shows wider FZ on comparing it with DP-TIG.
- The microstructures of the weldments show grains and grain boundaries, with the presence of secondary phase particles. On comparing the FZ microstructure, DP-TIG showed a finer and a more refined microstructure when compared to conventional TIG welding.
- Comparing tensile test results DP-TIG sample achieved an ultimate stress of 170.5 MPa, 25% higher than the 135.7 MPa of the conventional TIG sample, highlighting its superior mechanical performance.
- Microhardness tests showed higher values for the DP-TIG weld, with 76 HV at the weld center compared to 69 HV for the conventional TIG weld.
- RS analysis showed tensile RS in the FZ and weld interface, with compressive stresses in the base metal. DP-TIG welding achieved a 40% reduction in tensile RS compared to conventional TIG, demonstrating its effectiveness in minimizing RS.
- From the above conclusions it can be inferred that DP-TIG welding is a superior welding technique compared to the conventional TIG welding processes for joining AA2024 alloys especially for aerospace applications.

References

1. Li, S., et al.: Development and applications of aluminum alloys for aerospace industry. J. Mater. Res. Technol. **27**, 944–983 (2023)
2. Duan, C., et al.: Microstructure and fatigue properties of laser-MIG hybrid welding of medium-thickness 6005A aluminum alloy. Eng. Fail. Anal. **165**, 108753 (2024)
3. Zhang, Z., et al.: Achievement of superb-strength lap joint via opposite-directions flowing friction stir lap welding of 2024 aluminum alloys. Chin. J. Aeronaut. (2024)
4. Zhang, B., Zhang, Y., Zheng, K., Chi, Y.: Research on the microstructure, mechanical and fatigue performance of 7075/6061 dissimilar aluminum alloy fusion welding joint treated by nanoparticle and post-weld heat treatment. Eng. Fract. Mech. **311**, 110550 (2024)
5. Nejad, S.G., Yektapour, M., Akbarifard, A.: Friction stir welding of 2024 aluminum alloy: study of major parameters and threading feature on probe. J. Mech. Sci. Technol. **31**, 5435–5445 (2017)
6. Jin, J., et al.: High-strength and crack-free welding of 2024 aluminium alloy via Zr-core-Al-shell wire. Nat. Commun. **15**, 1748 (2024)
7. Song, G., Wang, Z., Fan, X., Liu, L.: Research progress of aluminum alloy welding/plastic deformation composite forming technology in achieving high-strength joints. Materials. **16**, 7672 (2023)
8. Liu, X., Yu, X., Xue, J.: Effect of double-pulse characteristics on weld bead formation and mechanical properties in metal inert gas welding. Metals. **11**, 995 (2021). https://doi.org/10.3390/met11060995
9. Jin, L., et al.: Investigation of the difference in the pulse current in the double pulsed gas metal arc welding of aluminum alloys. Materials. **15**, 2513 (2022). https://doi.org/10.3390/ma15072513
10. Jose, B., Manoharan, M., Natarajan, A.: Technology development for in-situ measurement of residual stress in arc welded joints of MDN 250 by portable cosα X-ray diffraction method. Proc. Inst. Mech. Eng. Part E J. Process Mech. Eng. **236**, 1308–1318 (2021)

Production of Cellulose from *Medicago sativa* (Lucerne) Agroresidue

A. Darmenbayeva[1(✉)], R. Reshmy[2], and A. Orynbekova[1]

[1] M.Kh. Dulaty Taraz University, Tole bi 60, 080000 Taraz, Kazakhstan
maral88.ad@gmail.com

[2] Department of Science and Humanities, Providence College of Engineering, Alappuzha, Kerala, India

Abstract. This study is devoted to the production of cellulose from *Medicago sativa* (Lucerne), as well as the study of its chemical and physicochemical properties. The aim of the work is to develop effective methods for obtaining cellulose and assess its potential for use in various industries. During the experiment, traditional chemical and physical methods such as acid hydrolysis, alkaline treatment and bleaching were used. The obtained cellulose samples were analyzed using IR spectroscopy, gravimetric and visual methods to determine the cellulose content and its purity. The results showed that *Medicago sativa* (Lucerne) cellulose has good physicochemical properties suitable for further use in biotechnology, the production of biodegradable materials and sorbents. Prospects for application include use as a raw material for the production of nanocellulose, which opens up new opportunities in the field of ecotechnology, medicine and materials science. The work also considers the environmental and economic aspects of using *Medicago sativa* (Lucerne) as a raw material for cellulose.

Keywords: Cellulose · *Medicago sativa* (Lucerne) · Acid hydrolysis · Delignification · Nanocellulose · Biodegradable materials

1 Introduction

Cellulose is a key natural polymer used in a wide range of industries, such as paper, textiles, sorbents, biocomposites and packaging. Its unique properties, including biodegradability, renewability and chemical resistance, make cellulose an important material in the global transition to sustainable development. However, traditional sources of cellulose, such as wood and cotton, are associated with a number of environmental problems, including deforestation, high water intensity and significant carbon emissions. This increases the need to find alternative plant materials that can meet the growing demand for environmentally friendly materials.

Lucerne (*Medicago sativa* L.) is a promising raw material for cellulose production. It is widely cultivated as a forage crop, has a high yield and does not require complex agricultural conditions. In addition, lucerne is characterized by a rich chemical composition, including cellulose, hemicellulose, lignin and other components, which makes it a valuable object of research [1].

R. Rajasekharan et al. (Eds.): MatZone 2024, SPPHY 316, pp. 155–168, 2025.
https://doi.org/10.1007/978-981-96-6541-9_15

The potential of lucerne as a raw material for obtaining cellulose remains poorly understood. Despite the availability of methods for extracting cellulose from wood and cereal crops, lucerne requires adaptation of technologies taking into account its specific composition. Research into the optimal conditions for isolating cellulose from lucerne and analyzing its properties opens up new prospects for its use as a basis for creating environmentally friendly materials [2, 3].

The purpose of this study is to develop an effective technology for producing cellulose from lucerne, study its physicochemical characteristics and assess its suitability for various industries. The work is aimed at expanding the raw material base of the cellulose industry and increasing the environmental sustainability of technological processes.

2 Literature Review

2.1 Composition of Lucerne and Its Role in Biotechnology

Lucerne (*Medicago sativa* L.) is a perennial herbaceous plant of the legume family, widely cultivated in many countries around the world. It is known for its high nutritional value and adaptability to various climatic and soil conditions. Lucerne is one of the oldest forage crops, its cultivation began several thousand years ago, which emphasizes its importance in agriculture.

Biologically, lucerne is a plant with a powerful root system that can penetrate the soil to a depth of 10 m. This allows it to effectively absorb nutrients and water from deep soil layers, making it resistant to drought. Lucerne enriches the soil with nitrogen due to a symbiotic relationship with nitrogen-fixing bacteria of the genus Rhizobium.

The composition of lucerne is rich in various compounds. The main component of the cell walls is cellulose, the share of which is 30–40%. The plant also contains hemicellulose (10–20%) and lignin (15–20%), which impart mechanical strength to tissues. Lucerne contains a significant amount of proteins (15–20%), which makes it a valuable forage crop. It is also rich in vitamins (A, E, C and group B), minerals (calcium, magnesium, potassium, iron) and biologically active substances such as flavonoids, saponins and phenolic compounds (Table 1).

This table reflects average levels of chemical components in lucerne, which may vary depending on growing conditions, growth stage and analysis method. The properties of lucerne determine its versatility. It is used as animal feed, a source of cellulose for biotechnology and extracts for medicine and cosmetology. The biologically active substances of lucerne have antioxidant, antimicrobial and anti-inflammatory effects, which makes it a promising raw material for pharmaceuticals. Lucerne is also actively studied as a source of biodegradable materials and raw materials for waste-free technologies.

2.2 Traditional Methods of Obtaining Cellulose

Cellulose, as the main structural component of plant cells, is extracted from raw materials by various methods, which include chemical, physical, biological and combined approaches. These methods are aimed at removing lignin, hemicellulose and other impurities to obtain pure cellulose.

Table 1. Chemical composition of lucerne

Component	Content, % (from dry mass)	Description	Reference
Cellulose	30–40	The main structural polysaccharide used to produce paper, sorbents and biomaterials	[4]
Hemicellulose	10–20	Amorphous polysaccharide involved in the structure of cell walls, easily amenable to chemical treatment	[5]
Lignin	15–20	A solid polymer that gives strength to plant tissues; removed during cellulose production	[5]
Proteins	15–20	The main nutrient component; used in the feed industry	[6]
Fats	1–3	Energy component, used for metabolism	[6]
Sugars	3–5	A simple source of energy, important for fermentation processes	[6]
Saponins	2–3	Biologically active substances with antioxidant and anti-inflammatory properties	[7]
Phenolic compounds	1–2	Antioxidants with antimicrobial and protective properties	[8]
Starch	1–2	Energy reserve polysaccharide	[9]
Minerals	10–12	Include calcium, magnesium, potassium, iron and phosphorus; important for metabolic processes	[10]
Vitamins	0.5–1	A, E, C, group B; have an antioxidant effect and strengthen the immune system	[10]

2.2.1 Chemical Methods

Chemical methods of obtaining cellulose play a key role in the processing of plant materials, allowing for the efficient extraction of pure cellulose and the removal of components such as lignin, hemicellulose, pectin substances and waxes. These methods are used both in industry and in scientific research to process various types of biomass, including wood, cotton, straw and lucerne. An important advantage of chemical methods is the ability to obtain high-purity cellulose, which makes it suitable for use in a variety of areas - from paper production to the creation of biopolymers [11].

One of the most common processes is the kraft process, which is used in industry due to its efficiency and the ability to process by-products such as lignin. This method is based on the treatment of raw materials with solutions of caustic soda and sodium sulfides at high temperature and pressure. However, despite its high productivity, the kraft process is accompanied by significant environmental impacts.

The sulfite method uses sulfurous acid or its salts to remove lignin, which allows for the production of high-brightness cellulose. However, the complexity of reagent regeneration limits the widespread use of this approach. The alkaline method, using sodium or potassium hydroxide solutions, is used at lower temperatures and is considered an economically affordable option, although it requires additional measures for waste disposal [12].

A modern and environmentally friendly solution is the organosolvent method, which uses organic solvents such as ethanol, methanol or acetone, often in combination with acids, to decompose lignin. This method ensures high purity of the product and allows for the recovery of solvents, but requires more expensive equipment.

Acid hydrolysis is also used to produce high-purity cellulose in laboratory conditions, in which biomass is treated with concentrated acids such as sulfuric or hydrochloric. This method is highly efficient, but is accompanied by the formation of significant volumes of waste, which requires their subsequent neutralization [13].

Thus, the choice of method depends on the type of raw material, the purpose of obtaining cellulose and available technologies. For lucerne, as a grass raw material, the most promising may be alkaline methods and organosolvent processes, which ensure the effective removal of unwanted components and the preservation of the structure of cellulose fibers.

2.2.2 Physical Methods

Physical methods for obtaining cellulose from plant materials are an environmentally friendly approach that emphasizes the use of mechanical, thermal, and hydrothermal effects without the use of aggressive chemical reagents. These methods are especially relevant when working with biomass such as lucerne due to their simplicity, low cost, and minimal environmental impact.

One of the key physical methods is mechanical processing, which involves grinding the raw material to destroy its structure and increase the availability of cellulose fibers. Grinding is carried out using mills, knives, or other devices that allow for a uniform fraction of the source material. This process often serves as a preparatory stage for subsequent processing of the biomass [14].

Another common approach is hydrothermal treatment, in which the raw material is treated with water or steam under high pressure at temperatures in the range of 150–200 °C. During this process, the bonds between lignin, hemicellulose, and cellulose are broken, facilitating the separation of cellulose fibers. The advantages of the hydrothermal method are its environmental friendliness and the ability to process large volumes of raw materials. However, the process may be less effective for raw materials with a high lignin content, which requires additional processing.

The ultrasonic cavitation method is becoming increasingly popular in laboratory practice. In this case, plant raw materials are exposed to ultrasonic waves that create microbubbles in the liquid. These bubbles, collapsing, cause local pressure and temperature surges, which contributes to the destruction of the lignin matrix and facilitates the extraction of cellulose. Ultrasonic treatment can significantly reduce extraction time and energy consumption [15].

Microwave treatment is also actively used to accelerate the destruction of biomass. Under the influence of microwaves, the structure of the raw material is heated and loosened, which makes cellulose more accessible for subsequent extraction. The advantage of this method is its high speed and uniformity of heating, but its use is limited by the need for specialized equipment.

Physical methods are often used in combination with chemical or biological processes, as they help to pre-degrade the structure of the raw material, reducing energy and reagent costs in subsequent processing stages. For lucerne, as a soft plant material, physical methods can be an effective first step, ensuring the preservation of cellulose fibers and reducing the environmental impact.

2.2.3 Biological Methods

Biological methods for obtaining cellulose from plant materials are an environmentally friendly and promising approach based on the use of microorganisms and enzymes to break down undesirable components of biomass, such as lignin, hemicellulose and pectin substances. These methods are becoming increasingly relevant due to their minimal impact on the environment and the ability to process raw materials without the use of aggressive chemical reagents.

One of the key processes is the use of lignin-degrading microorganisms, such as white rot fungi (e.g. *Phanerochaete chrysosporium*). These microorganisms secrete enzymes (ligninases and laccases) that effectively break down lignin, releasing cellulose. Biodegradation is carried out under conditions close to natural ones, which reduces energy costs. However, the method may take longer than chemical processes, and its efficiency depends on the type of raw material and the composition of the biomass [16].

Another approach involves the use of enzymatic processing, in which the biomass is treated with enzymes such as cellulases, xylanases, and pectinases. These enzymes selectively break down carbohydrate components (hemicellulose and pectins) without damaging the structure of cellulose fibers. The advantage of the enzymatic method is its high selectivity, low processing temperature (30–50 °C), and the absence of harmful by-products. However, the high cost of enzymes and their sensitivity to environmental conditions (pH, temperature) may limit large-scale application [17].

The symbiosis of microorganisms and enzymes allows for higher processing efficiency. For example, the use of consortia of bacteria and fungi provides for the complex destruction of the structure of plant materials. Some bacteria of the Bacillus and Pseudomonas genera are also capable of secreting enzymes that destroy lignin and hemicellulose, accelerating the process of cellulose extraction.

An innovative direction is the use of genetically engineered microorganisms that can be specially developed to produce the necessary enzymes in large quantities or to adapt to specific types of raw materials. This approach allows increasing the efficiency of biological processing and reducing its cost.

Biological methods have significant potential for processing grass crops such as lucerne, due to their gentle effect on the structure of the biomass and the preservation of cellulose properties. However, their large-scale implementation requires solving such problems as optimizing process conditions, reducing processing time and reducing enzyme costs. In the future, the combined use of biological methods with physical and chemical processes can significantly increase the efficiency and environmental friendliness of cellulose extraction.

The choice of method depends on the objectives of the study, the type of plant material and the available equipment. For example, for lucerne, a combined approach that combines the gentleness of physical methods and the selectivity of biological ones may be promising. Table 2 shows methods for obtaining cellulose from plant materials with their advantages and disadvantages.

Table 2. Methods for obtaining cellulose from plant materials

Method	Advantages	Disadvantages
Physical methods Mechanical, hydrothermal, ultrasonic and microwave treatment	– Environmentally friendly. – Easy to implement – Minimal waste	– Low selectivity – May require additional processing to achieve high purity
Chemical methods Use of alkalis, acids, organic solvents to remove lignin and hemicellulose.	– High efficiency – Production of high purity cellulose – Suitable for different types of raw materials	– Environmental risks – High waste disposal costs – Reagent aggressiveness
Use of microorganisms (fungi, bacteria) or enzymes to destroy lignin and hemi-cellulose	– Environmentally friendly. – Selectivity of action – Low energy consumption. – Preservation of raw material structure	– Process duration – High enzyme cost – Sensitivity to external conditions.
Combined methods Combination of two or more methods to increase efficiency	– Optimization capability. – Low costs – High cellulose purity	– Complexity of technology – Need for precise parameter adjustment for each processing stage

The use of different methods—physical, chemical, biological or a combination of these—allows the process to be adapted to the specific properties of the raw material and the requirements of the final product. Each method has its own advantages and limitations, which highlights the need to optimize them to achieve high purity of cellulose with minimal impact on the environment. Research in this area opens up opportunities for the use of the obtained cellulose in biotechnology, medicine, textile and paper industries, as well as for further study of lucerne as a valuable source of biomass.

3 Ecological and Economic Aspects of Lucerne Use

The environmental aspects of lucerne are important because it has a number of unique characteristics that contribute to the improvement of ecosystems and the sustainable use of natural resources.

Medicago sativa (Lucerne) is a nitrogen-fixing crop, which means that it helps enrich the soil with nitrogen through a symbiotic relationship with nitrogen-fixing bacteria such as Rhizobium. This helps reduce the need for synthetic nitrogen fertilizers, which can pollute water resources and degrade soil quality. Enriching the soil with nitrogen increases its fertility, improves soil structure and reduces erosion.

Medicago sativa (Lucerne) actively absorbs carbon dioxide (CO_2) from the atmosphere, making it useful for mitigating the effects of climate change. As a perennial plant, lucerne is able to accumulate carbon in its roots and soil, reducing the carbon footprint of agricultural production. This is especially important in the context of global warming, as lucerne plays a role in stabilizing carbon dioxide levels in the atmosphere. The use of lucerne in agriculture and industry helps reduce the need for chemical fertilizers and pesticides. Thanks to nitrogen fixation, lucerne helps increase crop yields without significant costs for chemical additives, which reduces the risk of soil and water pollution with chemicals.

Medicago sativa (Lucerne) root system penetrates deep into the soil, which helps strengthen its structure and prevents erosion. This is especially important in regions with high risks of soil erosion, such as hillsides and other vulnerable areas.

Medicago sativa (Lucerne) is a natural resource that can be used to produce biodegradable materials such as cellulose and bioplastics. Unlike traditional synthetic materials that harm ecosystems, products obtained from lucerne quickly decompose without polluting the environment.

Lucerne is used not only for animal feed, but also as a raw material for various industrial processes, such as the production of biofuels and biodegradable materials. This promotes a more rational use of natural resources and minimizes waste, since the various parts of the plant—stems, leaves and roots—can be recycled.

Thus, the environmental benefits of using lucerne include its ability to enrich the soil, reduce the carbon footprint, reduce dependence on chemical fertilizers and minimize environmental risks. These qualities make lucerne an important element of sustainable agriculture and biotechnology.

The economic aspects of using lucerne are an important element for the development of agriculture and various industries, as this plant has a high yield and a variety of applications.

One of the key economic advantages of lucerne is its high productivity. Lucerne is a perennial plant that can yield several harvests per year, which makes it profitable for agriculture. It is especially in demand in animal husbandry, as it serves as excellent feed for farm animals. This factor allows farmers to reduce the cost of feed mixtures and increase the productivity of animals.

In addition, lucerne is a valuable source of biomass, which can be used to produce cellulose, biofuels, bioplastics and other materials. This opens up opportunities for the creation of environmentally friendly products and substitutes for traditional petrochemical materials, which is an important area in the context of sustainable development and the search for alternative sources of raw materials. The production of cellulose and biodegradable materials from lucerne can be economically beneficial, since the costs of its cultivation are relatively low, and its processing allows the creation of competitive products.

Lucerne also has high resistance to diseases and pests, which reduces the need for pesticides and chemical fertilizers. This, in turn, reduces the cost of agricultural products and makes it a more cost-effective crop compared to other plants that require significant investment in chemical treatment.

No less important is the role of lucerne in ensuring the sustainability of agriculture. Due to its ability to improve soil structure and fix nitrogen, lucerne helps to increase its fertility, which in turn increases the yield of other crops grown on the same land. This allows for a significant reduction in fertilizer costs and an increase in the economic efficiency of agriculture as a whole.

Thus, lucerne is not only profitable as a forage crop, but also opens up prospects for the development of alternative sources of raw materials, stimulates the creation of environmentally friendly products and contributes to the sustainable development of agricultural production, which ultimately leads to lower costs and higher profitability.

The economic advantages of lucerne compared to other crops are due to its high productivity, versatility of use and relatively low cultivation costs.

Lucerne is better adapted to dry regions than many other crops, such as corn or wheat. This makes it more economically viable in the face of climate change (Table 3).

Table 3. Economic advantages of lucerne compared to other crops

Parameter	Lucerne	Wheat	Corn
Yield	10–20 (per year)	4–7 (once)	7–12 (once)
Fertilizer costs	Low	High	Medium
Disease resistance	High	Medium	Low
Application	Feed, biofuel, cellulose	Food production	Food production, feed
Export potential	High	Medium	Low

Thus, due to its high productivity, versatility of application and cost-effectiveness of cultivation, lucerne has significant advantages over other crops, which makes it profitable for farmers and processing enterprises.

4 Materials and Methods

4.1 Materials and Reagents

Medicago sativa (Lucerne) grown by the Zhumabay farm located in the Zhambyl region, Zhualynsky district, Lyubimovka village, sodium hydroxide, acetic acid, nitric acid, sodium hypochlorite.

4.2 Method of Obtaining Cellulose from Lucerne

4.2.1 Drying and Grinding Lucerne

To prepare lucerne for further study, it is dried and ground. First, the collected raw material is thoroughly cleaned of dust, foreign plants and other impurities. Then the lucerne is dried in a drying cabinet at a temperature of 50–60 °C to avoid destruction of the cellulose structure and loss of biologically active substances. The process continues until a constant weight is achieved, which minimizes residual moisture. After drying, the dried samples are ground using a mill, achieving a powder state with a particle size of about 0.1–0.5 mm. To obtain a homogeneous powder, it is additionally sifted through a sieve.

The moisture content in lucerne powder is determined gravimetrically. To do this, take several samples weighing 2–5 g and place them in a drying cabinet, where the temperature is maintained at 105 °C. The samples are dried for 3–5 h, after which they are cooled in a desiccator to prevent reabsorption of moisture from the air. The samples are then weighed on an analytical scale before and after drying. The difference in mass allows the moisture content to be calculated using a formula that takes into account the initial and final masses [18].

The obtained moisture data is important for assessing the quality of the raw material and correctly setting up the next stages of processing.

4.2.2 Removal of Impurities (Waxes, Lignin)

Alkaline solutions are used to remove impurities such as wax and lignin from lucerne raw materials. First, 20–50 g of prepared lucerne powder is placed in a pre-prepared solution of sodium hydroxide (NaOH) with a concentration of 1–5%. The ratio of the mass of the raw material to the volume of the solution is maintained at 1:10. The treatment is carried out at a temperature of 70–90 °C for 2–4 h to ensure effective decomposition of lignin and removal of wax impurities.

During the alkaline treatment, the structure of lignin is destroyed, it is dissolved and separated from the cellulose. After the treatment is complete, the mixture is filtered, and the remaining solid raw material is repeatedly washed with hot distilled water until a neutral pH is achieved. This allows the removal of residual alkali and dissolved impurities [19].

The final stage is drying the treated raw material at a temperature of 50–60 °C to a constant weight. The result is a purified mass that is virtually free of lignin and waxes, which increases its suitability for further processing. To improve the degree of purification, bleaching agents such as hydrogen peroxide or sodium hypochlorite can be additionally used, which remove residual colored components.

4.2.3 Acid Hydrolysis

Acid hydrolysis allows the destruction of hemicellulose and residual impurities, ensuring the isolation of cellulose with a high degree of purity. This process also improves the physical and chemical properties of the material, making it suitable for subsequent use in biotechnology or other applied areas.

To carry out acid hydrolysis, the pre-cleaned lucerne material was filtered and washed with hot distilled water to a neutral pH. This step is necessary to completely remove any alkali and dissolved substances remaining after the previous treatment.

Then, a mixture of acids for hydrolysis was prepared. It consisted of acetic acid, water, and nitric acid mixed in a ratio of 2:2:1 by volume. The prepared solution was used immediately to preserve its activity. The purified raw material was immersed in this mixture, after which the container was placed in a water bath with a temperature of 90 °C.

The hydrolysis process continued for 30 min with periodic stirring to ensure uniform processing of the material. The acid mixture had a destructive effect on residual impurities, such as hemicellulose, facilitating the release of pure cellulose.

After hydrolysis was complete, the material was again washed with distilled water until a neutral pH was reached. This allowed all acid residues to be removed and possible side reactions to be prevented. The processed material was then dried in a drying oven at a temperature of 50–60 °C to constant weight [20, 21].

As a result of hydrolysis, cellulose of a high degree of purity was obtained, which makes it suitable for subsequent use in biotechnological processes or other applied areas.

4.2.4 Bleaching of Obtained Cellulose

To improve the purity and whiteness of lucerne pulp, a bleaching process was carried out using sodium hypochlorite. For this, a 5–10% NaClO solution was prepared in advance, which was used immediately after preparation to ensure maximum activity.

The pulp was placed in a sodium hypochlorite solution in a ratio of 1:10 (material weight to solution volume) and kept at a temperature of 40–50 °C for 1–2 h. During the bleaching process, the mixture was periodically stirred to ensure uniform action of the reagent on the material. After processing, the pulp was filtered and washed with a large amount of distilled water until a neutral pH was achieved in order to completely remove residual chemical reagent.

The degree of bleaching was assessed visually: the material acquired a significantly lighter and more uniform white shade, indicating successful removal of residual impurities. The results showed that sodium hypochlorite treatment effectively improved the whiteness and purity of pulp, making it suitable for use in high-quality industries such as pharmaceuticals, textiles and biotechnology.

5 Characterisation Methods

The following analytical methods are used to characterise and confirm the cellulose produced, which allow a deeper understanding of their morphology, chemical composition, structure and other key characteristics.

The gravimetric method for determining cellulose content involves accurately measuring the mass of cellulose before and after sample processing. To do this, the starting material is first prepared — lucerne, which is dried to a constant mass in a drying oven at a temperature of 105 °C to remove moisture. The raw material is then treated to remove impurities such as lignin using alkaline solutions, after which the material is washed to a neutral pH and dried again. After processing, the mass of the resulting cellulose is measured with high accuracy using an analytical balance. The cellulose content is calculated as the ratio of the mass of pure cellulose to the mass of the starting material. This method allows for an accurate determination of the percentage of cellulose in a sample before and after its processing, which is important for assessing the efficiency of cellulose extraction and quality control of the material at different stages.

IR spectroscopy (infrared spectroscopy) is used to analyze the molecular structure of cellulose and identify its functional groups. This method allows you to study the chemical composition of the material, which is especially important for assessing the purity of cellulose and identifying changes that occur in its structure during various treatments. IR spectroscopy can identify key functional groups of cellulose, such as hydroxyl groups (-OH) and ether bonds (C–O–C). These groups are characteristic of cellulose and help determine its composition. This method is also effective for assessing the degree of delignification, that is, the process of removing lignin and other impurities, since changes in the spectrum can indicate the remains of lignin and other contaminants. In addition, IR spectroscopy allows to track changes in the structure of cellulose after treatment, for example, acid hydrolysis or bleaching. Changes in the intensity of peaks in the spectrum can indicate structural transformations of the polysaccharide. This method is also used for qualitative and quantitative analysis, allowing not only to determine the composition, but also to assess the content of various components, such as moisture or hemicellulose.

6 Results and Discussion

The scheme for obtaining cellulose is shown in Fig. 1.

Fig. 1. The procedure of obtaining cellulose from lucerne

Based on the gravimetric analysis, it was found that the cellulose content in lucerne ranged from 30 to 40% of the total dry plant material weight. Weight losses associated with the removal of impurities were observed at different stages of processing. Thus, about 5–10% of the original weight was lost at the wax removal stage, and the delignification process aimed at removing lignin reduced the weight by another 15–20%. An additional 5–8% of the weight was lost during bleaching, which was associated with the removal of the remaining color impurities. The final yield of pure cellulose was 25–30% of the original lucerne weight. The resulting material was light white, indicating the successful removal of lignin and other undesirable components.

IR (infrared spectroscopy) data of isolated cellulose can provide information about the characteristic functional groups and structural changes in the cellulose molecule. Typical peaks and their corresponding functional groups that can be identified in the cellulose spectrum are presented below (Table 4).

Table 4. IR spectroscopy results

Wavenumber (cm^{-1})	Peak characteristic	Notes
3200–3500	Sodium-O-group (–OH)	A broad peak characteristic of hydroxyl groups located on cellulose molecules
2900–2800	C–H vibrations (CH_2, CH_3)	Associated with CH_2 groups present in cellulose
1600–1650	C=C vibrations (aromatic rings, if any)	May appear in the presence of impurities, such as lignin
1500–1450	Deviation of CH_2- and CH_3-groups (strain)	Indicates the presence of a polysaccharide structure
1420–1410	C–O–C vibrations (ether bonds)	Characteristic of cellulose chains
1300–1200	C–O vibrations (internal asymmetric strain)	Indication of the presence of C–O bonds in cellulose
1150–1080	C–O vibrations (hemicelluloses)	Associated with the presence of hemicellulose in the sample
1050–1030	C–O vibrations (cellulose)	Direct indication of the cellulose structure
890–850	Vibrations for the β-glucosidic bond group	Indicates the β-form of cellulose, characteristic of natural cellulose

The IR spectroscopy data of the isolated cellulose provide valuable information about its chemical composition and structure. Several characteristic peaks can be observed in the spectrum of cellulose. The broad peak in the region of 3200–3500 cm^{-1} indicates the presence of hydroxyl groups (–OH), which is characteristic of cellulose. The peaks in the region of 2900–2800 cm^{-1} indicate C–H vibrations associated with CH_2 and CH_3 groups that are present in cellulose molecules. Other peak areas, such as 1420–1410 cm^{-1}, 1300–1200 cm^{-1} and 1050–1030 cm^{-1}, reflect C–O and C–O–C vibrations, which confirms the presence of ether and alcohol bonds characteristic of the cellulose structure. A peak at 890–850 cm^{-1} can also be seen, which is associated with β-glucosidic bonds, indicating the natural β-form of cellulose.

These peaks together help confirm the composition and structure of cellulose, and also allow us to assess its purity. IR spectroscopy is an important tool for cellulose analysis, as it allows us to identify the functional groups present in the sample and determine changes that have occurred as a result of processing.

7 Conclusions

In the study, experiments were successfully conducted to produce pulp from lucerne using various processing methods, including acid hydrolysis, alkali treatment and bleaching. The results showed that the pulp obtained from lucerne has high quality and can be used for further applications in biotechnology and other industries. Analysis of the chemical composition and IR spectroscopy confirmed the presence of the main structural components of cellulose, such as cellulose, hemicellulose and lignin. Optimum conditions for obtaining pure cellulose were determined, which opens up prospects for its application in various fields.

Lucerne-derived cellulose is a promising raw material for various industries. It can be used to produce biodegradable materials such as bioplastics, water purification sorbents, as well as in medicine and pharmaceuticals. The environmental friendliness of lucerne, its availability and high fiber content make it an attractive source for obtaining materials with improved properties, which contributes to sustainable development. In the future, lucerne cellulose can be used to produce nanocellulose, which has unique mechanical and chemical properties such as high strength, low density and the ability to be modified. This opens up new opportunities for creating lightweight, durable and environmentally friendly materials.

Medicago sativa (Lucerne) is a promising raw material for the production of nanocellulose due to its high cellulose content, availability and environmental safety. Nanocellulose obtained from lucerne has unique properties, such as high mechanical strength, lightness, biocompatibility and the ability to be modified for various technical needs. It can be used in the production of environmentally friendly materials, such as biodegradable plastics, water filters, textiles, and as a filler for composite materials.

In addition, nanocellulose from lucerne can find application in medicine, for example, for the creation of biocompatible implants, medicines and carriers for drug delivery. The potential of lucerne for the production of nanocellulose opens up new opportunities for sustainable development and the creation of environmentally friendly technologies.

To improve the efficiency of lucerne cellulose production, it is recommended to conduct additional research aimed at optimizing the processes of impurity removal and improving delignification. This will help increase the yield of cellulose and improve its physicochemical properties. It is also worth paying attention to the extraction of cellulose with minimal energy and chemical consumption, which will ensure the environmental safety of the process. It is necessary to continue working on improving the methods for obtaining nanocellulose from *Medicago sativa* (Lucerne), exploring various ways of modifying it and using it in high-tech areas such as nanomaterials, medicine and electronics.

Acknowledgments. This research is funded by the Committee of Science of the Ministry of Science and Higher Education of the Republic of Kazakhstan (Grant No. AP 22686347).

References

1. Katić, S., et al.: Variation of protein, cellulose and mineral contents of lucerne as influenced by cultivar and cut. Biotechnol. Anim. Husb. **25**(6), 1189–1195 (2009)
2. Bahreini, Z., Abedi, M., Sadeghi Fateh, D.: Preparation and characterization of microcrystalline cellulose from Lucerne (Medicago sativa) waste fibers as food additive. Innov. Food Technol. **10**(1), 87–99 (2022)
3. Bahreini, Z., Abedi, M., Fateh, D.S.: A study on preparation and characterization of microcrystalline cellulose from Lucerne (alfalfa). Waste Fibers. (2022)

4. Gaweł, E.: Chemical composition of lucerne leaf extract (EFL) and its applications as a phytobiotic in human nutrition. Acta Sci. Pol. Technol. Aliment. **11**(3), 303–309 (2012)
5. Nordkvist, E., Åman, P.: Changes during growth in anatomical and chemical composition and in-vitro degradability of lucerne. J. Sci. Food Agric. **37**(1), 1–7 (1986)
6. Yolcu, H., et al.: Nutrient value of some lucerne cultivars based on chemical composition for livestock. Asian J. Chem. **20**(8), 4110 (2008)
7. Christian, K.R., Jones, D.B., Freer, M.: Digestibility and chemical composition of fractions of lucerne during spring and summer. J. Agric. Sci. **75**(2), 213–222 (1970)
8. Repetto, J.L., et al.: Relationship between ruminal degradability and chemical composition of dehydrated lucerne. Anim. Res. **52**(1), 27–36 (2003)
9. Katic, S. et al.: Yield, Morphology and Chemical Composition of Five Lucerne Genotypes as Affected by Growth Stage and the Environment (2003)
10. Vasileva, V.: Effect of mineral nitrogen fertilization and water-deficiency stress on chemical composition of lucerne (*Medicago sativa* L.). In: Grassland–A European Resource, pp. 391–393 (2012)
11. Kuznetsov, B.N., et al.: Optimized methods for obtaining cellulose and cellulose sulfates from birch wood. Wood Sci. Technol. **49**, 825–843 (2015)
12. Kuznetsov, B.N., Tarabanko, V.E., Kuznetsova, S.A.: New catalytic methods for obtaining cellulose and other chemical products from vegetable biomass. Kinet. Catal. **49**, 517–526 (2008)
13. Heinze, T., Liebert, T.: Unconventional methods in cellulose functionalization. Prog. Polym. Sci. **26**(9), 1689–1762 (2001)
14. Spence, K.L., et al.: A comparative study of energy consumption and physical properties of microfibrillated cellulose produced by different processing methods. Cellul. **18**, 1097–1111 (2011)
15. French, A.D., et al.: Cellulose. In: Encyclopedia of polymer science and technology, vol. 1838, 4th edn, pp. 1–69 (2002)
16. Kalia, S., et al.: Cellulose-based bio-and nanocomposites: a review. Int. J. Polym. Sci. **2011**(1), 837875 (2011)
17. Jonoobi, M., et al.: Different preparation methods and properties of nanostructured cellulose from various natural resources and residues: a review. Cellul. **22**, 935–969 (2015)
18. Rinaldi, R., Schüth, F.: Acid hydrolysis of cellulose as the entry point into biorefinery schemes. ChemSusChem. **2**(12), 1096–1107 (2009)
19. Dussan, K.J., et al.: Dilute-acid hydrolysis of cellulose to glucose from sugarcane bagasse. Chem. Eng. Trans. **38**, 433 (2014)
20. Huang, Y.B., Fu, Y.: Hydrolysis of cellulose to glucose by solid acid catalysts. Green Chem. **15**(5), 1095–1111 (2013)
21. Mahmud, M.M., et al.: Preparation of different polymorphs of cellulose from different acid hydrolysis medium. Int. J. Biol. Macromol. **130**, 969–976 (2019)

Modern Achievements in the Field of Bentonite Clays for the Removal of Anionic Contaminants in Water Purification Processes

B. M. Kussainova[1], G. K. Tazhkenova[1], I. A. Kazarinov[2], A. N. Nurlybayeva[3](✉), and A. Bazarkhankyzy[4,5]

[1] L.N. Gumilyov Eurasian National University, 010000 Astana, Kazakhstan
[2] Department of Physical Chemistry, Saratov State University, 410000 Saratov, Russian Federation
[3] M.Kh. Dulaty Taraz Regional University, 080000 Taraz, Kazakhstan
rustem_ergali@mail.ru
[4] Research Institute of New Chemical Technologies, L.N. Gumilyov Eurasian National University, 010000 Astana, Kazakhstan
[5] Department of General and Biological Chemistry, Astana Medical University, 010000 Astana, Kazakhstan

Abstract. The importance of this review article is that it provides a comprehensive overview of the latest developments in the field of adsorbents derived from bentonite clays for the removal of chromium and arsenic contaminants from various sources. The review provides a mini-review of the latest achievements in the field of synthesis and development of adsorbents obtained from bentonite clays. By summarizing various modification methods, structural characteristics, and methods for evaluating performance characteristics used in the development of adsorbents, the review provides an idea of modern approaches to improving the efficiency, selectivity, and environmental friendliness of adsorption. By clarifying the physicochemical processes governing adsorption, the review expands our understanding of the interaction of pollutants with adsorbents and provides a theoretical basis for optimizing the characteristics of adsorption and technical design. Finally, the review identifies gaps, challenges, and opportunities for further research and innovation in the field of adsorbents derived from bentonite clays. By identifying knowledge gaps, methodological limitations, and unresolved issues, the review facilitates further research aimed at solving critical research issues, improving the characteristics of adsorbents, and promoting sustainable remediation technologies. As a result, we can say that this review article plays a key role in promoting scientific knowledge, informing about technological innovations, and making scientifically sound decisions in the field of environmental restoration. By summarizing recent achievements, existing knowledge, and identifying gaps and opportunities for further research, the review contributes to ongoing efforts to reduce chromium and arsenic pollution and protect human health and the environment.

Keywords: Bentonite clay · Modification · Anionic contaminant · Chromium · Arsenic

R. Rajasekharan et al. (Eds.): MatZone 2024, SPPHY 316, pp. 169–179, 2025.
https://doi.org/10.1007/978-981-96-6541-9_16

1 Introduction

Water pollution is one of the most urgent environmental challenges of the twenty-first century [1]. Human activities have led to the introduction of many harmful substances into natural water sources [2], including anionic pollutants such as nitrates [3–5], phosphates [6], sulfates [7], chromates [7], and other toxic compounds [8]. These substances not only degrade water quality but also pose serious threats to human health and ecosystems. Although these compounds often exist in low concentrations, they can still cause significant toxic effects, especially when they coexist in complex mixtures. In response to these concerns, the World Health Organization (WHO) [9] has established acceptable standards for the presence of hazardous pollutants in water. Consequently, researchers around the world are focused on developing effective, affordable, and reliable water purification technologies to address this pressing issue.

Purifying water from anionic pollutants like sulfates, nitrates, phosphates, and chromates is a crucial task for modern ecology and industry [10]. Dangerous concentrations of these anions have been identified in many water sources, where they negatively impact the quality of natural reservoirs and pose risks to human health. The low concentrations of many of these anions complicate detection, potentially increasing health risks.

Water purification encompasses a variety of methods, including reverse osmosis, membrane techniques, coagulation, electrodialysis, biological methods, and adsorption. Among these, sorption technologies are particularly notable for their high efficiency in removing pollutants.

The advantages of these sorption methods include their effectiveness and ability to treat wastewater containing multiple pollutants. Adsorption purification can achieve efficiencies of approximately 90%, depending on factors such as the chemical nature of the sorbent, the size and accessibility of the sorption surface, the chemical structure of the pollutants, and their chemical forms in the aqueous medium.

A range of materials can serve as sorbents, but from an economic perspective, natural clay minerals and aluminosilicates are particularly advantageous [11]. Bentonites, which are inexpensive and environmentally friendly, stand out among natural aluminosilicates. Bentonite clays are widely used for the adsorption of various pollutants, including anions, due to their availability, low cost, and potential for modification. Ion adsorption on clay minerals is influenced by several mechanisms, including ion exchange, electrostatic interactions, and complexation processes. Factors such as the solvate shell, pH of the medium, and the ability of simple ions to form complex ions affect their electrostatic interactions with the charged surface of the adsorbent. Thanks to their unique layered structure, large specific surface area, and ion exchange capabilities, bentonites are highly effective in removing various anions. Additionally, modifying bentonite can significantly enhance its sorption characteristics, expanding its application in water treatment processes.

The purpose of this article is to review recent advancements in the use of bentonite clays for the removal of anionic contaminants over the past decade, identify the primary mechanisms of action, analyze the effectiveness of modified materials, and evaluate prospects for future research.

2 Anions and Water Pollution

Contamination with chromium and arsenic anions is a common environmental problem with significant consequences for human health and ecosystem integrity. Both pollutants are widespread in the natural environment and are usually associated with industrial activities, mining, agricultural practices, and natural geological processes. Understanding the sources, distribution, and environmental and human health impacts of chromium and arsenic pollution is essential for developing effective strategies to restore and protect public health and environmental quality.

Chromium exists in various oxidation states, with the most common forms of environmental pollution being chromium (III) and chromium (VI). Chromium (III) occurs naturally and is necessary for human nutrition in small quantities, while chromium (VI) is predominantly of anthropogenic origin and is highly toxic. Sources of chromium pollution include industrial emissions: chromium (VI) is widely used in processes such as electroplating, leather tanning, and metalworking, leading to the release of chromium-contaminated wastewater; mining; the development of chromite and coal deposits can result in the release of hexavalent chromium into soil and water sources; and precipitation: chromium emissions from industrial plants and combustion processes can settle on soil and surface water bodies, contributing to environmental pollution.

Arsenic is a metalloid element of natural and anthropogenic origin, occurring in inorganic and organic forms [12]. Sources of arsenic pollution include its natural presence in the Earth's crust, where it can leach into groundwater and surface waters due to the weathering of arsenic-containing minerals. In the mining and smelting industry, arsenic is a common by-product, especially in regions with deposits of metallic ores such as gold, copper, and lead. Additionally, pesticides, herbicides, and fertilizers containing arsenic have historically been used in agriculture, leading to soil and water pollution. The main pollutants are the industrial processes of arsenic synthesis, which are used in various industries, including wood preservation, semiconductor production, and pesticide manufacturing, resulting in emissions into the environment. Chromium can undergo oxidation-reduction transformations between its trivalent and hexavalent forms, with chromium (VI) being more mobile and toxic than chromium (III). Arsenic exists in various oxidation states (-3, 0, $+3$, $+5$), with inorganic forms of arsenic (arsenate and arsenite) being the most common in the environment [13].

Chromium and arsenic are known to cause a number of adverse effects on human and wildlife health, including: Chromium (VI) compounds are classified by international health agencies as carcinogenic substances for humans, with prolonged exposure leading to lung, skin, and gastrointestinal cancers [14]. Neurotoxicity: Arsenic exposure is associated with neurological disorders, cognitive impairment, and developmental delays, especially in children exposed at critical stages of growth and development. Effects on the cardiovascular system: Chronic exposure to chromium is associated with cardiovascular diseases, including hypertension, atherosclerosis, and peripheral vascular diseases. Effects on the reproductive system and development: Both chromium and arsenic are linked to reproductive toxicity, including infertility, miscarriage, and birth defects, highlighting the importance of protecting vulnerable populations such as pregnant women and children. Global importance and the urgent need to address the consequences: Chromium and arsenic pollution pose significant risks to human health, ecosystems, and socio-economic development worldwide [15]. In many regions, access to safe drinking water and uncontaminated food products is challenging due to high levels of chromium and arsenic, particularly in low-income and marginalized communities. The global scale of pollution problems underscores the urgent need to implement mitigation strategies to reduce risks and protect public health and environmental quality. Effective remediation efforts require interdisciplinary collaboration, innovative technologies, and proactive policy measures to address the complex challenges of chromium and arsenic contamination.

The existing literature on chromium and arsenic reduction based on adsorption using bentonite clays highlights the potential of these materials as effective and stable adsorbents for pollutant removal.

2.1 Adsorption Capacity and Efficiency

Numerous studies have demonstrated the high adsorption capacity of bentonite clays for removing chromium and arsenic from aqueous solutions [16]. The unique properties of bentonite clays, such as high surface area, porosity, and cation exchange ability, ensure effective adsorption of metal ions through electrostatic interactions, the formation of surface complexes, and ion exchange mechanisms.

2.2 Modification Methods

Researchers have explored various modification methods to improve the adsorption properties of bentonite clays, including acid activation, modification with surfactants, treatment with metal oxides, and the introduction of nanomaterials. These modifications enhance the surface properties, structural stability, and selectivity of bentonite-based adsorbents, resulting in increased efficiency of pollutant removal and accelerated kinetics.

During the research, the features of the competitive adsorption of chromium and arsenic in multicomponent systems containing other coexisting ions such as sulfate, nitrate, and chloride were studied. Understanding the competitive interactions between various pollutants and bentonite clays is important for predicting adsorption kinetics, equilibrium behavior, and the efficiency of adsorption systems in real conditions.

2.3 Regeneration and Reusability

Researchers have developed methods for the regeneration of spent bentonite-based adsorbents to extract adsorbed contaminants and restore adsorption capacity. Common regeneration methods include desorption with acids, alkalis, or chelating agents, heat treatment, and immobilization of pollutants for safe disposal or reuse for other purposes. Assessment of the feasibility and effectiveness of regeneration processes is crucial for evaluating the long-term sustainability and economic feasibility of adsorption technologies.

In the course of several field studies and experimental demonstrations, the effectiveness of bentonite-based adsorbents for the removal of chromium and arsenic in real environmental conditions was evaluated. These studies provide valuable information on practical problems, operational considerations, and performance limitations of adsorption systems in the field, allowing for the scaling up and implementation of recovery technologies.

Limitations of traditional methods of purification from chromium and arsenic, such as chemical precipitation, membrane filtration, and ion exchange, have led to interest in the study of alternative adsorbents such as bentonite clays. Traditional methods often suffer from disadvantages such as high operating costs, the formation of toxic sludge or brine, limited selectivity, and the need for complex infrastructure and maintenance. On the contrary, bentonite clays have a number of advantages, including their abundance, low cost, environmental compatibility, and ease of modification, which makes them attractive candidates for use in sustainable restoration work.

In addition, the development of new adsorbents derived from bentonite clays satisfies the need for more efficient, selective, and environmentally friendly reclamation technologies. Using the properties inherent in bentonite clays and innovative modification methods, researchers strive to overcome the limitations of traditional processing methods and achieve higher adsorption capacity, accelerated kinetics, and improved selectivity for the removal of chromium and arsenic. The rationale for exploring alternative adsorbents lies in their potential to improve treatment efficiency, reduce environmental impacts, and promote sustainable management of contaminated sites and water resources.

3 Methods for Cleaning Polluted Waters

Modern technologies for purifying water from anionic pollutants include physico-chemical, biological, chemical, and hybrid methods. Each method offers distinct advantages and limitations, depending on factors such as the type of pollutant, its concentration, environmental conditions, and the desired quality of the purified water (Table 1).

Natural clay materials exhibit exceptional adsorption properties compared to other adsorbents used for heavy metal removal and are widely available. Their popularity stems from their high cation exchange capacity, significant surface area, and structural stability. The substitution of aluminum ions (Al^{3+}) for silicon ions (Si^{4+}) imparts a negative charge to the clay matrix, which facilitates the formation of Brønsted acid centers on its surface through hydration [17].

To create polymer-clay nanocomposites, techniques such as polymer intercalation from a melt or solution, as well as monomer intercalation followed by in-situ polymerization within the matrix, are employed [18].

The modification of clay materials for wastewater treatment involves methods such as acid activation, heat treatment, and mechanical pressing. Acid activation enhances the physico-chemical properties of clay, including surface area and porosity, thereby improving its adsorption effectiveness. Heat treatment alters the clay structure, increasing its adsorption capacity by modifying particle size, porosity, crystal structure, and shape. This process also enhances the mechanical and thermal stability, as well as the catalytic properties of the clay, optimizing its use for purification and adsorption applications.

Table 1. Comparative analysis of known methods of water purification

Method	Effectiveness	Disadvantages	Applicability
Physico-chemical	Are high for most anions, especially with membrane technologies	The limited service life of materials (membranes, sorbents) High energy consumption for membrane technologies	Suitable for low and medium concentrations of anions; effective in the purification of drinking and wastewater
Chemical methods	They remove nitrates, phosphates, and sulfates well	Disposal of precipitation Side reactions and products are possible Dependence on reagents	They are widely used on an industrial scale for processing large volumes of water
Biological methods	High efficiency for nitrates (denitrification), organic compounds.	Dependence on environmental conditions (pH, temperature). Long processing time	They are limited by low concentrations of pollutants; they are used for natural and wastewater treatment
Hybrid methods	Maximum efficiency through a combination of methods; reduction of waste volume	High cost of equipment The complexity of design and operation Technical maintenance	They are suitable for cleaning complex mixtures of pollutants; they are used in high-tech industrial systems

3.1 Adsorption Methods for Purification

When developing adsorption methods, three primary mechanisms are considered: ion exchange, chemical bonding through the formation of chemical bonds, and physical bonding driven by Van der Waals forces. Isothermal models are commonly employed to analyze adsorption equilibrium, as they help describe and understand adsorption processes. Notable models include the isotherms of Henry, Freundlich, Langmuir, Temkin, Hill-De Boer, and Dubinin-Kaganer-Radushkevich [18]. Various materials, such as hydrogels, clay compounds, activated carbon, zeolites, silica gel, biocarbon, activated alumina, polymers, and magnetic adsorbents, are being researched as potential adsorbents for wastewater treatment. Adsorption kinetics influence the speed of the process and are dependent on physical and chemical conditions. This knowledge not only aids in predicting optimal models and process parameters but also allows for estimating the absorption rate of sorbents, thereby enabling control over the interaction time at the interface [19]. Removing dyes and heavy metal ions from wastewater continues to be a significant research focus. New adsorbents based on nanomaterials greatly enhance purification efficiency due to their unique structural and morphological characteristics, making them promising solutions for environmental challenges [20].

Clay is actively employed in water purification due to its unique physico- chemical properties. Recent research has concentrated on creating clay nanocomposites for removing pollutants from water sources (Table 2). A promising avenue involves developing clay-based composites with modified structures, which enhances their effectiveness as adsorbents [21].

The efficacy of clay as an adsorbent can be attributed to its high surface area, permeability, charge, cation exchange capacity, acidity, and diverse active centers. To improve adsorption properties, clay can be modified, making the choice of modifier a critical step in creating effective materials. Several types of clays are specifically utilized for wastewater treatment [22].

Researchers have employed various clay-based adsorbents for water purification [23]. One such material is the clay–poly(methoxyethyl) composite acrylamide (PMEA), which has demonstrated effectiveness in removing lead (II) ions from aqueous solutions, achieving an adsorption capacity of 81.02 mg/g after 60 min of interaction [24]. The chitosan-g-poly(acrylicacid)/ Montmorillonite (CTS-g-PAA/MMT) nanocomposite has shown the ability to absorb methylene blue from water with an efficiency of 1859 mg/g and can be reused after saturation [25]. Activated attapulgite has been utilized both as a bleaching agent and as an adsorbent for purifying soybean oil.

The adsorption processes within the clay layer are influenced by its composition, which affects how it interacts with pollutants. Mechanisms for the absorption of metal ions by clay nanoparticles include chemical adsorption, physical adhesion, micro deposition (including ion transfer to charged sites), and complex formation on hydroxyl groups present on the clay surface. Consequently, the fixation of pollutants on the clay surface occurs through various mechanisms, allowing for the adaptation of materials for specific applications [23].

4 Recommendations and Future Direction

Clay-based adsorbents have garnered global attention due to their versatile properties, particularly in wastewater treatment, owing to their enhanced adsorption capacity. Various types of clay, including montmorillonite, sepiolite, palygorskite, and bentonite, have been extensively studied for their effectiveness in removing dyes and heavy metals from wastewater. These adsorbents possess unique characteristics, such as a high surface area, a layered structure, and the ability to exchange cations. Additionally, they are environmentally friendly and cost-effective, as they are non-toxic and generally less expensive than other adsorbents.

Ongoing research and development aim to further improve the adsorption efficiency of clay-based adsorbents in wastewater treatment. This includes exploring methods to alter their adsorption capacity and selectivity concerning different dye molecules and heavy metals.

The effectiveness of clay-based adsorbents can be enhanced by integrating nanoparticles, modifying polymers, or introducing functional groups onto their surfaces. Additionally, employing a hybrid adsorption system can optimize recyclability and reuse,

Table 2. Clay-based adsorbents for wastewater treatment

Removed anions	Adsorbent	Modification	Reference
Nitrate	PDDA-Mt nanocomposite	Montmorillonites (Mt) and a polymer, polydiallyldimethylammonim (PDDA) chloride	[26]
As(III) and as(V)	Granulated attapulgite (GAP)	Fe-impegnated charred GAP is syntesized by mixing 10 g og chared GAP into 40 ml soluyion of ($FeCl_3$*$6H_2O$)	[27]
Bichromate and arsenate	Bentonite clay	Modified with polyhydroxocomplexes of aluminum and iron (III) by co-precipitation	[28]
Phosphate	Halloysite nanotubes (HNTs)	Halloysite nanotubes (HNTs) are modified with nano-scale iron oxide (Fe_2O_3)	[29]
Cr(VI)	Natural and modified attapulgite	Modification is carried out acis leaching	[30]
As (V)	Iranian natural clay ans clay/Fe-Mn composite	Prepared in water for 20 min, then $KMnO_4$ and $FeCl_3$ added	[31]
Anionic surfactants and polysphates	Saponite, Palygorskite and saponite clay	Magnetized by magnetic modifiier Fe_3O_4 used in the form of magnetic fluid	[32]
Phosphate	Sand-clay mixture	Calcination	[33]
Phosphate	Magnetic diatomite and illite clay	Modified by loading Fe3O4 NPs onto raw diatomite or illite clay surface	[34]
Chromuim	Colombian clay	Modified with hexadecyltrimetilammonium bromide (HDTMA-Br)	[35]
Chromuim	Kaolinite	Clay-crystalline cellulose adsorbent synthesized by dissolving cellulose in an alkali solution with an additive	[7]
Sulfate	Organo-nano-clay	Organo-nano-clay prepared by natural zeolite and cationic surfactant cetyltrimethylammonium bromide (CTAB)	[36]

thereby reducing overall processing costs and environmental impacts. Moreover, modified clay-based adsorbents can also be utilized to eliminate pathogenic microorganisms and pharmaceutical waste. In conclusion, clay-based adsorbents hold significant potential for wastewater purification.

5 Conclusions

Clay-based materials are regarded as promising adsorbents for water treatment. This review provides a detailed analysis of various types of clay, their properties, mechanisms, and potential for wastewater treatment. It begins by discussing the different clay types primarily used for this purpose. Each type of clay has a unique composition, and the review explores how pollutants attach to the surfaces of clay-based adsorbents. It also addresses the modification of clay-based materials for wastewater treatment, highlighting their minimal environmental and economic impact. Furthermore, examples of both raw and modified clay-based materials used in previous studies for the removal of specific dyes and heavy metals are discussed. The review concludes by exploring potential advancements to enhance the capabilities of clay-based adsorbents for wastewater treatment, positioning them as one of the most suitable options for this application.

References

1. Balali-Mood, M., Naseri, K., Tahergorabi, Z., Khazdair, M.R., Sadeghi, M.: Toxic mechanisms of five heavy metals: mercury, lead, chromium, cadmium, and arsenic. Front. Pharmacol. **12**, 643972 (2021). https://doi.org/10.3389/FPHAR.2021.643972/BIBTEX
2. Babel, S., Kurniawan, T.A.: Low-cost adsorbents for heavy metals uptake from contaminated water: a review. J. Hazard. Mater. **B97**(1–3), 219–243 (2003). https://doi.org/10.1016/s0304-3894(02)00263-7
3. Sabri, A., Abbood, N.: Adsorption study of nitrate anions by different materials using fixed bed column. Eng. Technol. J. **37**(1C), 156–162 (2019). https://doi.org/10.30684/ETJ.37.1C.25
4. Zhou, H., Tan, Y., Gao, W., Zhang, Y., Yang, Y.: Selective nitrate removal from aqueous solutions by a hydrotalcite-like absorbent FeMgMn-LDH. Sci. Rep. **10**(1), 1–10 (2020). https://doi.org/10.1038/s41598-020-72845-3
5. Khatamian, M., Derakhshan, S.K., Nami, S.H., Fazli-Shokouhi, S.: Nitrate removal study of synthesized nano γ-alumina and magnetite-alumina nanocomposite adsorbents prepared by various methods and precursors. Sci. Rep. **14**(1), 7673 (2024). https://doi.org/10.1038/S41598-024-58459-Z
6. Yin, H., Yang, C., Jia, Y., Chen, H., Gu, X.: Dual removal of phosphate and ammonium from high concentrations of aquaculture wastewaters using an efficient two-stage infiltration system. Sci. Total Environ. **635**, 936–946 (2018). https://doi.org/10.1016/j.scitotenv.2018.04.218
7. Islam, M.M., et al.: Preparation and characterization of bijoypur clay-crystalline cellulose composite for application as an adsorbent. Adv. Mater. Sci. **2**(3), 1–7 (2017). https://doi.org/10.15761/AMS.1000126
8. Mudgal, V., Madaan, N., Mudgal, A., Singh, R.B., Mishra, S.: Effect of toxic metals on human health. Open Nutraceut. J. **3**(1), 94–99 (2010) Accessed 6 Oct 2024
9. WHO: Guidelines for drinking-water quality: fourth edition incorporating the first addendum. World Heal. Organ. **4**, 608 (2017) Accessed 25 Jun 2024. [Online]

10. Yılmazoğlu, M.: Organic-Inorganic Ion Exchange Materials for Heavy Metal Removal from Water, pp. 179–198 (2021). https://doi.org/10.1007/978-3-030-80334-6_7
11. Kussainova, B., et al.: Comparative study of the physico-chemical properties of sorbents based on natural bentonites modified with Iron (III) and Aluminium (III) polyhydroxocations. Coatings. **13**, 1974 (2023). https://doi.org/10.3390/COATINGS13111974
12. Kong, S., Wang, Y., Zhan, H., Yuan, S., Liu, M., Zhou, C.: Arsenite and arsenate removal from contaminated groundwater by nanoscale iron-manganese binary oxides: column studies. Environ. Eng. Sci. **30**(11), 689–696 (2013). https://doi.org/10.1089/EES.2013.0037
13. Mohapatra, D., Mishra, D., Chaudhury, G.R., Das, R.P.: Arsenic(V) adsorption mechanism using kaolinite, montmorillonite and illite from aqueous medium. J. Environ. Sci. Heal. Part A Toxic/Hazardous Subst. Environ. Eng. **42**(4), 463–469 (2007). https://doi.org/10.1080/10934520601187666
14. Ahmed, S., Khan, S.H., Hashem, M.A.: Chromium from tannery waste in poultry feed: a potential cradle to transport human food chain. Cogent. Environ. Sci. **3**(1), 1312767 (2017). https://doi.org/10.1080/23311843.2017.1312767
15. Joya-Cárdenas, D.R., Rodríguez-Caicedo, J.P., Corona-Rivera, M.A., Saldaña-Robles, N., Damián-Ascencio, C.E., Saldaña-Robles, A.: Removal of as(V) in the presence of Cr(VI) in contaminated water from the Bajio region of Mexico using ferrihydrite-functionalized graphene oxide (GOFH): a case study. Emerg. Contam. **10**(3), 100312 (2024). https://doi.org/10.1016/J.EMCON.2024.100312
16. Liu, F., Yang, W., Li, W., Zhao, G.C.: Simultaneous oxidation and sequestration of arsenic(III) from aqueous solution by copper aluminate with Peroxymonosulfate: a fast and efficient heterogeneous process. ACS Omega. **6**(2), 1477–1487 (2021). https://doi.org/10.1021/ACSOMEGA.0C05203
17. Tiruneh, A.T., Debessai, T.Y., Bwembya, G.C., Nkambule, S.J.: Combined clay adsorption-coagulation process for the removal of some heavy metals from water and wastewater. Am. J. Environ. Eng. **8**(2), 25–35 (2018). https://doi.org/10.5923/J.AJEE.20180802.02
18. Nguyen, Q.T., Baird, D.G.: Preparation of polymer–clay nanocomposites and their properties. Adv. Polym. Technol. **25**(4), 270–285 (2006). https://doi.org/10.1002/ADV.20079
19. Awad, A.M., et al.: Adsorption of organic pollutants by natural and modified clays: a comprehensive review. Sep. Purif. Technol. **228**, 115719 (2019). https://doi.org/10.1016/J.SEPPUR.2019.115719
20. Sadegh, H., et al.: The role of nanomaterials as effective adsorbents and their applications in wastewater treatment. J. Nanostruct. Chem. **7**(1), 1–14 (2017). https://doi.org/10.1007/S40097-017-0219-4
21. Ayalew, A.A.: A critical review on clay-based nanocomposite particles for application of wastewater treatment. Water Sci. Technol. **85**(10), 3002–3022 (2022). https://doi.org/10.2166/WST.2022.150
22. Rashid, R., Shafiq, I., Akhter, P., Iqbal, M.J., Hussain, M.: A state-of-the-art review on wastewater treatment techniques: the effectiveness of adsorption method. Environ. Sci. Pollut. Res. **28**(8), 9050–9066 (2021). https://doi.org/10.1007/S11356-021-12395-X
23. Awasthi, A., Jadhao, P., Kumari, K.: Clay nano-adsorbent: structures, applications and mechanism for water treatment. SN Appl. Sci. **1**(9) (2019). https://doi.org/10.1007/S42452-019-0858-9
24. Şölener, M., Tunali, S., Özcan, A.S., Özcan, A., Gedikbey, T.: Adsorption characteristics of lead(II) ions onto the clay/poly(methoxyethyl)acrylamide (PMEA) composite from aqueous solutions. Desalination. **223**(1–3), 308–322 (2008). https://doi.org/10.1016/J.DESAL.2007.01.221
25. Wang, X.S., Zhou, Y., Jiang, Y.: Removal of methylene blue from aqueous solution by non-living biomass of marine algae and freshwater macrophyte. Adsorpt. Sci. Technol. **26**(10), 853–863 (2008). https://doi.org/10.1260/026361708788708234

26. Zhang, H., Huang, W., Komarneni, S.: Polymer-clay nanocomposites for the uptake of hazardous anions. Nanomaterials. **14**, 467 (2024). https://doi.org/10.3390/NANO14050467
27. Yin, H., Kong, M., Gu, X., Chen, H.: Removal of arsenic from water by porous charred granulated attapulgite-supported hydrated iron oxide in bath and column modes. J. Clean. Prod. **166**, 88–97 (2017). https://doi.org/10.1016/J.JCLEPRO.2017.08.026
28. Kussainova, B., et al.: Adsorption of bichromate and arsenate anions by a sorbent based on bentonite clay modified with polyhydroxocations of iron and aluminum by the 'Co-Precipitation' method. Molecules. **29**, 3709 (2024). https://doi.org/10.3390/MOLECULES29153709
29. Almasri, D.A., Saleh, N.B., Atieh, M.A., McKay, G., Ahzi, S.: Adsorption of phosphate on iron oxide doped halloysite nanotubes. Sci. Rep. **9**(1), 1–13 (2019). https://doi.org/10.1038/s41598-019-39035-2
30. Tichapondwa, S.M., van Biljon, J.B.: Adsorption of Cr (VI) pollutants in water using natural and modified attapulgite clay. Chem. Eng. Trans. **74**, 355 (2019). https://doi.org/10.3303/CET1974060
31. Foroutan, F., Iorio, A., Thabane, L., Guyatt, G.: Calculation of absolute risk for important outcomes in patients with and without a prognostic factor of interest. J. Clin. Epidemiol. **117**, 46–51 (2020). https://doi.org/10.1016/j.jclinepi.2019.08.012
32. Makarchuk, O., Dontsova, T., Perekos, A., Skoblik, A., Svystunov, Y.: Magnetic mineral nanocomposite sorbents for wastewater treatment. J. Nanomater. **2017**, 1–7 (2017). https://doi.org/10.1155/2017/8579598
33. Almilly, R.F., Hasan, M.H.: Waste water treatment by liquid-solid adsorption using calcined sand-clay mixture adsorbent. Iraq. J. Chem. Pet. Eng. **18**(2), 69–81 (2017). https://doi.org/10.31699/IJCPE.2017.2.6
34. Chen, J., et al.: Efficient removal of phosphate by facile prepared magnetic diatomite and illite clay from aqueous solution. Chem. Eng. J. **287**, 162–172 (2016). https://doi.org/10.1016/J.CEJ.2015.11.028
35. MacÍas-Quiroga, I.F., Giraldo-Gómez, G.I., Sanabria-González, N.R.: Characterization of Colombian clay and its potential use as adsorbent. ScientificWorldJournal. **2018**, 1–11 (2018). https://doi.org/10.1155/2018/5969178
36. Chen, W., Liu, H.C.: Adsorption of sulfate in aqueous solutions by organo-nano-clay: adsorption equilibrium and kinetic studies. J. Cent. South Univ. **21**(5), 1974–1981 (2014). https://doi.org/10.1007/S11771-014-2145-7/METRICS

Soil Cleaning from Heavy Metals Using Activated Carbon

A. U. Khabashova[1], A. N. Nurlybayeva[1], Helder T. Gomes[2,3], and M. S. Kalmakhanova[1](✉)

[1] Department of Chemistry and Chemical Technology, M.Kh. Dulaty Taraz Regional University, Taraz 5300-253, Kazakhstan
marjanseitovna@mail.ru

[2] Centro de Investigação de Montanha (CIMO), Instituto Politécnico de Bragança, Bragança 5300-253, Portugal

[3] Laboratório Associado para a Sustentabilidade e Tecnologia em Regiões de Montanha (SusTEC), Instituto Politécnico de Bragança, Bragança, Portugal

Abstract. Rapid industrialization, expanding urbanization, and unrestrained use of natural resources contribute to environmental issues, particularly the contamination of water bodies by a variety of chemical compounds due to inadequate wastewater treatment. These contaminants, referred to as contaminants of emerging concern (CECs), pose global challenges due to their widespread occurrence, persistence, and bioaccumulation, stemming from anthropogenic activities. Notable examples of CECs include pharmaceuticals like paracetamol and synthetic chemicals such as perfluorooctanoic acid (PFOA), underscoring the imperative for effective wastewater treatment. Furthermore, phenolic species originating from pharmaceutical and dye industries endure in wastewater, posing threats to human health and ecosystems. To address these challenges, permeable reactive barriers (PRBs) have been developed to treat polluted waters, utilizing various materials. However, concerns about their manufacturing complexity processes have prompted the exploration of simpler and eco-friendly alternatives. Promising options include geopolymers, activated carbons derived from waste, and carbon nanotubes sourced from plastic solid waste. Integrating these materials into permeable barriers and adopting the concept of PRBs, with a focus on circular economy principles, presents a sustainable approach to mitigate population exposure to contaminated water. This innovative strategy not only addresses the environmental and health risks associated with water pollution but also aligns with the broader purpose of this work, which is to produce environmentally friendly permeable barriers for application in systems post wastewater treatment plants.

Keywords: Contaminants · Heavy metal · Permeable barriers · Wastewater treatment

1 Introduction

Many toxic substances such as heavy metal ions, harmful dyes, micropollutants, personal care products, pharmaceuticals and pesticides have been found in drinking water at dangerous levels. Soil pollution with heavy metals (HM), especially in large cities

R. Rajasekharan et al. (Eds.): MatZone 2024, SPPHY 316, pp. 180–191, 2025.
https://doi.org/10.1007/978-981-96-6541-9_17

and industrial centers, has become one of the pressing environmental problems for our country. Significant foci of anthropogenic impact and soil pollution are widespread in the industrial regions of the country. There is an urgent need to develop reliable, cost-effective and environmentally friendly smart materials using an eco-friendly approach [1–3]. In this regard, biomass-derived activated carbon is a valuable and effective product that acts as an excellent adsorbent for the removal of various micro, inorganic and organic pollutants from water and atmosphere. Activated carbon has high surface area, large porosity, excellent thermal stability, adjustable surface morphology and low acid-base activity [4].

We hope that the new knowledge gained from this research will aid in the development and production of highly efficient biomass-derived activated carbon for energy and environmental purposes (water, air and soil purification).

2 Materials and Research Methods

Soil samples with a violation of their structure, activated carbon, $Cr(SO_4)_3 \cdot 6H_2O$, KBr, zeolite and brown coal, sampler-manual soil drill model "AM-16", vacuum furnace, IR spectroscopy, APP spectroscopy, AES spectroscopy.

2.1 Soil Cleaning Methods

Soil samples were collected from two locations near the Asa River in Zhamyl Oblast. The first sample was collected near the discharge, and the second near where agricultural crops are grown for livestock feed (Fig. 1).

Fig. 1. Soil sampling process

Samples were taken at a depth of 0–15; 15–30 in the first place and 0–10; 10–20; 20–30 in the second place. All samples were taken with special soil sampler equipment. We had the most popular type of soil sampler - a manual soil drill, model "AM-16", which is designed to take soil samples with a violation of their structure, as well as to measure the depth of soil freezing, as well as for drilling wells for permafrost meters.

2.2 Efficiency of the Conductive Reactive Barrier

Permeable adsorption barriers consisting of individual (compost, zeolite and lignite) and composite (lignite compost and zeolite-compost) adsorbents were evaluated for their hydraulic performance and aqueous benzene removal efficiency using batch and column experiments [5, 6].

Our results generally demonstrate how useful such studies can be in evaluating potential reactive barrier materials. We investigated the optimal composition of permeable reactive barrier (PRB) materials for the treatment of groundwater heavily contaminated with landfill leachate by testing different zero-valent iron columns. The zeolite (Zeo) and activated carbon (AC) mixtures had particle sizes of 0.01–0.25, 3.0–5.0, and 0.7–1.0 mm, respectively. Inorganic contaminants and AC were the main contributors to organic matter removal, and the optimal weight ratio of the three FIRE materials to remove contaminants and maintain adequate hydraulic conductivity was 5:1:4 (Fig. 2) [7–10].

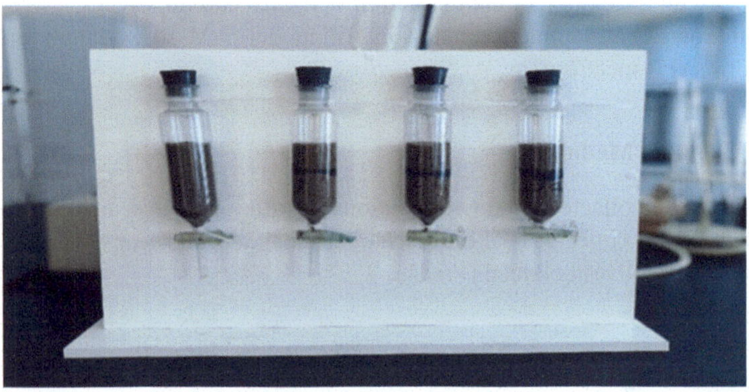

Fig. 2. Model of a reactive barrier created in laboratory conditions

The average reductions in chemical oxygen demand and total nitrogen, ammonium, Ni, Pb and 16 polycyclic aromatic hydrocarbons in the test samples using this mixture were 55.8%, 70.8%, 89.2%, 70.7%, 92.7% and 94.2%, respectively.

2.3 Preparation of Activated Carbon

Biochar can be activated both physically and chemically. The activation process can consist of one or two stages. One stage of the process is the direct production of activated carbon from biomass; the other stage is the production of biochar followed by activation [11]. The properties of activated carbon largely depend on such parameters as biomass, temperature, pressure, flow rate. First of all, the sawdust was left in distilled water for 12 h, as shown in Fig. 3. After 12 h, the sawdust was filtered. It was left at room temperature until the water evaporated. After the water evaporated, it was dried for 3 h at 70 °C in a vacuum furnace. At this stage, the material was burned in an oven at 320 °C for 1 h. The nitrogen oven program was brought to 320 °C, and then the suspended sawdust was burned for 1 h. The resulting carbon was compared before and after placing it in the oven, as shown in Table 1, the mass of coal lost during combustion was calculated and subtracted from the original mass. And I received about 3 g of material. After receiving the coal at 320 °C, it was activated in two ratios - 1:1, 2:1.

Fig. 3. Biomass preparation

Table 1. Main agrochemical characteristics

Place of selection and depth	General humus, %	Nitrogen mg/kg	Phosphorus mg/kg	Potassium mg/kg	pH
P-1; 0–15		162.4	134	290	6.26
P-2; 15–30		142.8	106	210	5.86
P-1; 0–10	0.70	33.6	10	190	8.33
P-2; 10–20	0.83	33.6	8	180	8.46
P-3; 20–30	0.59	36.4	6	170	8.57

To obtain activated carbon with a concentration of 1:1, 1 g of KOH and 1 g of carbon burned at 320 °C were used. To carry out the activation process, water was used, which was first boiled for 30 min and then cooled to room temperature, since CO could be released in the reaction 2 or other gases. Then 20 mL of water cooled to room temperature was measured and poured into a graduated cylinder, then the measured water was poured into an Erlenmeyer flask, on top of which 1 g of KOH and 1 g of charcoal were added and stirred. The prepared solution was left for 1 night (Fig. 3).

When activating the second concentration of 2:1, 20 mL of water was used, boiled and cooled for 30 min, as with the previous concentration, measured in a graduated cylinder and poured into an Erlenmeyer flask. 0.5 g of KOH and 1 g of carbon were added to 20 mL of the measured water, mixed well and left overnight. Thus, the activation process was carried out in the ratio of 1:1 and 2:1. The next process is washing the activated carbon with a KOH solution of two concentrations. Boiled water at room temperature was taken, 1 liter of both concentrations was poured in. It was collected using a Bunsen flask and a Buchner funnel in a vacuum pump. Before the first washing of the activated carbon, filter paper was placed on the Buchner filter and fixed with distilled water (UP). Then the activated carbon was first placed in the prepared KOH solution of a concentration of 1:1 and washed with 1 L of water. In the same order, activated carbon with a 2:1 KOH solution was washed with 1 liter of boiled water at room temperature. After washing, two concentrations of activated carbon on filter paper were left in a vacuum oven for 1 day and dried. After drying, the resulting materials of 1:1 and 2:1 concentrations were scraped off the filter paper to the last granules on a porcelain grater. Activated carbon in these two concentrations was reweighed on a scale, the percentage of losses was calculated and recorded in a table.

Then 0.0030 g of activated carbon in a concentration of 2:1 was tested in IR spectroscopy. When tableting in IR spectroscopy, a KBr solution, activated carbon was added and mixed on an agate grater until a completely homogeneous mass, as shown in the figure. 0.2 g of 100 mL of this concentration of activated carbon 2:1 was left to stir on a magnetic stirrer for 24 h by time. Take 0.4 g of a model solution of 50 mg/L $Cr(SO_4)_3 \cdot 6H_2O$, place it in a 1 L flask and mix to the line with water. The prepared solution was diluted to 25 mL, determined by the APP spectroscopy method, a calibration table was built, checked and it was established that the model solution is purified. At different time intervals, 100 mL of the model solution were poured into a 250 mL flask at 0 min and mixed. During this time, 2 mL were collected and poured into a 25 mL flask. After 0 min of sampling, 0.2 g of activated carbon was added to the flask filled with 250 mL of the model solution in a ratio of 2:1. Every minute specified in the table, 2 mL of the solution were collected and poured into a 25 mL flask. That is, small filters were installed in a 25 mL flask, filter paper was placed on the surface, 2 mL was poured into it, water was added to the measuring line and 1.25 mL of 5% nitric acid was added to it. Having checked everything every minute, we checked it in AES spectroscopy.

3 Results and Discussion

3.1 Agrochemical Soil Analysis

Results of agrochemical soil analysis. Agrochemical parameters: Location and depth of sampling - P-1; 0–15, mobile: nitrogen = 162.4 mg/kg, phosphorus = 134 mg/kg, potassium = 290 mg/kg, pH = 6.26. Soil samples were taken from two different places. The first sample was collected at a depth of 0–15 cm near the septic tank in the village of Asa, Zhambyl district, and the second sample - at a depth of 0–10 cm from the place of sowing for animal feed. All samples were taken with special equipment - a soil sampler (MemST 26205-91). Determination of mobile compounds of phosphorus and potassium by the Matsevina method modified by Tsinao (MemST 26423-85). Methods for determining specific conductivity, pH and water absorption of solid residue (MemST 26423-85-26428-85). Methods for determining calcium and magnesium in an aqueous extract.

The results showed that the average soil pH in Asa village was 7.5, which is slightly above the optimal range for crops. This may be due to anthropogenic pressure, which may interfere with the conversion of organic matter into plant-available mineral compounds. The analysis of the amount of humus in the soil was carried out only on the second plot. The results showed a very low humus content in the upper soil layers, which indicates unfavourable conditions for the development of plant roots at certain depths. This is typical for some types of soil, such as podzolic or clay-podzolic. pH = 6.26 is the level of soil acidity that affects the availability of nutrients to plants. The optimum pH for most crops is considered to be 6–7 (Table 1).

The results of the agrochemical analysis of the soil for the sampling site P-1 at a depth of 0–10 cm are shown in Table 1. Analysis of these data allows us to assess the fertility of the soil and its suitability for growing crops. High pH levels can affect the availability of certain nutrients to plants, such as phosphorus. Low humus content may indicate the need to improve the soil with organic fertilizers or green manure.

The data show the concentration of various elements and ions in the aqueous extract of the soil (Table 2). Based on the results of the analysis of the aqueous extract of the soil, the following conclusions can be made:

Table 2. Water extract

No p/p	Water extract inon absolutely dry soil							
	Total salts, %	Alkalinity	Cl′	SO_4''	Ca++	Mg++	Na+	K+
3	0.122	0.027	0.001	0.066	0.008	0.015	0.005	0.001
		0.44	0.04	1.37	0.38	1.24	0.21	0.02
4	0.177	0.027	0.000	0.109	0.011	0.022	0.007	0.001
		0.44	0.00	2.27	0.57	1.81	0.30	0.02
5	0.179	0.029	0.000	0.110	0.008	0.026	0.006	0.001
		0.48	0.00	2.28	0.38	2.10	0.27	0.02

Salts: The total soil salinity is 0.122%, which is an important indicator for assessing soil fertility and its suitability for agriculture.

Alkalinity: The amount of bicarbonate ion (HCO_3^-) varies from 0.027% to 0.44%, indicating the alkalinity of the soil. This parameter is important for optimal plant growth.

Chlorides, sulfates, calcium, magnesium, sodium, potassium: The content of these elements is also varied and varies widely. These elements play an important role in plant nutrition and metabolism. Soil pH 8.33 indicates an alkaline environment. This parameter is important for the availability of nutrients to plants. Total humus: the humus content is 0.70%, which is important for improving the structure and fertility of the soil.

Based on this data, it is possible to draw conclusions about the chemical composition of the soil and its suitability for farming. Further research and analysis will help determine the optimal measures to increase soil fertility and ensure optimal conditions for plant growth and development. As we already know, the value of the cation exchange capacity depends on the granulometric composition of the soil, or more precisely on the silt content (Table 3).

Table 3. Content of soil granulometric composition

No. P/P	Place of selection	Depth	A.S.N. % H_2O	Fraction content in % on absolute dry soil						
				Fraction sizes in mm						
				Sand		Dust			Silt	3-x
				1.0–0.25	0.25–0.05	0.05–0.01	0.01–0.005	0.005–0.001	<0.001	Factions < 0.01
1	P-1	0–15	1.72	19,434	45,157	18,722	4477	11,396	0.814	16,687
2	P-2	15–30	1.44	20,901	36,080	27,192	2029	7305	6494	15,828
3	P-3	0–10	0.98	7049	47,304	16,966	2020	17,370	9291	28,681
4	P-4	10–20	0.82	6917	50,333	13,712	8469	18,552	2017	29,038
5	P-5	20–30	0.94	5492	36,362	17,363	13,325	23,420	4038	40,783

According to our data, the silt content varies from 0.8–4.038%, and the physical sand is 19.434–5.492, which means that the sand content decreases with depth. Classification of soils by granulometric composition for stony soils should be accompanied by indications of the degree of their stonyness. According to our data, the first sampling location is dominated by a greater sand content, as in other soil layers.

Based on the results of the study, the indicators of total soil contamination were calculated taking into account toxicity in all variants of the studied variants and in the filtration area. Assessing the obtained contamination indicators according to the scheme for assessing agricultural soils, these soils can be characterized as having an acceptable degree of contamination (less than 16.0) (Table 4).

Table 4. Microelement content

No. p/p	Place of selection	Depth	Mobile forms of microelements, mg/kg			
			Zn	Cu	Cd	Pb
1	P-1	0–15	131.8	16:30	1.10	1.60
2	P-2	15–30	71.6	8.40	0.50	1.40
3	P-3	0–10	3.10	0.80	0.70	1.70
4	P-4	10–20	3.80	1.30	0.60	1.50
5	P-5	20–30	2.80	1.20	0.90	1.60

At the sampling site near the canal is 6048.07 and is classified as acceptable. The maximum environmental indicator of total pollution at the first sampling site is 131.8 due to the concentration of cadmium compounds and zinc compounds, which are elements of hazard class 1. At the filtration site, the indicator is 3.10. In all the studied options, during the growing season, there is a fluctuation in the content of mobile copper compounds and a decrease in depth, no excess of MPC is noted. The highest copper content in the soil is observed at the first site near the canal 16.30 mg/kg, on the filtration field 0.70 mg/kg. Zn mobility is associated with acidification of hydrolytic acidity due to the use of mineral fertilizers. Cadmium is not among the elements necessary for plants, moreover, it is considered to be especially dangerous for plants and the vital activity of animals and humans. Therefore, constant monitoring of the cadmium content in the soil and plants is required. The level of soil pollution with heavy metals is established based on the content of mobile compounds and their compliance with maximum permissible concentrations. It was established that the content of mobile cadmium compounds in the soil was 1.0–0.9 mg/kg, and is decreasing with depth, which exceeds the MAC value.

Thus, the content of total cadmium and zinc compounds exceeds the Clarke values, but the calculated environmental indicator of total pollution, taking into account toxicity, is acceptable for the use of land in agriculture. The studied variants revealed a low content of mobile copper and zinc. Contents in the mobile form of cadmium exceeding the maximum permissible concentration of soil on fallow land under no-till technology. The obtained results indicate the need for constant close attention to the problem of optimizing the microelement balance of arable soils, the importance of which increases in soil health management.

3.2 Scanning Electron Microscopy

Figure 4 shows SEM images of activated carbon obtained from sawdust at a concentration of 2:1 at 720 °C. The samples in their original form, as shown in the picture, are very dense and there are no holes in their pieces.

Fig. 4. SEM images of activated carbon obtained from sawdust at a concentration of 2:1, obtained at 720 °C

This Fig. 4 shows the surface of activated carbon particles obtained from sawdust at a concentration of 2:1. During the carbonization process, the structure of the shell changes. As shown in this image, as a result of heat treatment, holes and irregularities are visible on the surface of the activated carbon particles, which were previously closed and did not connect with the pore space. The increase in temperature can be explained by the fact that wood chips cause thermal decomposition of organic matter, so the number of pores and irregularities increases. The internal structure and thick cell walls are shown.

3.3 Infrared Spectroscopy

The IR spectra of activated carbon with concentrations of 1:1 and 2:1 were studied and discussed. The spectra are presented in Fig. 5. The IR spectra of activated carbon were

Table 5. IR spectra of concentrated activated carbon 1:1

Activated carbon	Ammonium molybdate (cm^{-1})	Ammonium sulfate (cm^{-1})	Arsenic oxide (cm^{-1})	Ammonium iodide (cm^{-1})	Aluminum potassium sulfate (cm^{-1})
Concentration 1:1	516	622	1039	687	689

determined based on the results of detection of inorganic elements in them. These data are shown in Table 5.

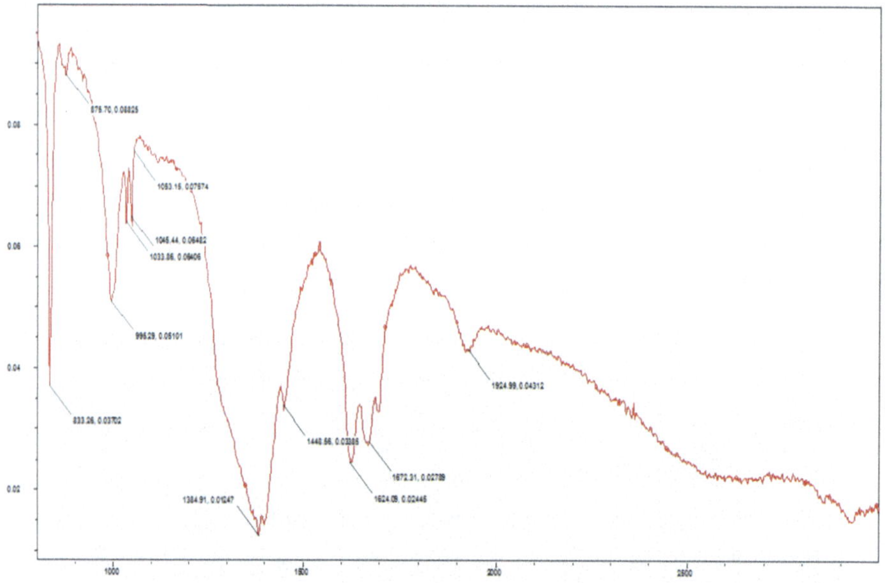

Fig. 5. Activated carbon concentration 1:1 at 720 °C

The IR spectra of activated carbon with a concentration of 1:1 correspond to the absorption lines of ammonium molybdate 516, ammonium sulfate 622, arsenic oxide 1039, ammonium iodide 687, aluminum potassium sulfate 689 (Table 6).

Table 6. IR spectra of concentrated activated carbon 2:1

Activated carbon	Ammonium carbonate (cm^{-1})	Ammonium peroxydisulfate (cm^{-1})	Ammonium chloride (cm^{-1})	Aluminum chloride (cm^{-1})	Aluminum potassium sulfate (cm^{-1})
Concentration 2:1	831	592	616	598	832

The IR spectra of activated carbon with a concentration of 1:1 correspond to the absorption lines of ammonium carbonate 831, ammonium peroxydisulfate 592, ammonium chloride 616, aluminum chloride 598, aluminum potassium sulfate 832 (Fig. 6).

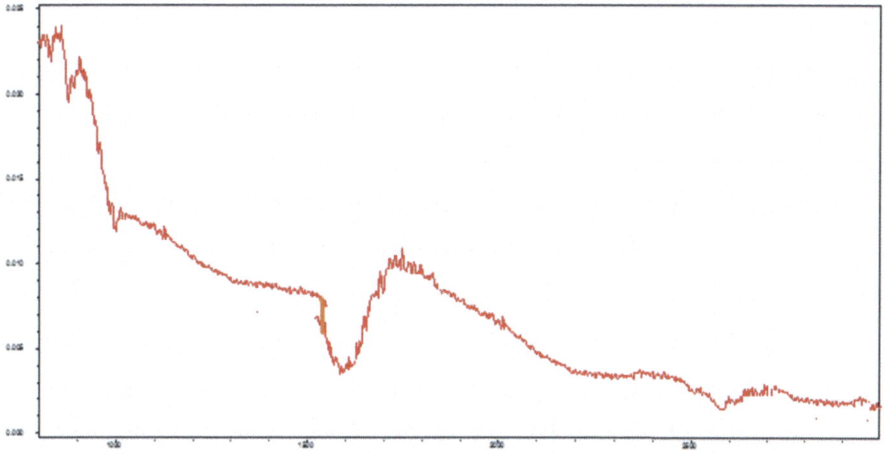

Fig. 6. Activated carbon at 720 °C has a concentration of 2:1

Activated carbon with concentration of 1:1 and 2:1 contains 516 (cm^{-1}) ammonium molybdate, 622 (cm^{-1}) ammonium sulfate, 1039 (cm^{-1}) arsenic oxide, 687 (cm^{-1}) ammonium iodide, 689 (cm^{-1}) 1) refers to aluminum potassium sulfate. The composition corresponds to the results of X-ray structural analysis, 831 (cm^{-1}) ammonium carbonate, 592 (cm^{-1}) ammonium peroxydisulfate, 616 (cm^{-1}) ammonium chloride, 598 (cm^{-1}) aluminum chloride, 832 (cm^{-1}) corresponds to the absorption lines of aluminum and potassium sulfate groups.

3.4 Adsorption of Heavy Metal with the Activated Carbon

To determine the effect of the adsorbent mass on the adsorption process, 10 different weighed portions of each sample (m(adsorbent) = 0.2; 0.4; 0.6; 0.8; 0.10 g; 0.12 g; 0.14 g; 0.16 g; 0.18 g; 0.20 g) were prepared and studied. An adsorbent with a concentration of 2:1 was chosen as a sample, that is, taken from sawdust burned at 720 °C (0 min, 15 min, 30 min, 60 min, 1.5 h, 2 h, 4 h), 6 h, 8 h, 24 h). According to the research results, the adsorption of heavy metal ions Cr^{3+} increased due to the increase in mass. The yield of the adsorbent for the heavy metal ion Cr^{3+} 0.2 g/L–49.375; 0.4 g/L–15.25; 0.6 g/L–16,625; 0.8 g/L–13,125; 0.10–12,125; 0.12–7.5; 0.14–5625; 0.16–5625; 0.18–2.5; It achieved low values of 0.20–0.125. These results were due to the 2:1 increase in surface area and Cr^{3+} performed moderately well compared to other adsorbents. The results obtained are shown in Fig. 7 below.

The graph shows the results of the adsorbent analysis by atomic emission spectroscopy. According to the results obtained, it was noted that over time it was well purified.

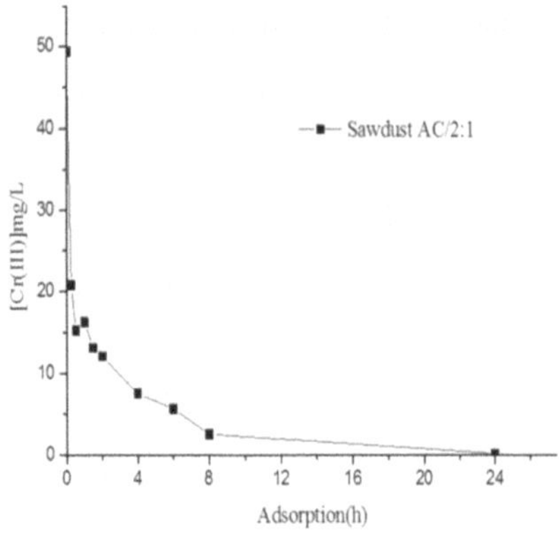

Fig. 7. Graph of ion adsorption studies

4 Conclusion

In modern conditions, biomass waste is an accessible, renewable and cost-effective resource that can be used to produce popular and environmentally friendly materials. The study assessed the potential of activated carbon made from natural wood chips as an adsorbent for purifying water contaminated with chromium ions (Cr). The purification efficiency within 24 h was 99.52%, which confirms the high efficiency of this material as a reactive barrier for water purification. The results of the experiment demonstrated that activated carbon from wood biomass is an effective means for removing heavy metals from aquatic environments. The two tested adsorbents showed a high degree of purification, which proves their promise for further use in water purification from heavy metals. Thus, the developed adsorbent has great potential for large-scale use in environmental technologies.

Acknowledgements. This research is funded by the Science Committee of the Ministry of Education and Science of the Republic of Kazakhstan (Grant No. BR24992867).

References

1. Hamad, H.: Modification and Performance of Activated Carbon for CO2 Sequestration in Pervious Concrete. Doctoral dissertation, University of Cambridge (2019). https://doi.org/10.17863/CAM.42881
2. Devi, R., Kumar, V., Kumar, S., Jatrana, A., Agrawal, A., Singh, P.: Development of biochar-based functional materials for electrochemical supercapacitor applications. ECS Trans. **107**(1), 7979–7986 (2022). https://doi.org/10.1149/10701.7979ecst
3. Applied Mathematics Issues "Extraction of Heavy Metals from Aqueous Solutions on Carbon-Containing Sorbents". p. 6
4. Wong, S., Ngadi, N., Inuwa, I.M., Hassan, O.: Recent advances in applications of activated carbon from biowaste for wastewater treatment: a short review. J. Clean. Prod. **175**, 361–375 (2018). https://doi.org/10.1016/j.jclepro.2017.12.059

5. Zhang, Q., Ma, M.Q., Chen, T.T., Zhang, H., Hu, D.F., Wu, B.H., Ji, J., Xu, Z.Q.: One-step dopamine-triggered polymerization and coprecipitation of acrylate monomers for functional coatings. ACS Suppl. Mater. Interfaces. **9**, 34356–34366 (2017)
6. Sawood, D.G., Gupta, S.K.: Kinetic equilibrium and thermodynamic analysis of As(V) removal from aqueous solution using iron-impregnated carbon of Azadirachta indica. Appl. Water Sci. **10**, 131 (2020)
7. Musyoka, N.M., Mutuma, B.K., Manyala, N.: Onion-derived activated carbon with increased surface area for improved hydrogen storage and electrochemical energy applications. RSC Adv. **10**, 26928–26936 (2020)
8. Xing, R., Wang, W., Jiao, T., Ma, Q., Zhang, Q., Hong, W., Qiu, H., Zhou, J., Zhang, L., Peng, Q.: Bioinspired polydopamine-shelled nanofibers containing graphene oxide carboxylate nanosheet for highly efficient dye scavenger. ACS Sustain. Chem. Eng. **5**, 4948–4956 (2017)
9. Khorasgani, N.C., Sengul, A.B., Asmatulu, E.: Briquetting grass and tree leaf biomass for sustainable future fuel production. Biomass. Conv. Bioref. **10**, 915–924 (2020)
10. Yang, H.M., Zhang, D.H., Chen, Y., Ran, M.J., Gu, J.Q.: Study on the application of KOH to produce activated carbon for the realization of distillate grain recycling. In: Proceedings of the 3rd International Conference on Advances in Energy, Environment and Chemical Engineering, Chengdu, China, May 26–28, p. 012051 (2017)
11. Blose, D.W., Ptacek, K.J., Bain, J.G., Waybrant, K.R., Robertson, W.D.: Mine drainage water treatment using in situ permeable reaction walls. In: Sudbury proc. 95, Mining and Environment, CANMET, Ottawa, Ontario, vol. 3, pp. 979–987 (1995)

A Docking Study of Quercetin Derivatives Against Covid 19 Spike Protein (Pdb Id 7 Cab) Target

K. S. Jayalekshmi[1,2] and M. S. Latha[2,3(✉)]

[1] Department of Chemistry, M S M College, Kayamkulam, Kerala, India
[2] Department of Chemistry, S. N. College, Kollam, Kerala, India
lathams2014@gmail.com
[3] Department of Chemistry, Sree Narayana Guru College of Advanced Studies, Nangiarkulangara, Haripad, Kerala, India

Abstract. Quercetin is a flavonol present in most fruits and vegetables. It has immense potential for treating various diseases. It also has antibacterial and antiviral effects. The present study focussed on In silico analysis of Quercetin derivatives against Covid spike protease PDB ID 7CAB.It is evident that Quercetin derivatives have anti-covid effects with high binding affinity ranging from −6.5 Kcal/mol to −11.8 Kcal/mol and can be highly useful for future drug development for Covid.

Keywords: Docking · Anti-viral · Quercetin · Covid 19 · Binding affinity

1 Introduction

Quercetin is a polyphenol under the class of flavonoids present in all fruits and vegetables. Flavonoid-containing fruits and vegetables have antimicrobial, anti-viral, anti-fungal activities. Flavonoids show various pharmacological activities including anti-oxidant, antimicrobial, anti-inflammatory, antimutagenic, antitumour, antidiabetic Vaso-relaxant, immunomodulatory and both oestrogenic and anti-oestrogenic etc. Computational docking methodology for finding drugs for treating various harmful diseases is very fruitful nowadays. Before applying drug in to animals including human, it should confirm many times and in many ways that it is less hazardous. This method of testing drugs computationally helps to decrease wastage of time, wastage of money, wastage of manpower and handling carcinogenic or toxic materials in laboratories. Covid 19 is a major threat that ever faced the whole world and people facing many health problems due to variant formation and ineffective use of antibiotics. Whole people above 15 years got vaccinated but still it is a threat for children below 15 years all over the world. Covid also a threat that affect the health of next generation in the whole world. In this paper Docking of Quercetin derivatives against covid spike protein with PDB ID 7CAB.

R. Rajasekharan et al. (Eds.): MatZone 2024, SPPHY 316, pp. 192–201, 2025.
https://doi.org/10.1007/978-981-96-6541-9_18

2 Materials and Methods

2.1 Computational Tools

Docking is highly useful tool for drug discovery. It is a computational programme for determining binding affinity of ligand molecules (small molecules) to the target proteins explaining ligand interact with a protein through various forces like hydrogen bonding, van der Waals forces etc. It gives conformational effects. It helps to determine good docking pose of ligands within the pocket of target protein.

2.1.1 PDB

RCSB PDB (Protein Data Bank)—It is a database used for the 3D structural data for large molecules such as nucleic acids, proteins etc. Protein Data Bank is a useful key in the structural genomic areas. It was established as the first accessible digital data source in the area of bio medicinal and pharmaceutical research. Knowing the three-dimensional structure of a biomolecule is essential to know its role in human health, its function in plants and other animals for the sustainability From RCS PDB, 3D structure of protein is obtained. The Covid Spike protein PDB ID 7CAB selected for docking and saved it in pdb format.

2.1.2 Pubchem

Downloaded the 2D structure of ligand molecules from PubChem database and saved it in sdf format.

2.1.3 PyMol

PyMol is an open-source molecular visualization program, for the creation of images of biomolecules such as proteins. The protein needed to be docked should be prepared by removing the additional molecules attached to it like water molecule, i.e., the protein should contain only amino acids. It is done by using PyMol.

2.1.4 Autodock Vina

It is the fastest docking engine. Docking is performed using Auto Dock 1.5.6.

2.1.5 Command Prompt

The Command Prompt, CMD, is the command line interface for Windows operating systems and is a way of interacting with a computer directly using text commands.

2.1.6 Biovia Discovery Studio

The protein selected for docking was minimised using BIOVIA software. It gives the 3D diagram of protein–ligand interaction.

2.1.7 Protox II

It is used for predicting ligand toxicity.

2.2 Structure of Covid 19 SPIKE PROTEIN (ID 7CAB)

See Fig. 1.

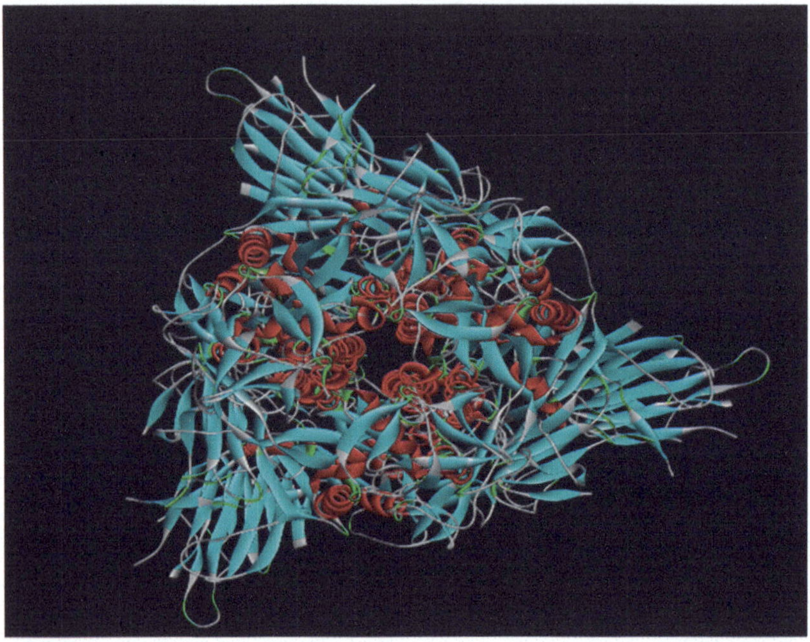

Fig. 1. Structure of Covid Spike protein ID 7CAB with Resolution: 3.52 A°

3 Results and Discussion

3.1 Molecular Docking Result Analysis

Quercetin (3,3′,4′,5,7-pentahydroxylflavone) is a secondary metabolite among plant derived flavonoid. It is a group of polyphenolic compounds (Murakami et al. 2008). A basic skeleton of diphenylpropane (C6–C3–C6) present. The flavonoids containing Quercetin are commonly found as O-glycosides with one of its hydroxyl groups substituted by sugars of various types. Here molecular docking of various Quercetin derivatives against Covid spike protein target (PDB ID 7CAB) results showing good binding affinity values. Toxicity prediction of ligand shows quercetin dihydrate is a good one because toxicity of ligand under certain circumstances may present a danger to vulnerable populations, ligand classified in high toxicity category may have toxic effect in humans indicate a concern for human health or an acute nature. Recognising the need to protect animal welfare, testing in Category 5 ranges is discouraged and should only be considered when there is a strong likelihood that results of such a test would have a direct relevance for protecting human health (Table 1).

Table 1. Docking scores various Quercetin derivatives against 7CAB

Sl.No.	Ligand	Structure	Docking Score/Affinity(kcal/mol)	Oral Toxicity class
1	**Quercetin pentamethyl ether**	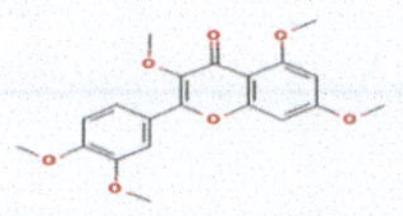	**-6.5**	**5**
2	**Quercetin dihydrate**	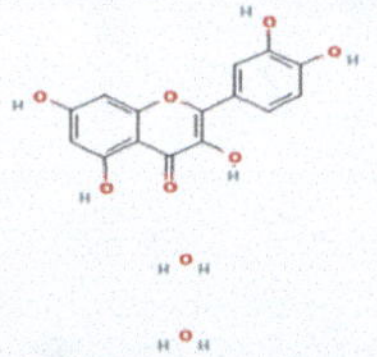	**-9.2**	**3**
3	**Quercetin pentaacetate**	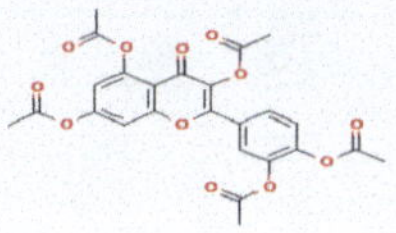	**-9.8**	**5**
4	**Quercetin-3-monoglucoside**	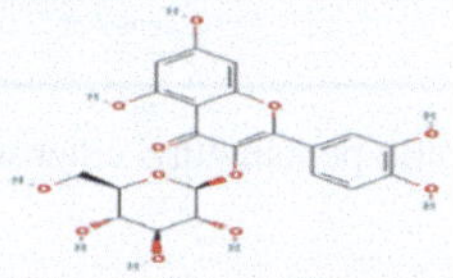	**-11.4**	**5**
5	**Quercetin-3-rutinoside**	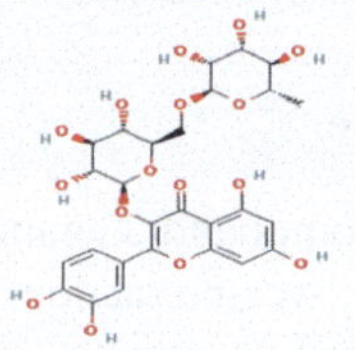	**-11.8**	**5**

3.2 Docking with Quercetin Pentamethyl Ether

The results of the docking study of the compound 1 revealed that the ligand binds to target protein spike glycoprotein receptor having PDB ID 7CAB with the binding affinity of −6.5 kcal/mol. It contains only two main interactions; one is weak van der Waals forces and the other is strong Conventional Hydrogen Bonds interactions. Here, ligand forms a conventional hydrogen bond with an amino acid Serine, which is in the 59th position of chain A of the protein. And van der Waals interaction is formed between amino acids Cystine (Chain A of 538 position), Cystine (Chain A of 590 position) Valine (Chain A of position 615), Valine (Chain A of position 551), Valine (Chain A of position 320) Threonine(Chain A of position 588), Threonine(Chain A of position 553) and Phenyl alanine(Chain A of 318) which are in different positions of the chain A of the protein (Fig. 2). The predicted toxicity class is 5 which has relatively low acute toxicity where the average similarity in toxicity prediction is 97.6% which has relatively low acute toxicity hazard.

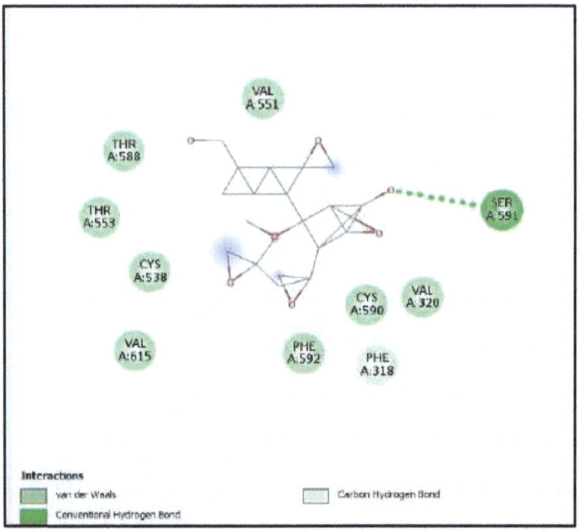

Fig. 2. 2D interaction of Quercetin pentamethyl ether with target protein

3.3 Docking with Quercetin Dihydrate

The results of the docking study of the compound 2 against 7CAB the binding affinity is −9.2 kcal/mol. It also contains only two interactions, van der Waals interaction, Conventional Hydrogen Bonding and Alkyl interaction.

Here the weak van der Waals interaction is between amino acids Valine (Position 991), Threonine (Position 998, C Chain), Threonine (Position 998, B Chain) Phenyl alanine (Position 970, A chain), Glycine (Position 971), Arginine (position 995), Aspartic acid (Position 994) which are in the different positions of chain A, B and C of the protein and ligand forms Conventional Hydrogen Bond with Arginine (Chain B, 995) and Phenyl alanine (Chain B, 970), which is the Chain B of the protein respectively. The predicted toxicity class is 3 where the average similarity in toxicity prediction is 100%, which means slightly toxic and slightly irritating (Fig. 3).

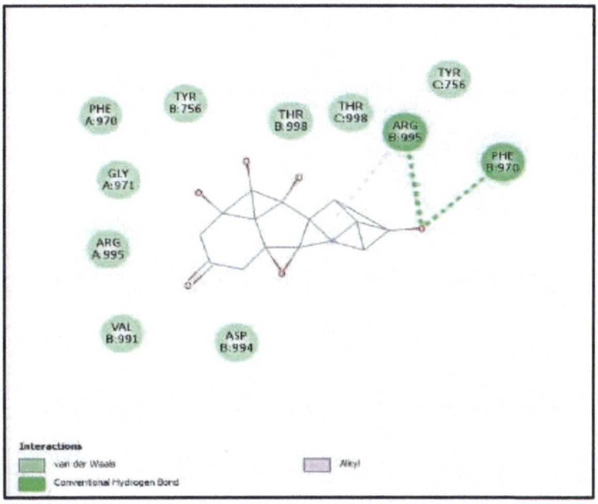

Fig. 3. 2D interaction of Quercetin dihydrate with target protein

3.4 Docking with Quercetin Pentaacetate

The results of the docking study of the compound 3, against 7CAB, the binding affinity is −9.8 kcal/mol. It also contains two interactions, van der Waals and Conventional Hydrogen bond. Here ligand forms conventional hydrogen bond with amino acids Arginine, Aspartic acid, Threonine, Phenyl Alanine which are in the different positions of chain A, B and C of the protein. And van der Waals interaction is between amino acids Glycine, Arginine, and Phenyl alanine of the protein. The predicted toxicity class is 5 where the average similarity in the prediction is 70.97% which means relatively low acute toxicity hazard (Fig. 4).

3.5 Docking with Quercetin-3-Monoglucoside

The results of the docking study of the compound 4, against 7CAB, the binding affinity is −11.4 kcal/mol. This has 4 possible interactions they are, van der Waals, Alkyl, Pi-Alkyl and Covalent bond. van der Waals interaction arises between ligand and amino acids like Alanine in 1087th position of chain A, Phenylalanine in the 1089th position of chain B, Glutamine in the 1113th position of chain B and Serine on the 1123th position of chain B of target protein. Covalent bond is between Phenylalanine on the 1121th and Serine on the 1123th position of chain A of spike protein. Alkyl and Pi-alkyl interaction arises between amino acids Lysine on the 1086th position and Histidine on the 1088th position of chain A of the protein respectively. The predicted toxicity class is 5 where the average similarity in the prediction is 72.9%, which means relatively low acute toxicity hazard (Fig. 5).

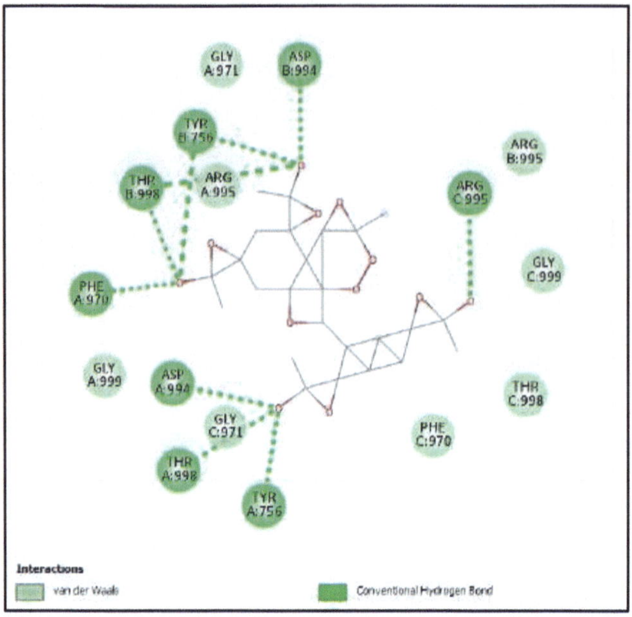

Fig. 4. 2D interaction of Quercetin pentaacetate with target protein

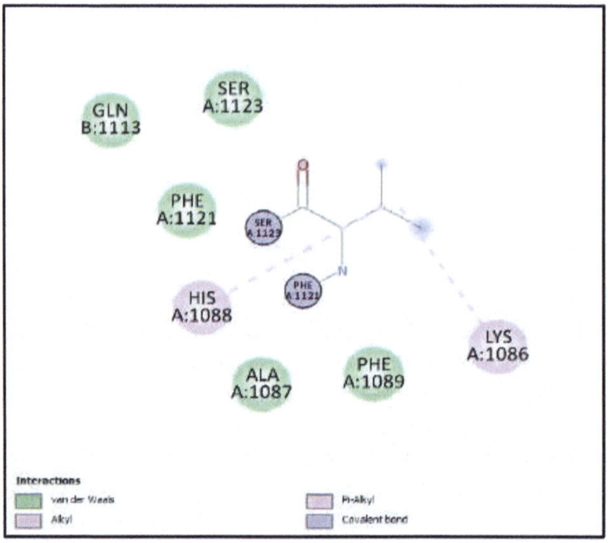

Fig. 5. 2D interaction of Quercetin-3-monoglucoside with target protein

3.6 Docking with Quercetin-3-Rutinoside

The results of the docking study of the compound 5, Quercetin-3-rutinoside, here the binding affinity is −11.8 kcal/mol. The interactions in this case are quite different. There are 5 possible interactions they are, van der Waals, Pi-sigma, Pi-Pi stacked, Pi-alkyl and Covalent bond. Van der Waals interaction arise between ligand and amino acids like Lysine, Proline, Leucine, Glutamine, Glycine, Arginine which are on different positions of Chain C of protein. The Pi-sigma interaction is between amino acids Leucine and Lysine of Chain C on the 56th and 55th position of the protein. The Pi-Pi stacked interaction is between ligand and Phenyl alanine on the 43rd position on Chain A of the protein. The Pi-alkyl interaction is between ligand and amino acid Alanine on the on the 57th position of Chain C of the protein. And the Covalent bond is between ligand and amino acid Isoleucine which is on the 58th position of Chain C of the protein. It belongs to toxicity class 5 with 100% average similarity in prediction which is a plus (Fig. 6).

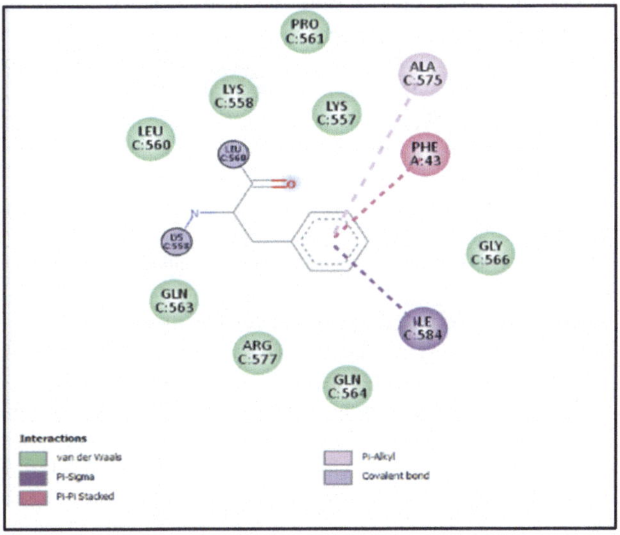

Fig. 6. 2D interaction of Quercetin-3-rutinoside with target protein

4 Conclusions

The 2D interaction diagrams of the best docking poses of the ligand with protein shows that the ligand is best fitted within the active site of the protein. For the energetically favoured ligand's intermolecular interactions such as hydrogen bonding, covalent bonding, van der Waals forces etc. plays an important role in stabilizing, in an open conformational environment of the protein structure. It will give the best docking pose of ligand in target protein. Docking of Quercetin derivatives against Covid Spike protease with PDB ID 7CAB gives good result of docking score. In the case of protein 7CAB, the highest binding affinity is shown by Quercetin-3-rutinoside (−11.8 kcal/mol) and Quercetin-3-monoglucoside (−11.4 kcal/mol) among 5 studied Quercetin derivatives. The lowest affinity shown by Quercetin pentamethyl ether(−6.5 kcal/mol). All the others show good binding affinities (above -9 kcal/mol). In the case of oral toxicity studies, all four ligands except Quercetin dihydrate (toxicity class 3) fall under toxicity class 5. That is low toxic. The study will be helpful for drug development in future.

References

1. Young, D.C.: Computational Drug Design—A Guide for Computational and Medicinal Chemists. Wiley, Hoboken, NJ
2. Meng, X.-Y., Zhang, H.-X., Mezei, M., Cui, M.: Molecular Docking: A powerful approach for structure-based drug discovery. Curr. Comput. Aided Drug Des. **7**(2), 146–157 (2011). https://doi.org/10.2174/157340911795677602
3. Coumar, M.S.: Molecular Docking for Computer-Aided Drug Design Fundamentals, Techniques, Resources and Applications. Centre for Bioinformatics School of Life Sciences Pondicherry University, Kalapet, Pondicherry
4. Pagadala, N.S., Syed, K., Tuszynski, J.: Software for molecular docking: a review. Biophys. Rev. **9**(2), 91–102 (2017). https://doi.org/10.1007/s12551-016-0247-1
5. Duan, L., Zheng, Q., Zhang, H., Niu, Y., Lou, Y., Wang, H.: The SARS-CoV-2 spike glycoprotein biosynthesis, structure, function, and antigenicity: implications for the design of spike-based vaccine immunogens. Front. Immunol. **11**, 576622 (2020)
6. Coronavirus History: How did coronavirus start?—WebMD. https://www.webmd.com/lung/coronavirus-history
7. Coronavirus disease (COVID-19)—World Health Organization. https://www.who.int/health-topics/coronavirus
8. Gupta, R.C.: Nutraceuticals Efficacy, Safety and Toxicity (2016)
9. Li, F.: Structure, function and evolution of coronavirus spike proteins. Annu. Rev. Virol. **3**(1), 237–261 (2016). https://doi.org/10.1146/annurev-virology-110615-042301
10. Noman, A., Aqeel, M., Khalid, N., Hashem, M., Alamari, S., Zafar, S., Qasim, M., Irshad, M.K., Qari, S.H.: Spike glycoproteins: Their significance for corona viruses and receptor binding activities for pathogenesis and viral survival. Microb. Pathog. **150**, 104719 (2021). https://doi.org/10.1016/j.micpath.2020.104719
11. Martinez-Flores, D., Zepeda-Cervantes, J., Cruz-Resendiz, A., Aguirre-Sampieri, S., Sampieri, A., Vaca, L.: SARS-CoV-2 vaccines based on the spike glycoprotein and implications of new viral variants. Front. Immunol. **12**, 701501 (2021). https://doi.org/10.3389/fimmu.2021.701501
12. Formica, J.V., Regelson, W.: Review of the biology of Quercetin and related bioflavonoids. Food Chem. Toxicol. **33**(12), 1061
13. Nam, J.S., Sharma, A.R., Nguyen, L.T., Chakraborty, C., Sharma, G., Lee, S.S.: Application of bioactive quercetin in oncotherapy: from nutrition to nanomedicine. Molecules. **21**(1), E108 (2016). https://doi.org/10.3390/molecules21010108
14. Yang, D., Wang, T., Long, M., Li, P.: Quercetin: its main pharmacological activity and potential application in clinical medicine. Oxid. Med. Cell. Longev. **2020**, 8825387 (2020). https://doi.org/10.1155/2020/8825387
15. Robertford, M.B.: Inulin type fructans: functional flood ingredients. J. Nutr. **137**(11 Suppl), 2493S–2502S (2007). https://doi.org/10.1093/jn/137.11.2493S
16. Pernak, J., Syguda, A., Janiszewska, D., Materna, K., Praczyk, T.: Ionic liquids with herbicidal onions. Tetrahedron. **67**(26), 4838–4844 (2011)
17. Singh, S.P., Konwar, B.K.: Molecular docking studies of quercetin and its analogues against human inducible nitric oxide synthase. Springerplus. **1**, 69 (2012)

18. Alsaifa, N.A., Wani, T.A., Bakheita, A.H., Zargarb, S.: Multi-spectroscopic investigation, molecular docking, and molecular dynamic simulation of competitive interactions between flavonoids (quercetin and rutin) and sorafenib for binding to human serum albumin. Int. J. Biol. Macromol. **165**(Pt B), 2451–2461 (2020). https://doi.org/10.1016/j.ijbiomac.2020.10.098
19. Bhuiya, S., Haque, L., Pradhan, A.B., Das, S.: Inhibitory effects of the dietary flavonoid quercetin on the enzyme activity of zinc (II)- dependent yeast alcohol dehydrogenase by spectroscopic and molecular docking studies. Int. J. Biol. Macromol. **95**, 177–184 (2017). https://doi.org/10.1016/j.ijbiomac.2016.11.047
20. Boyina, H.K., Geethakhrishnan, S.L., Panuganti, S., Gangarapu, K., Devarakonda, K.P., Bakshi, V., Guggilla, S.R.: In silico and in vivo studies on Quercetin as potential anti-Parkinson agent. Adv. Exp. Med. Biol. **1195**, 1–11 (2020)
21. Hashemzaei, M., Delarami Far, A., Yari, A., Heravi, R.E., Tabrizian, K., Taghdisi, S.M., Sadegh, S.E., Tsarouhas, K., Kouretas, D., Tzanakakis, G., Nikitovic, D., Anisimov, N.Y., Spandidos, D.A., Tsatsakis, A.M., Rezaee, R.: Anticancer and apoptosis-inducing effects of quercetin in vitro and in vivo. Oncol. Rep. **38**(2), 819–828 (2017). https://doi.org/10.3892/or.2017.5766
22. Xu, W., Xie, S., Chen, X., Pan, S., Qian, H., Zhu, X.: Effects of quercetin on the Efficacy of various chemotherapeutic drugs in cervical cancer cells. Drug Des. Dev. Ther. **15**(15), 577–588 (2021 Feb). https://doi.org/10.2147/DDDT.S291865
23. Imran, M., Thabet, H.K., Alaqel, S.I., Alzahrani, A.R., Abida, A., Alshammari, M.K., Kamal, M., Diwan, A., Asdaq, S.M.B., Alshehri, S.: The therapeutic and prophylactic potential of quercetin against COVID-19: an outlook on the clinical studies, inventive compositions, and patent literature. Antioxidants. **11**(5), 876 (2022). https://doi.org/10.3390/antiox11050876
24. Raval, K., Ganatra, T.: Review Article; Basics, types, and applications of molecular docking: a review. IP Int. J. Comp. Adv. Pharmacol. **7**(1), 12–16 (2022)
25. Seeliger, D., de Groot, B.L.: Ligand docking and binding site analysis with PyMOL and Autodock/Vina. J. Comput. Aided Mol. Des. **24**(5), 417–422 (2010). https://doi.org/10.1007/s10822-010-9352-6

Synthesis of Aluminum-Cobalt-Molybdenum Nanocomposites

A. Orynbekova[1], D. Aubakirova[2], B. Massalimova[3], and A. Darmenbayeva[1(✉)]

[1] M.Kh. Dulaty Taraz University, Tole bi 60, 080000 Taraz, Kazakhstan
maral88.ad@gmail.com

[2] D. Serikbayev East Kazakhstan Technical University, Serikbayev 17, 070004 Ust-Kamenogorsk, Kazakhstan

[3] M. Kozybayev North Kazakhstan University, Zhumabayev 114, 150000 Petropavlovsk, Kazakhstan

Abstract. This paper presents the synthesis and characterization of aluminum, cobalt, and molybdenum (Al-Co-Mo) nanocomposites for various applications in materials science and catalysts. The Al-Co-Mo nanoparticles were prepared using the salt synthesis method, which ensured their high stability and homogeneity. The main physicochemical properties of the nanocomposite, such as particle size, morphology, elemental composition, and their distribution at the microlevel, are investigated in this work. Scanning electron microscopy (SEM), infrared spectroscopy (IR), and elemental analysis (EDX) were used for the analysis. SEM studies showed that the obtained nanoparticles have a spherical or oval shape with a size of about 60–80 nm and are uniformly distributed over the sample, with little particle aggregation. IR spectroscopy confirmed the presence of oxide bonds between the metals, and elemental analysis showed the correct ratio of aluminum, cobalt, and molybdenum in the nanocomposite. The use of such nanomaterials as catalysts and in electronics promises significant advantages due to their high activity, stability and potential for further functionalization. The work provides important data for the development of new efficient materials with improved catalytic properties and potential in various fields, including energy technologies and ecology.

Keywords: Nanocomposites · Aluminum · Cobalt · Molybdenum · Scanning Electron · Synthesis

1 Introduction

The development of nanotechnology opens up new prospects for the creation of materials with unique properties, making them one of the most important areas of modern science. Among such materials, special attention is paid to nanocomposites, which combine the properties of several components and ensure their synergistic interaction. Nanocomposites containing aluminum, cobalt, and molybdenum are particularly promising objects of research due to their high strength, heat resistance, and catalytic activity. These properties make Al-Co-Mo-based materials in demand in such areas as catalyst production, the creation of materials for high-temperature processes, and the development of sensors.

R. Rajasekharan et al. (Eds.): MatZone 2024, SPPHY 316, pp. 202–215, 2025.
https://doi.org/10.1007/978-981-96-6541-9_19

The relevance of the study of Al-Co-Mo nanocomposites is due to their multifunctional characteristics and wide range of applications. Cobalt and molybdenum have already proven themselves as key components of catalysts used in petrochemistry for hydrodesulfurization and hydrogenation processes. Adding aluminum to the composition of such materials can improve their dispersion, increase stability and durability. In addition, Al-Co-Mo nanocomposites demonstrate high activity in reactions associated with hydrogen production, which is especially important in the context of the development of hydrogen energy.

No less significant is the use of such materials in environmental and industrial applications. For example, cobalt-molybdenum nanomaterials are used in wastewater treatment and the removal of heavy metals, which helps to solve urgent environmental problems. At the same time, the lightness and mechanical strength of aluminum allow the use of the resulting nanocomposites in the aviation and automotive industries.

The aim of this paper is to develop an effective method for synthesizing aluminum, cobalt, and molybdenum-based nanocomposites, as well as to study their structural and functional characteristics. Particular attention will be paid to the influence of synthesis conditions on the properties of materials and their application in catalysis, energy, and ecology.

The work is aimed at developing fundamental knowledge about Al-Co-Mo nanocomposites and their practical potential. The results of the study can be useful both for improving existing technologies and for creating new materials that meet modern challenges of science and industry.

2 Literature Review

Modern approaches to the synthesis of nanocomposites based on transition metals are based on the wide application of chemical, physical and combined methods that allow controlling the structure and properties of the resulting materials [1–3].

Chemical methods include approaches such as coprecipitation, salt-gel technology, hydrothermal and solvothermal synthesis, as well as reductive chemical deposition. For example, the coprecipitation method allows obtaining materials with a high degree of homogeneity due to the joint precipitation of metal salts from a solution, and the salt-gel method is used to form nanostructures with high purity and precise control of their morphology. Hydrothermal synthesis, carried out at high temperatures and pressures, is especially effective for creating crystalline structures [4].

Among physical methods, mechanochemical synthesis, thermal spraying, and pyrolytic synthesis are popular. The mechanochemical approach involves intensive grinding and mixing of components, which contributes to the formation of nanocomposites in the solid phase. Thermal spraying, including plasma and laser, allows for the creation of thin coatings with specified characteristics, and pyrolytic synthesis provides precise control over the crystal structure and particle size [5, 6].

Combined methods combine the advantages of chemical and physical approaches. For example, template synthesis uses templates to form a given structure, which are then removed, leaving a nanocomposite. An innovative direction is the use of environmentally friendly methods, such as synthesis using green chemistry, where plant extracts replace toxic reagents.

Synthesis of nanocomposites is a key step in the creation of new materials with unique properties, such as high catalytic activity, thermal stability and strength. The influence of different synthesis methods on the characteristics of the final nanocomposites, such as aluminum-cobalt-molybdenum materials (Al-Co-Mo), is crucial to optimize their application in various industries, such as catalysis, electronics and energy.

One of the most common synthesis methods is chemical precipitation. This method involves the deposition of metal or oxide particles from a solution. Chemical precipitation allows for precise control of the composition, particle size and their uniform distribution. For example, in the synthesis of Al-Co-Mo nanocomposites, this method can ensure a stable distribution of components, which is important for maintaining high catalytic activity and strength of the material. However, to prevent aggregation and obtain uniform particles, it is important to carefully control the reaction parameters, such as temperature, pH and reagent concentration.

Another promising method is salt synthesis, in which metals are mixed with salts and reduced to their metallic or oxide forms. This approach allows for the production of materials with high thermal stability and resistance to external influences. For aluminum-cobalt-molybdenum nanocomposites, salt synthesis can be particularly effective, as it helps to obtain high-quality materials with good structural characteristics. An important aspect of this method is the ability to precisely control the composition of the material, which ensures the stability and efficiency of the nanocomposite.

The aerosol deposition method is also used in the synthesis of nanocomposites. This process involves the deposition of nanoparticles on a substrate from aerosol solutions, which allows for the creation of high-quality coatings. Aerosol deposition can lead to the production of nanoparticles with good monodispersity, but its complexity lies in the need for careful control of temperature and pressure to avoid particle aggregation.

Another method, although less common, for the synthesis of Al-Co-Mo nanocomposites is biotechnology. Biosynthesis using microorganisms or biopolymers has the advantage of being environmentally friendly, but for the synthesis of complex materials such as Al-Co-Mo, this method may be less suitable, as it is difficult to obtain the desired proportions and shape of the materials.

All these methods have a significant impact on the properties of the resulting nanocomposites. For example, chemical precipitation and salt synthesis can provide nanoparticles with a size of 10 to 50 nanometers, which is important for their high activity in chemical reactions. Aerosol deposition makes it possible to obtain small particles, but there is a risk of their aggregation, which can reduce the stability and catalytic activity of the material. At the same time, salt synthesis provides a better distribution of elements and a more stable structure, which is especially important for use in catalysis and other high-temperature processes.

Thus, the choice of synthesis method should depend on the specific requirements for the material, including its size, shape, stability and catalytic activity. For the synthesis of aluminum-cobalt-molybdenum nanocomposites, chemical precipitation and salt synthesis methods are of greatest interest due to their ability to control the composition and structure of materials, which helps to obtain highly efficient and stable nanocomposites for various applications.

An important addition is the use of digital technologies, such as machine learning, to optimize the synthesis processes and predict the properties of nanocomposites. This makes it possible to create materials with predetermined characteristics for their use in catalysis, energy, medicine and ecology. Thus, modern science offers diverse and effective approaches to the synthesis of nanocomposites, which allows not only to study the fundamental properties of materials, but also to find practical applications for them.

Aluminum, cobalt, and molybdenum are important components of nanocomposites due to their unique physicochemical properties, which determine a wide range of their applications [7, 8].

Aluminum attracts the attention of researchers due to its lightness, high strength, and excellent corrosion resistance. It is widely used in the creation of nanocomposites due to its high thermal and electrical conductivity, which makes it an ideal material for heat-conducting and electrically conductive systems [9]. Moreover, its high reactivity allows for the successful integration of aluminum into the structure of nanocomposites, improving their physicochemical characteristics.

Cobalt, in turn, stands out for its magnetic properties, since it is a ferromagnetic metal. This makes it an indispensable component for the creation of nanocomposites with magnetic activity, which are used in hard drives, medical sensors, and magnetic fluids. In addition, cobalt exhibits high catalytic activity, which opens up possibilities for its use in chemical synthesis, as well as in the processes of hydrogenation and oxidation of organic compounds [10]. Its resistance to corrosion and heat resistance also contribute to increased durability of materials.

Molybdenum, having one of the highest melting points among metals, is used to create heat-resistant nanocomposites. Its mechanical strength and resistance to deformation make molybdenum an important element in materials operating under high loads. Molybdenum is known for its unique catalytic properties and is widely used in oil refining, hydrogen production and hydrotreating. Of particular note is molybdenum sulfide (MoS_2), which has a low coefficient of friction and is used in tribological coatings [11].

The combination of aluminum, cobalt and molybdenum in nanocomposites allows for a synergistic effect. These materials combine the lightness and strength of aluminum, the magnetic and catalytic properties of cobalt, as well as the heat resistance and stability of molybdenum, which makes them ideal for use in energy, electronics, ecology and other high-tech industries. This approach opens up new prospects for the development of highly efficient nanocomposites capable of meeting the growing needs of modern science and industry.

Nanocomposites based on aluminum (Al), cobalt (Co) and molybdenum (Mo) attract much attention in various fields of science and industry due to their unique physical and chemical properties. Each of these metals plays an important role in the formation of a material with improved characteristics, such as high strength, heat resistance, magnetic properties and catalytic activity.

Al-Co-Mo nanocomposites are widely used in the energy sector, especially in processes that require high temperatures and thermal stability. Molybdenum, due to its high melting point, helps improve the thermal stability of such materials, making them suitable for use in reactors operating at high temperatures. Cobalt and aluminum are also involved in catalysis processes, where molybdenum can improve the efficiency of catalysis in petrochemistry, hydrogen synthesis, and gas purification. For example, cobalt and molybdenum are used in catalysts for exhaust gas purification [12].

In electronics, Al-Co-Mo nanocomposites are used as magnetic materials. Cobalt, being a ferromagnetic material, with the addition of aluminum and molybdenum can be used to create highly efficient magnetic nanocomposites. These materials are used in hard drives, magnetic sensors, and other magnetic devices. For example, cobalt-aluminum nanocomposites are used in magnetic recording and memory [13]. In addition, aluminum and molybdenum, due to their conductivity, are used in conductive layers in electronics [14].

Magnetic nanoparticles based on Al-Co-Mo are used in medicine, in particular to create biocompatible materials used in magnetic resonance imaging (MRI), as well as for targeted drug delivery. Cobalt, as a magnetic material, enhances these properties, and aluminum and molybdenum add strength and stability to the material [15]. One example is the use of magnetic nanocomposites to create antibacterial coatings on medical devices such as implants and catheters, which reduces the risk of infections [16–19].

In ecology, Al-Co-Mo nanocomposites are widely used in water purification. Molybdenum and aluminum, due to their properties to adsorb pollutants, are used as filtering and cleaning materials. These nanocomposites can effectively remove heavy metals and organic pollutants from wastewater, which makes them promising for eco-technologies. Studies have shown that molybdenum sulfide, obtained on the basis of Mo, effectively adsorbs toxic substances [20, 21].

Thus, nanocomposites based on aluminum, cobalt and molybdenum have many promising applications in a variety of fields, from energy and ecology to medicine and the aerospace industry. Their unique physical and chemical properties make them indispensable for the creation of highly effective materials and technologies that meet modern requirements.

3 Materials and Methods

3.1 Materials and Reagents

Aluminumhydroxide $Al(OH)_3$(pure for analysis), ammonium molybdate $(NH_4)_2MoO_4$(pure for analysis), cobalt nitrate $Co(NO_3)_2$ * $6H_2O$(pure for analysis), distilled water.

3.1.1 Methods for the Synthesis of Al-Co-Mo Nanocomposites

The sol-gel method is one of the most effective and widely used methods for the synthesis of nanoparticles and catalysts. The process produces materials with high homogeneity and controlled particle sizes.

The following is a step-by-step procedure for preparing an aluminum-cobalt-molybdenum (Al-Co-Mo) nanocomposite by the sol-gel method. Cobalt aluminate or a mixture of aluminum and cobalt salts soluble in an organic solvent is added to a pre-prepared glass vessel. A solution with a metal concentration in the range of 0.1–1.0 M is used, the solution should be thoroughly mixed with a magnetic stirrer until the reagents are completely dissolved. A pre-calculated and weighed mass of ammonium molybdate is added to the resulting solution. The ratio of the components may vary depending on the desired nanocomposite composition. The mixture is stirred until a homogeneous solution is obtained. The resulting sol is kept at room temperature for 12–24 h for complete gelation. During the gelation process, a three-dimensional network structure of the gel is formed. Heating at a temperature of 50–70 °C can be used to accelerate the process. This promotes faster evaporation of the solvent and the formation of a gel. The formed gel is thoroughly dried at 100–120 °C for 12–24 h. This allows the removal of residual solvent and water. The dried gel is then heat treated at 300–500 °C to remove organic residues and condense the material. The dried gel is heated in a muffle furnace at 500–800 °C for 2–4 h. Calcination promotes the crystallization of oxide phases and the formation of an active nanocomposite surface [22, 23].

3.1.2 Research Methods

The following analytical methods are used to study the synthesized Al-Co-Mo nanoparticles, which allow a deeper understanding of their morphology, chemical composition, structure and other key characteristics.

Scanning Electron Microscopy (SEM) is used to obtain high-resolution images of the surface of nanoparticles. This method allows one to study the morphology of particles, their shape, size, and to identify possible defects or agglomerates in the material. SEM can accurately measure the particle size, as well as study their distribution and shape (e.g., spherical or elongated). This method can also be used to evaluate the surface of the particles, the presence of pores or other defects that can affect their physicochemical properties.

Infrared spectroscopy is used to study the chemical composition of nanoparticles and to analyze the functional groups that may be present in their structure. The IR spectrum helps to determine the vibrations of chemical bonds between atoms, as well as to provide information on the interactions between components such as aluminum, cobalt, and molybdenum, as well as their possible oxide forms. Based on the IR spectrum, it is possible to study the characteristic peaks corresponding to different chemical groups, which allows us to identify the elements in the nanoparticles and find out what chemical reactions occur between them.

Elemental analysis methods are used to accurately determine the composition of nanoparticles and determine the ratio of elements in a sample. One such method is energy-dispersive X-ray spectroscopy (EDX), which is used in conjunction with SEM and allows one to determine the chemical composition of the material based on X-ray radiation.

All these methods together allow one to conduct a comprehensive analysis of Al-Co-Mo nanoparticles, determining their morphological characteristics, chemical composition and physicochemical properties. These data are key to understanding the potential of using such nanoparticles in various applications, for example, in catalysis, electronics or materials science.

4 Results and Discussion

IR spectroscopy is a powerful method for studying the structural characteristics of materials, in particular, the chemical composition, the presence of functional groups, and the nature of the interaction of nanocomposite components. The method for studying the aluminum-cobalt-molybdenum (ACM) composite material by IR spectroscopy is described below (Table 1).

Table 1. IR spectroscopy results

Name of the compound	υ (H_2O)	δ (H_2O)	υ (M-OH)	Interlayer -OH bonds	υ (C-OH)	υ (Al-O)	υ(M-NO_3)	Mo-O
$Al(OH)_3$	3000–4000	1642	3620	2335	–	–	–	–
$Co(NO_3)_2$ * $6H_2O$	3000–3150	1620	–	–	–	–	805	–
Al-Co-Mo nanocomposites	3520	1629	3618	–	1030	945	–	669

According to the literature [24–27], the presence of a broad peak around 3400 cm^{-1} indicates the presence of hydroxyl groups or water adsorbed on the surface of nanoparticles. This may indicate the effect of moisture on the samples or the presence of hydroxyl groups formed as a result of the reaction of oxides with moisture in the environment. The peak with a maximum intensity of about 1600 cm^{-1} may be associated with carbon pollutants or residual organic matter, which is also confirmed by the use of organic solvents or additives in the synthesis process.

In the region of 3200–3600 cm^{-1}, bands corresponding to stretching vibrations of hydroxyl groups (O-H) can be observed, which may indicate the presence of hydrated compounds or adsorbed water.

The bands in the region of 1400–1600 cm^{-1} and 800–1100 cm^{-1} may be associated with vibrations of nitrate groups (NO_3^{2-}), indicating the presence of nitrates on the surface of the nanocomposite or traces of complexing agents used in the synthesis. In the region of 400–1000 cm^{-1}, vibrations associated with metal oxide bonds such as Co-O, Mo-O and Al-O should be expected. These bands can provide information about the metal coordination and the structure of the oxide phases of the nanocomposites. The results of the IR spectra of the metal complexes are presented in Fig. 1.

Particular attention should be paid to the peaks in the 600–800 cm^{-1} region, which confirm the presence of metal oxides such as Al-O, Co-O and Mo-O, which is an expected result for nanocomposites consisting of these elements. These peaks reflect the vibrations of metal-oxide bonds, indicating that the metals are indeed in the form of oxides forming strong structures in the nanoparticles.

Another important point is the presence of peaks in the 1000–1100 cm^{-1} region, which can also be associated with deformation vibrations of metal-oxide bonds. This confirms the presence of strong chemical bonds between aluminum, cobalt and molybdenum in the nanoparticles, which is important for their stability and functionality in further application.

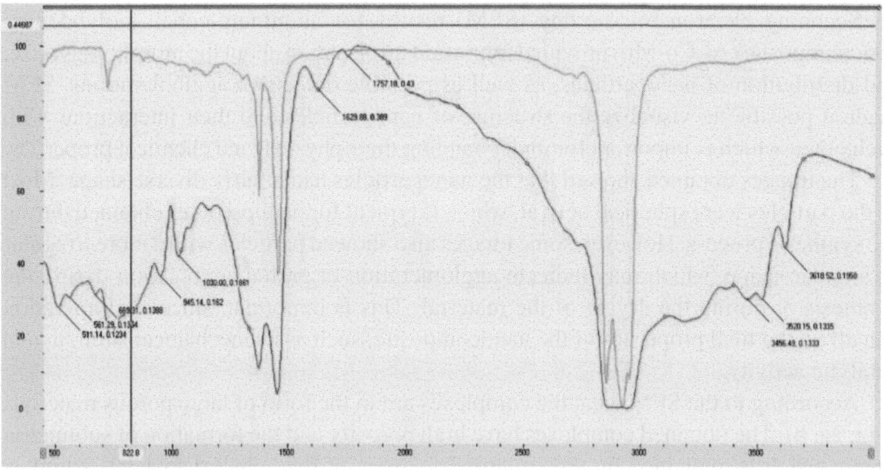

Fig. 1. Vibrational frequencies in IR spectra of metal complexes (cm^{-1})

In the IR spectra of the $Co(NO_3)_2{\cdot}6H_2O$ salt, intense absorption bands of stretching vibrations of water are observed in the range of 3000–3150 cm^{-1}. In the lower frequency region of the spectrum, absorption bands of deformation vibrations of water are located at 1620 + 10 cm^{-1}. The intense absorption band with a frequency of 800–860 cm^{-1} is attributed to the absorption of the NO_3^- anion.

In the spectra of the nanocomposites, absorption bands at 3520 cm^{-1} (stretching vibrations) and 1629 cm^{-1} (bending vibrations) are observed and belong to –OH groups. Some metal cations were converted into metal particles, and some remained in the form of $Co(OH)^-$, which was confirmed by the absorption band at 3618 cm^{-1} in the IR spectrum.

Changes in the position and intensity of the bands associated with Mo-O and Co-O vibrations may indicate the interaction of the nanocomposite components, such as the formation of mixed oxides or complexes on the surface. Comparison of the spectra of the initial precursors and the resulting nanocomposite allows us to identify changes during synthesis, such as the decomposition of precursors, the formation of new phases, and changes in metal coordination.

Thus, based on the obtained IR spectrum, it is possible to conclude about the presence or absence of certain functional groups and phases in the nanocomposite. Observation of characteristic bands for Co-O and Mo-O confirms the successful incorporation of cobalt and molybdenum into the nanocomposites structure. If unexpected bands are detected, it is possible to conclude about possible impurities or by-products of the synthesis, which may be important for the subsequent optimization of the nanocomposite preparation method. The results of IR analysis confirm the presence of oxide forms of all three metals in the nanoparticles and indicate possible impurities, such as carbon pollutants. These data help to better understand the synthesis process and characteristics of the obtained nanocomposites, and also provide an idea of their stability and chemical activity under various conditions.

Scanning electron microscopy (SEM) results for aluminum-cobalt-molybdenum nanocomposite (Al-Co-Mo) provided important information about the morphology, size, and distribution of nanoparticles, as well as possible defects or agglomerations. SEM made it possible to visualize the structure of nanoparticles and their interactions with each other, which is important for understanding their physical and chemical properties.

The images obtained showed that the nanoparticles had a fairly diverse shape. Most of the particles were spherical or oval, which is typical for nanoparticles obtained during the synthesis process. However, some images also showed particles with a more irregular or angular shape, which may indicate agglomeration or partial aggregation during the synthesis or during the drying of the material. This is important, since agglomeration can affect the final properties of the nanocomposite, such as its mechanical strength and catalytic activity.

According to the SEM data, the complexes are in the form of large porous materials (Fig. 2a, b). The obtained complexes have high porosity and the formation of submicron pores, which is explained by the structural features of the cobalt octahedral complex stabilized by the polymer.

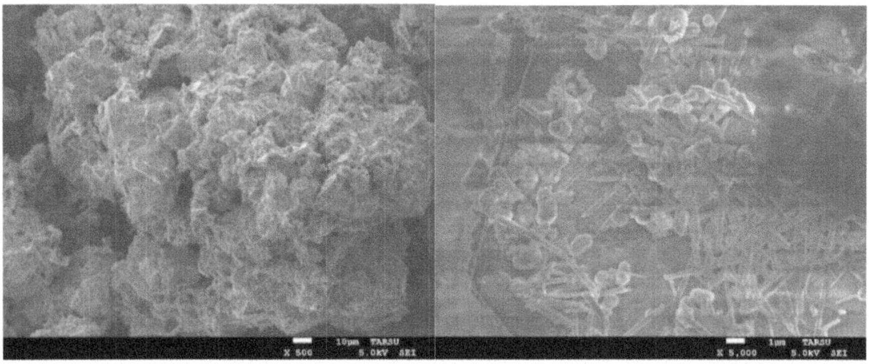

a) b)

Fig. 2. Microphotographs of Al-Co-Mo nanocomposites at different resolutions: (**a**) 500; (**b**) 5000.

SEM measurements showed that the average size of the nanoparticles was approximately 60–80 nm. However, the images also showed both smaller particles and larger agglomerates, which may indicate different growth rates of the nanoparticles during the synthesis process. The presence of large agglomerates may be due to insufficient control of the synthesis conditions, such as temperature, reagent concentration, or reaction time.

The Al-Co-Mo nanoparticles show a homogeneous distribution in the SEM images, indicating uniform mixing of the components during the synthesis process. However, some areas of the samples show a denser cluster of particles, which may indicate local aggregation. It is important to note that despite the presence of such areas, the overall trend indicates a fairly uniform distribution of the nanocomposite components.

The surface of the nanoparticles was examined and found to be rough, which is typical of metallic and metal-oxide nanoparticles. Such a surface could increase their active area for reactions, such as catalysis. The images also showed signs of surface defects, such as cracks or pores, which could be due to the synthesis process or storage conditions of the material.

Thus, the obtained data allow us to draw conclusions about the structure and morphology of the nanocomposite, such as the degree of agglomeration of the particles and their size. This may depend on the synthesis technology and processing conditions. The surface of the particles is rough, which may have a positive effect on their catalytic activity. These data suggest that the synthesized nanoparticles may have good physicochemical properties and are suitable for application in various fields such as catalysis and electronic materials.

The use of an X-ray microanalyzer in conjunction with SEM allows us to obtain a comprehensive understanding of the microstructure and composition of the aluminum-cobalt-molybdenum complex, which is an important step in assessing its efficiency and further optimizing the synthesis process. As a result of the analysis, it was found that the components are uniformly distributed over the sample, and their ratio corresponds to the calculated one, which confirms the correctness of the synthesis and the stability of the nanoparticle composition.

The presence of molybdenum in various oxidation states is indicated by X-ray spectra (Fig. 3).

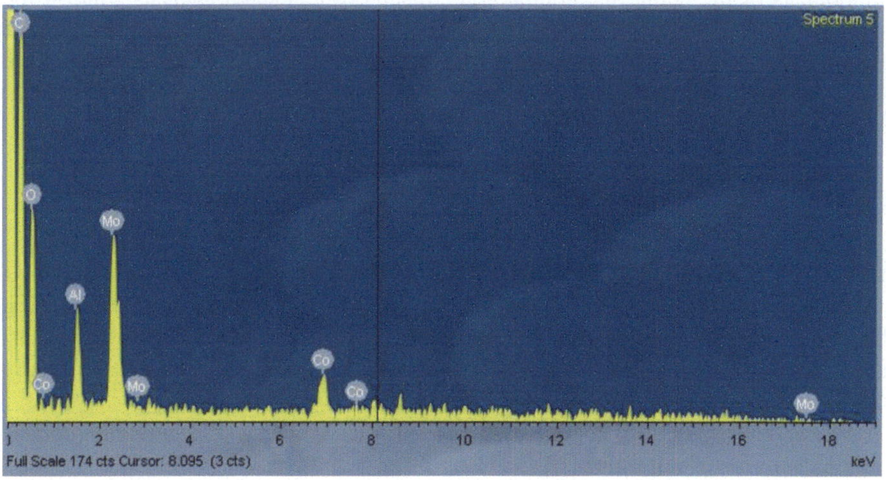

Fig. 3. X-ray spectrum of Al-Co-Mo nanocomposites

The analysis showed that the elements were distributed uniformly throughout the sample and their ratios corresponded to the predetermined proportions. The amounts of aluminum, cobalt, and molybdenum in the nanoparticles were close to the theoretical value, confirming the high accuracy of the synthesis. This is important because uniform distribution of the components affects the chemical activity and stability of the nanoparticles, which is necessary for their application in various fields such as catalysis, electronics, and battery materials.

The distribution of the elements throughout the sample was uniform, which also confirms the quality of the synthesis and the absence of significant agglomerations or inhomogeneities in the structure of the nanoparticles. This also correlates with the SEM results, where a uniform distribution of particles was noted, although there were small areas with a denser accumulation of nanoparticles. The uniformity of the distribution of the components confirms that the synthesis was carried out with sufficient control of the conditions.

In addition to the main elements (Al, Co, and Mo), the elemental analysis also showed the presence of trace amounts of other elements such as oxygen (O), which may be due to the formation of oxide forms of metals during the synthesis. These data are consistent with the results of IR spectroscopy, where a peak indicating the presence of oxide bonds was also recorded. It is important to note that traces of other elements did not significantly affect the overall composition and properties of the nanoparticles. Elemental analysis also confirms the presence of metallic and oxide forms of all three metals in the nanoparticles, which is consistent with the data obtained using other methods (IR spectroscopy and SEM). This confirms that the structure of the nanocomposite contains stabilized oxides of aluminum, cobalt and molybdenum, which is critical for their stability and activity.

Thus, element distribution maps help to determine the uniform distribution of cobalt and molybdenum on the surface of the aluminum carrier. Local differences in elemental composition may indicate inhomogeneity of the synthesis process or interaction of the components of the composite complex.The results of elemental analysis confirm the correctness of the synthesis of aluminum-cobalt-molybdenum nanocomposite, as well as the uniformity of the distribution of the components in the material. These data are consistent with the results of other analytical methods and provide a basis for further studies of the properties and applications of these nanoparticles.

5 Conclusions

In this work, aluminum-cobalt-molybdenum nanoparticles were synthesized and their structural properties were comprehensively studied. The use of synthesis methods using the sol-gel method allowed obtaining nanoparticles with different morphological characteristics and particle sizes, which was confirmed by scanning electron microscopy (SEM) data.

In the course of the conducted study of the synthesis of aluminum-cobalt-molybdenum nanocomposites (Al-Co-Mo), a significant amount of data was obtained that confirm the high prospects of these materials for application in various high-tech areas. Particular attention was paid to the comprehensive analysis of the obtained nanocomposites using various methods, such as infrared spectroscopy (IR), scanning electron microscopy (SEM) and elemental analysis, which allowed a deeper understanding of their structure, composition and functional properties.

The results of IR spectroscopy showed characteristic spectra reflecting the presence of specific bonds between the components of the nanocomposite. In particular, for Al-Co-Mo nanocomposites, absorption bands were recorded associated with vibrations of metallic bonds, as well as with functional groups present on the surface of the particles. This confirms the successful integration of each component into the structure of the nanocomposite and indicates possible interactions between elements within the material composition, which, in turn, affects its catalytic and mechanical properties.

Scanning electron microscopy (SEM) allowed us to study the morphology of the obtained nanoparticles. The results showed that chemical precipitation and salt synthesis can produce nanoparticles with a uniform size and uniform distribution, which is critical for their use in catalytic processes. The Al-Co-Mo nanoparticles had a size of 60 to 80 nm, confirming the possibility of size control using these synthesis methods. Additional attention was paid to the study of the particle surface, which demonstrates a high level of texturing, which improves their catalytic activity and increases the contact area with reagents in chemical reactions. Elemental analysis, carried out using energy-dispersive X-ray spectroscopy (EDX), confirmed that the nanocomposites contain all three components - aluminum (Al), cobalt (Co), and molybdenum (Mo) - in the required proportions. It was also noted that the components are uniformly distributed throughout the volume of the material, indicating the successful implementation of the synthesis process and the absence of aggregation or precipitation of individual phases. These data confirm that the nanocomposites have a stable chemical structure and can be used for further applications.

The obtained research results show that Al-Co-Mo nanocomposites have high physicochemical characteristics, which makes them suitable for use in catalytic reactions such as hydrogenation and oxidation. Uniform distribution of components and optimal morphology of nanoparticles provide high activity of the material, as well as its stability in aggressive chemical environments. The surface properties revealed by SEM allow us to state that these materials can be effectively used as catalysts in high-temperature processes, as well as for other industrial applications.

The analysis of literature sources conducted during the study also confirms that the use of aluminum-cobalt-molybdenum nanocomposites as catalysts in various reactions, such as hydrogenation, oxidation and reduction, can significantly increase the efficiency of these processes. The advantage of such materials lies not only in their high activity, but also in their durability, heat resistance and oxidation resistance, which makes them applicable for long industrial cycles.

Thus, the results of the study show that aluminum-cobalt-molybdenum nanocomposites are promising materials that can be used in various fields. It is important to note that further development of synthesis technologies and improvement of methods for monitoring their properties will significantly expand the scope of application of these nanocomposites. In particular, improvement of synthesis methods and characteristics of materials can contribute to their use in catalytic systems, as well as in other high-tech areas such as electronics, medicine and energy.

In the future, to achieve even better results, it is necessary to continue studying the influence of various synthesis parameters on the physicochemical properties of materials, as well as to conduct a comparative analysis with other types of nanocomposites to expand their functional capabilities.

References

1. Rana, A., Yadav, K., Jagadevan, S.: A comprehensive review on green synthesis of nature-inspired metal nanoparticles: mechanism, application and toxicity. J. Clean. Prod. **272**, 122880 (2020)
2. Ozin, G.A., Arsenault, A.: Nanochemistry: A Chemical Approach to Nanomaterials. Royal Society of Chemistry, London (2015)
3. Bhushan, B.: Introduction to nanotechnology. In: Springer Handbook of Nanotechnology, vol. 5, pp. 1–19. Springer, Berlin, Heidelberg (2017)
4. Rane, A.V., et al.: Methods for synthesis of nanoparticles and fabrication of nanocomposites. In: Synthesis of inorganic nanomaterials, pp. 121–139. Elsevier, Amsterdam (2018)
5. Ravichandran, K., et al.: Synthesis of nanocomposites. In: Synthesis of inorganic nanomaterials, pp. 141–168. Elsevier, Amsterdam (2018)
6. Camargo, P.H.C., Satyanarayana, K.G., Wypych, F.: Nanocomposites: synthesis, structure, properties and new application opportunities. Mater. Res. **12**, 1–39 (2009)
7. Pramanik, S., Das, P.: Metal-based nanomaterials and their polymer nanocomposites. In: Nanomaterials and polymer nanocomposites, pp. 91–121. Elsevier, Amsterdam (2019)
8. Yang, J., Liu, H.: Metal-Based Composite Nanomaterials. Springer International, Cham (2015)
9. Sadeghi, B., et al.: Architectural design of advanced aluminum matrix composites: a review of recent developments. Crit. Rev. Solid State Mater. Sci. **49**(1), 1–71 (2024)
10. Lai, H.K., et al.: Coordination polymer-derived cobalt nanoparticle-embedded carbon nanocomposite as a magnetic multi-functional catalyst for energy generation and biomass conversion. Chem. Eng. J. **332**, 717–726 (2018)
11. Morales-Guio, C.G., Hu, X.: Amorphous molybdenum sulfides as hydrogen evolution catalysts. Acc. Chem. Res. **47**(8), 2671–2681 (2014)
12. Chen, C., et al.: Selective catalytic reduction of nitrogen oxides with methanol over the (cobalt-molybdenum)/alumina dual catalysts under the diesel methanol dual fuel exhaust conditions. Chem. Eng. Sci. **211**, 115320 (2020)
13. Zhang, Y., et al.: Preparation and electrochemical properties of cobalt aluminum layered double hydroxide/carbon-based integrated composite electrode materials for supercapacitors. Electrochim. Acta. **442**, 141822 (2023)
14. Saji, V.S., Lee, C.W.: Molybdenum, molybdenum oxides, and their electrochemistry. ChemSusChem. **5**(7), 1146–1161 (2012)
15. Alavijeh, M.S., et al.: Cobalt ferrite decoration of molybdenum disulfide nanosheets; development of a nanocomposite-mediated hyperthermia method. J. Mech. Sci. Technol. **35**, 1319–1325 (2021)
16. Dhas, N., et al.: Molybdenum-based hetero-nanocomposites for cancer therapy, diagnosis and biosensing application: current advancement and future breakthroughs. J. Control. Release. **330**, 257–283 (2021)
17. Li, M., et al.: Hybrid structures of cobalt-molybdenum bimetallic oxide embedded in flower-like molybdenum disulfide for sensitive detection of the antibiotic drug nitrofurantoin. J. Hazard. Mater. **435**, 129059 (2022)
18. Kim, S., Park, J.: Catalytic hydrogenation using Co-Mo nanoparticles. Cat. Com. **98**, 58–62 (2017)
19. Lee, H., Kim, T.: Biomedical applications of Co-Mo nanoparticles: a review. Nano Res. **14**, 1563–1584 (2011)
20. Saleh, T.A., et al.: Ultra-deep adsorptive desulfurization of fuels on cobalt and molybdenum nanoparticles loaded on activated carbon derived from waste rubber. J. Colloid Interface Sci. **513**, 779–787 (2018)

21. Yu, S., et al.: Constructing sphere-like cobalt-molybdenum-nickel ternary hydroxide and calcined ternary oxide nanocomposites for efficient removal of U (VI) from aqueous solutions. Chem. Eng. J. **352**, 360–370 (2018)
22. Kwiatkowski, K.C., Lukehart, C.M.: Nanocomposites prepared by sol-gel methods: synthesis and characterization. In: Handbook of Nanostructured Materials and Nanotechnology, pp. 387–421. Elsevier, Amsterdam (2000)
23. Pomogailo, A.D.: Polymer sol-gel synthesis of hybrid nanocomposites. Colloid J. **67**, 658–677 (2005)
24. Okamoto, Y., et al.: Preparation of Co–Mo/Al_2O_3 model sulfide catalysts for hydrode sulfurization and their application to the study of the effects of catalyst preparation. J. Catal. **217**(1), 12–22 (2003)
25. Davydov, A.A., Goncharova, O.I.: Use of IR spectroscopy in studies of catalysts based on molybdenum heteropoly compounds supported on oxides. Russ. Chem. Rev. **62**(2), 105–120 (1993)
26. Huang, Y., et al.: Hierarchically macro−/mesoporous structured co–Mo–Ni/γ-Al_2O_3 catalyst for the hydrodesulfurization of thiophene. Chem. Eng. J. **172**(1), 444–451 (2011)
27. Wachowski, L., Łaniecki, M., Domka, F.: Activity of co−Mo catalysts supported on alumina modified with La_2O_3. React. Kinet. Catal. Lett. **67**, 123–127 (1999)

Development of Environmentally Friendly Packaging Material for Food Products Based on Biodegradable Polymer

Eleman Sabina, Nurlybayeva Aisha, and Rustem Ergali(✉)

M.Kh. Dulaty Taraz University, Taraz, Kazakhstan
rustem_ergali@mail.ru

Abstract. The main objective of the scientific work is to extract the pectin compound from the fruit peel and obtain bioplastic samples from it in laboratory conditions. The pectin compound was synthesized from the fruit peel and a biodegradable film was prepared from it. The composition and structure of the pectin compound were studied using the IR spectrum. A biodegradability test was conducted by preparing compost and microbial resistance was calculated. The degree of swelling of the pectin film was determined. The tensile strength of the pectin film was studied. Bioplastics are disposable packaging materials that do not decompose for a long time and are thrown away. Therefore, the relevance of the scientific work is the production of biodegradable bottles and eco-packaging from orange peel. Obtaining eco-packaging by processing pectin from waste streams is economically viable.

Keywords: Pectin · biodegradable film · IR spectroscopy · compost · microbial resistance · swelling degree

1 Introduction

The increasing global concern over environmental sustainability has led to a growing interest in the development and utilization of biodegradable polymers, which have the potential to address the pressing issue of plastic waste accumulation and its detrimental impact on ecosystems. Biopolymers, defined as polymers that are either biodegradable, bio-based, or possess both properties, offer a promising alternative to conventional petroleum-based plastics, providing a more sustainable solution for various applications, particularly in the packaging industry [1].

Polymers have become an integral part of our modern society, revolutionizing industries and transforming our daily lives. However, the growing concerns over the environmental impact of non-degradable polymers have prompted researchers to explore more sustainable alternatives [2]. Biodegradable polymers have emerged as a promising solution to tackle the issue of plastic waste. Biodegradable polymers are materials that can be broken down by the action of living organisms, such as bacteria, fungi, and enzymes, into simpler, non-toxic compounds, ultimately leading to their complete decomposition and

R. Rajasekharan et al. (Eds.): MatZone 2024, SPPHY 316, pp. 216–227, 2025.
https://doi.org/10.1007/978-981-96-6541-9_20

integration into the natural environment. Compared to traditional plastics, these materials offer several advantages, including reduced energy consumption and greenhouse gas emissions during production, as well as the ability to be recycled or composted, thereby minimizing their environmental footprint [1].

The use of biopolymers, which are either biodegradable, bio-based, or possess both properties, offers significant advantages over conventional polymers. Biopolymers provide a 65% energy savings and between 30% and 80% less greenhouse gas emissions during their production compared to conventional polymers [1]. This makes them a more environmentally-friendly option, addressing the pressing need to reduce the environmental impact of plastic waste.

Researchers have been actively exploring the development of bio-based fibers through various techniques, such as electrospinning and solution blowing, to create a wide range of applications for these biodegradable materials [3]. Combining several natural materials that are similar to petroleum-based polymers but biodegradable creates a cost-effective and [eco-friendly] composite.

Many materials in the food industry, such as leather and rubber, are recycled and discarded as waste. These components are a source of important biological compounds, such as pectins and polysaccharides, which are found in the primary cell walls of most plants. Pectin provides mechanical strength and flexibility for cell culture by interacting with other components of the cell wall [5]. Citrus fruit peels, especially orange peels, are rich in pectin and can be used as raw materials for the production of biodegradable polymers.

2 Literature Review

Pectin can be extracted from a variety of fruits, including bananas, guavas, mangoes, and citrus fruit peels [6, 7]. The extraction process significantly affects the yield and quality of pectin. Previous studies have reported the successful use of pectin and other bioactive compounds in various fruits such as mango, blueberry, watermelon, Poncan mandarin, grape, tomato, cucumber and carrot. to produce.

Pectin is an important compound found in the peels of many fruits, such as oranges. It is widely used as a gelling agent, emulsifier and thickener in the food and pharmaceutical industries. Pectin can be obtained from a variety of sources, with orange peel being the richest source [8]. The extraction process and the quality of the extracted pectin depend on many factors such as pH, solvent type, and extraction temperature [9].

The pectin extraction and characterization literature provides important insights into the effect of extraction and desorption processes on the properties of extracted pectin. There are many sources of starch that can be used to synthesize starch-based polymers. Examples include maize, avocado, rice, sugarcane, and fruit peels. However, these are primarily edible foods that serve as sources of energy for consumers, making it economically unfeasible to extract starch from them for industrial purposes. Consequently, there is potential to utilize starch-rich waste materials as substitutes. Among these, fruit peels stand out as a viable option, as they contain a high starch content of approximately 18.5%. Additionally, pectin compounds can be extracted from fruit peel waste, as summarized in Table 1.

Table 1. Amount of pectin in fruit waste

Fruit type	Calcium Pectate (%)	Pectin (%)
Citrus	24.5	25.5
Mango	7.5	8.8
Banana	2.5	2.8
Apple	10.8	12.5

Pectin is a type of starch known as a heteropolysaccharide, naturally found in plant cell walls where it provides structural support. This complex macromolecule comprises seventeen different monosaccharides linked by more than twenty distinct types of bonds. Pectin is one of the key components of plant cell walls, playing a vital role in their structural integrity. In general, the polymeric composition of the primary cell walls of dicotyledonous plants consists of approximately 30% pectin, 30% cellulose, 30% hemicellulose, and 5% protein.

Pectin can be obtained from a variety of sources, including fruits, vegetables, and agricultural wastes. Commonly used extraction methods include acid extraction, alkaline extraction, and enzymatic extraction. Acid extraction is the most commonly used method because it is performed under mild conditions and produces pectin with good gelling properties.

Some studies have investigated other extraction methods, such as microwave [10], ultrasound [11], and superhydrolysis [12], to improve the efficiency and stability of the pectin extraction process. Biodegradable products can be synthesized by obtaining esters of polyhydroxydicarboxylic acids and furandicarboxylic acids.

A review of the literature shows that pectin can be obtained from a variety of sources, including citrus fruits, apples, strawberries, plums, and oranges [11]. The physicochemical properties of the extracted pectin, such as molecular weight, degree of esterification, and acetylation, may vary depending on the source and extraction method [12]. Plant-based eco-packaging is the result of citrus peel recycling technology.

3 Materials and Methods

Fruit peel, glycerin (plasticizer), hydrochloric acid (HCl),sodium hydroxide (NaOH), filler ($CaCO_3$ calcium carbonate), кuanstaar, vacuum filtration, acetic acid (CH_3COOH), ethanol (C_2H_5OH), IR-spectroscopy, tensile strength, degree of swelling.

3.1 Extraction of Pectin from Fruit Peels

First of all, this selected fruit peel was cut into pieces, weighed 247 g, washed with distilled water from impurities and placed in a chemical beaker with a capacity of 800 ml. After that, I poured 600 ml of 2% acetic acid into the beaker and boiled for about 3 hours at a temperature of 100 ° C. A magnetic stirrer was used in the beaker so that the fruit peel was always immersed in the solution. The main purpose of such boiling is to make the cell walls of the shell soft, which helps to extract pectin compounds. Acetic acid was used as an acid solution, and I maintained the pH at 2.8, checking it with litmus paper. Because at a pH above 2.8, the pectin compound dissolves in the acetic acid solution, and at a low acidity, the pectin precipitate does not form.

Any organic or inorganic acid such as tartaric acid or HCl can be used to maintain the acidic pH level. After boiling, the solution is filtered through a Buchner funnel using filter paper and 70% ethanol and 30% water are added to the filtered solution. The solution is not clear for some time. After a while, the pectin compound becomes insoluble in alcohol and begins to precipitate in the solution. The resulting solution is concentrated in a vacuum filter. The filtration is repeated several times to reduce the amount of water in the pectin mixture. The crushed pectin mixtures are used to obtain biodegradable films.

The resulting pectin mixture is dried in an oven at 50 °C for 8 h. However, the pectin mixture is not completely dried at this time, so I dried it at the same temperature for 18 h. The resulting pectin mixture is ground into a fine powder using a mixer. I made a thin film with this pectin mixture. A water pectin paste was obtained [13].

3.2 Dry Pectin Powder

2 g of dried pectin and 0.1 g of wheat germ were added to the biomass. Then add 20 mL of water and heat the flask to 100° in the microwave, stirring occasionally with a magnetic stirrer.3 mL of 0.5 M HCl solution were added to this solution, and then 2.1 mL of 99% glycerol. To neutralize this reaction, 0.5 M NaOH solution was added. The pH value of the mixture was checked with litmus paper. The hydrolysis process in the solution was carried out for 20–30 min, as a result of which a thick gel-like structure of the mixture was formed. The resulting mixture was poured into a Petri dish and left to dry in a drying cabinet at a temperature of 50 °C for about 24 h [14].

3.3 Pectin Paste

The resulting 25 g of pectin paste was placed in a 50 mL beaker and heated on an electric hotplate at 100 °C. To the pectin paste mixture, 3 mL of 0.5 M HCl solution were added along with 2.1 mL of 99% glycerol. A magnetic stirrer was used for continuous stirring. 0.5 M HCl was added to neutralize the solution. The pH of this mixture was checked with litmus paper. After acid hydrolysis for 30 min, the mixture was poured into a Petri dish and left to dry in an oven at 50 °C for about 24 h. Two samples of pectin powder and two samples of pectin paste were prepared, the difference between them being that corn starch was added to the pectin powder [15].

3.4 Characterization of Materials and Analytical Methods

3.4.1 Biodegradation

The biodegradability of the polymers was tested by the standard method by composting. The residues are decomposed mainly by bacterial fungi, radiant fungi and earthworms with the help of many microorganisms. This helps to form a good compost soil for the decomposition of various products. The biodegradability of a product is directly proportional to the amount of carbon available for consumption by microbes. The biodegradation test was carried out using a home composting equipment shown in Fig. 1.

Fig. 1. Pectin film and polyethylene film in compost

Compost includes wood chips, biowaste, kitchen waste, etc. including household waste. Household compost was used to test the biodegradability of the film. The samples were labeled with tapes of different colors to make it easier to distinguish between them.

3.5 Infrared Spectroscopy (IR) Method

To determine the structure, composition and structure of synthesized polymers, to characterize polymers together with the NMR method, the IR spectrum method is used in the range of 4000–400 cm^{-1} in the program "Spectrum 65 FT". The "IR" device recorded and analyzed the spectrum.

4 Results and Discussion

4.1 Obtained Pectin Compound

After drying, the pectin was very difficult to separate because it solidified on the plate during drying. The total weight of the obtained dry pectin was 8 g, which corresponds to a yield of about 5%. The pectin was used to produce bioplastic film, as shown below.

Finally, the bioplastic films were taken out of the drying oven to see the results of the obtained samples. There were some obstacles in the formation of the film. After the drying process began, the oven temperature was gradually reduced. The thickness of the four obtained samples was measured. It was found that the thickness of the samples was uneven, which was caused by improper grinding and the presence of impurities. The films were made flexible and opaque. The film was cut into pieces for IR spectroscopy analysis and biodegradation tests.

4.2 Tensile Strength of Samples

Pectin paste and pectin powder were used to form the film from the samples, but pectin paste produced a thick film while pectin powder produced a thin film. The two samples prepared with pectin powder remained more unchanged than those with pectin paste. It was found that the tensile strength of the films made from pectin paste was low while the tensile strength of the films made from pectin powder was high. At least 4 intact films were needed for the tensile strength analysis [16].

The color of the obtained films was the same except that the films made from pectin paste were slightly lighter. In the early days of this project, the film synthesized using more corn starch (20%) showed good tensile strength. But since our project mainly aims to utilize waste materials, we reduced the amount of corn starch to 5%. As a result of measuring the strength of these films, the tensile strength of the samples prepared from pectin paste was low, that is, the tensile strength of the film made from pectin powder with 20% starch was 31.88, and the tensile strength of the film with 5% starch was higher. After all, the lower the starch content in the film, the higher its tensile strength. The film prepared from pectin paste without adding starch showed very low tensile strength, since it did not contain a plasticizer. A denser arrangement of hydrogen bonds in pectin solutions with polymers is confirmed by the value of the parameters of microhomogeneous bonds in the polymer shell (Fig. 2) [17].

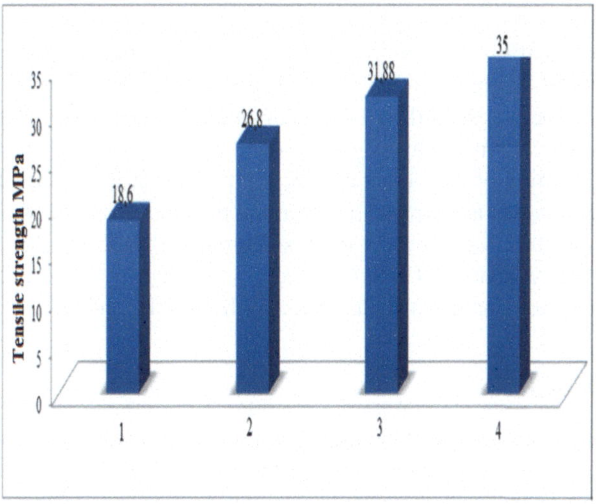

Fig. 2. Tensile strength 1-P:25, 2-P:35, 3-U2:S20, 4-U2:S5

The thickness of the films was not the same, so each film was measured ten times to calculate the average thickness. The thickness of the films is shown in Table 2 below.

Table 2. Samples and their thickness

No	Calcium pectate (%)	Pectin (%)
1	Pasta pectin (25 g), P:25	0.85
2	Pasta pectin (35 g), P:35	0.87
3	Untagy pectin (2 g) + starch 20%, U2:S20	0.45
4	Untagy pectin (2 g) + starch 5%, U2:S5	0.59

4.3 Investigation of Pectin Samples by IR Spectroscopy

The main purpose of obtaining the spectrum of biodegradable film is to determine the functional groups of its composition, which allows us to study the interaction of its ingredients during film formation. The absorption bands of the OH group in Fig. 3 are shown at 3374.14 cm^{-1}. These are the stretching peaks of the OH group hydrogen-bonded to the hydroxyl (OH) functional group, which arises from the complex band stretching that occurs naturally in carbohydrate structures. The skeletal vibrations of the pyranose ring of the glucose unit of starches include peaks below the 800 cm^{-1} region. The aliphatic vibration of CH shows small band peaks around 2900 cm^{-1}, while the absorption bands of pectin can be seen in small bands around 1600 cm^{-1} and 1700 cm^{-1}, which correspond to the bonding of the carbonyl groups C-O. In the IR spectrum, carbonyl excitation of CO and deformation vibration of CH between carbon and hydrogen were detected in the absorption bands of 1109.17 cm^{-1} and 1048.36 cm^{-1}, indicating that pectin contains hydroxyl groups OH, aliphatic groups CH and carbonyl groups CO (Fig. 3) [18].

Judging by the spectra, the shift to the right of the hydroxyl groups OH and the stretching of C-H between carbon and hydrogen are due to the action of the plasticizer, which confirms the weakening of the bonds connecting the starch molecules. During plasticization, hydrogen bonds are formed between the plasticizer and the starch, replacing the damaged molecules with stronger hydroxyl groups. Plasticizers have a significant effect on the hydroxyl groups of starch molecules, as they form hydrogen bonds with the hydroxyl groups of starch molecules, making the starch soluble.

4.4 Biodegradation

The samples were composted and the first visual inspection was carried out in 11 days. The maximum compost temperature in the first test was 32 °C, but the compost temperature dropped to 5 °C because the outside air temperature ranged from −7 to −20 °C, which proved that the biodegradation process was slower in the second test. Compared with the non-biodegradable samples, the samples obtained after the tests were too fragile. After the test, the samples became thinner and lost their transparency.

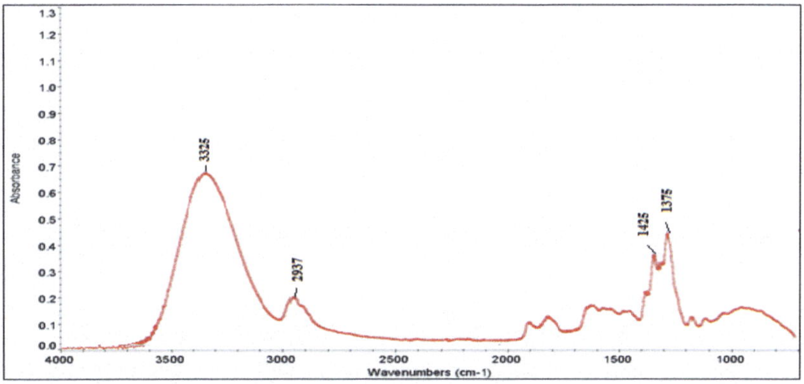

(a) Pectin paste

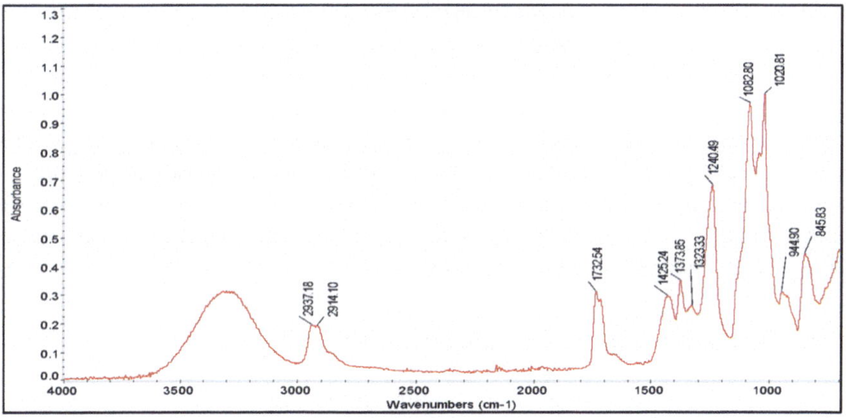

(b) Pectin powder

Fig. 3. IR spectrum of the obtained film

A biodegradation test of the composite was performed and the biodegradation rate was calculated. The initial weight of the film is measured before fertilization and measured again after one month of fertilization. The antimicrobial resistance of this sample was determined and calculated using the following formula.

$$Microorganism\ resistance\ (\%)(\%) = \frac{m_{initial} - m_{final}}{m_{initial}} * 100 \quad (1)$$

Antimicrobial resistance (%) was calculated for all samples (Table 3).

Table 3. Microbiological resistance of samples

Sample	Microbiological resistance	m_{final}	$m_{initial}$
1	1.01	0.02	98.01
2	1.53	0.07	95.42
3	0.36	0.03	91.67
4	1.06	0.08	93.39

Biodegradation is affected by many factors, including infrared radiation, temperature, concentration and metabolism. Bacteria, fungi and other microorganisms are involved in the decomposition process. This process produces carbon dioxide, heat, water and nutrients. The heat produced by these microbes increases the temperature of the compost to 70–80 °C. The degradation process is affected by the ambient temperature. During freezing, the temperature of the compost is maintained at −50 °C, preventing biodegradation. The optimum temperature for the decomposition of microorganisms is 32 °C, and during the first two weeks it is within the range of 23–32 °C. In addition, the external temperature is between °C4 and °C20. It can be said that at low temperatures the decomposition process slows down and frozen compost can be stored for longer. As the temperature increases, the rate of decomposition of organisms increases. However, at high temperatures, biodegradation decreases due to increased cohesion and intermolecular forces of starch in the fruit skin.

4.5 Determination of the Degree of Swelling

Then the samples were examined for swelling in distilled water (Fig. 4). As shown by the data obtained, it is clear that the swelling of the polymer film improves with increasing temperature, and is also observed in samples with a higher content of starch in relation to the pectin compound. Since these starch grains swell in warm water, fewer polysaccharides are released into the solution. It is known that it varies in starch from different plants and is due to the intense swelling of the starch grains, their bursting and the formation of a more or less homogeneous solution - a starch solution. The figure below shows that compounds containing corn starch (sample 3.4) have a higher degree of swelling. This process is due to the rupture of hydrogen bonds connecting the structural elements of the starch grains, in combination with swelling. It is clear that the nature of the anion affects the swelling of the gel [21].

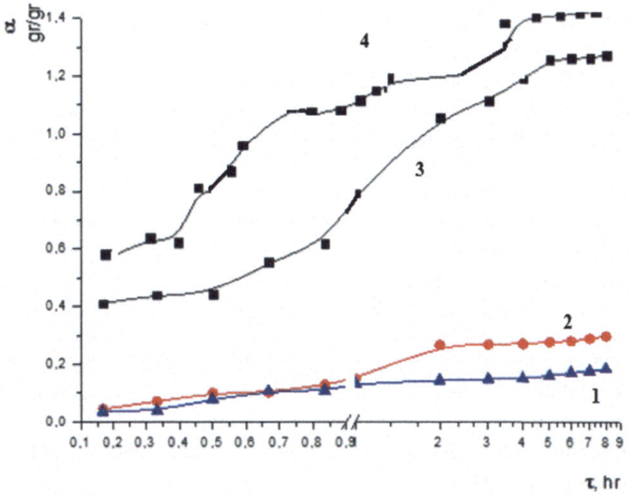

Fig. 4. Degree of swelling of the sample: 1-P:25, 2-P:35, 3-U2:S20, 4-U2:S5

5 Conclusion

In this study, a polymer film was synthesized by extracting fruit peel to extract pectin from the fruit peel. Pectin was extracted by boiling 2% acetic acid solution at 100 °C for 3 h. The pectin compound obtained by extraction was used to synthesize polymer films and various working conditions including drying temperature, HCl concentration, glycerol and hydrolysis residence time were adjusted to obtain good quality films with biodegradable and functional properties.

The obtained film was subjected to IR spectroscopy to determine the functional groups and structure, measure the biodegradability of the samples using compost for 30 days and determine the swelling ratio of tensile strength. The main objective of the study was to investigate the feasibility of obtaining biofilms of acceptable quality from waste or waste, the use of pectin from waste and its use in the production of biodegradable films. The obtained film can be used for packaging food and beverages. In addition, visual inspection of the films revealed shrinkage and loss of shape of the films obtained from pectin paste. It can be concluded that the pectin films obtained from fruit peels have high quality (about 90%) biodegradability. Further research is needed to improve the properties and quality of biofilm, which in the future can become an ideal process for sustainable synthesis. Using waste as a source of raw materials is an innovative method and a very effective part of the economy.

References

1. Dominguez-Candela, I., García-García, D., Perez-Nakai, A., Lerma-Canto, A., Lora-García, J., Fombuena, V.: Contribution to a circular economy model: from lignocellulosic wastes from the extraction of vegetable oils to the development of a new composite. Polymers. **13**, 2269 (2021). https://doi.org/10.3390/polym13142269

2. Abdullah, Z.W., Dong, Y., Davies, I.J., Barbhuiya, S.: PVA, PVA blends, and their nanocomposites for biodegradable packaging application. Polym.-Plast. Technol. Eng. **56**, 1307–1344 (2017). https://doi.org/10.1080/03602559.2016.1275684
3. Kakoria, A., Sinha-Ray, S.: A review on biopolymer-based fibers via electrospinning and solution blowing and their applications. Fibers. **6**(3), 45 (2018). https://doi.org/10.3390/fib6030045
4. Rendón-Villalobos, R., Ortíz-Sánchez, A., Tovar-Sánchez, E., Flores-Huicochea, E.: In: Rendón-Villalobos, R., Ortíz-Sánchez, A., Tovar-Sánchez, E., Flores-Huicochea, E. (eds.) The Role of Biopolymers in Obtaining Environmentally Friendly Materials. InTech eBooks (2016). https://doi.org/10.5772/65265
5. Khamsucharit, P., Laohaphatanalert, K., Gavinlertvatana, P., Sriroth, K., Sangseethong, K.: Characterization of pectin extracted from banana peels of different varieties. Food Sci. Biotechnol. **27**, 623–629 (2017). https://doi.org/10.1007/s10068-017-0302-0
6. Schneider, V.S., Bark, J.M., Winnischofer, S.M.B., dos Santos, E.F., Iacomini, M., Cordeiro, L.M.C.: Dietary fibres from guavira pomace, a co-product from fruit pulp industry: characterization and cellular antioxidant activity. Food Res. Int. **132**, 109065 (2020). https://doi.org/10.1016/j.foodres.2020.109065
7. Dias, P.G., Sajiwanie, J.W.A., Rathnayaka, R.M.U.S.K.: Chemical composition, physicochemical and technological properties of selected fruit peels as a potential food source. Int. J. Fruit Sci. **20**, S240 (2020). https://doi.org/10.1080/15538362.2020.1717402
8. Jin, Y., Yang, N.: Array-induced voltages assisted extraction of pectin from grapefruit (Citrus paradisi Macf.) peel and its characterization. Int. J. Biol. Macromol. **152**, 1205 (2019). https://doi.org/10.1016/j.ijbiomac.2019.10.215
9. Picauly, P., Tetelepta, G.: Characterization of pectin from tongka langit banana peels with various extraction temperature. Tetelepta. **883**, 12060 (2021). https://doi.org/10.1088/1755-1315/883/1/012060
10. Ciriminna, R., Fidalgo, A., Scurria, A., Ilharco, L.M., Pagliaro, M.: Pectin: new science and forthcoming applications of the most valued hydrocolloid. Food Hydrocoll. **127**, 107483 (2022). https://doi.org/10.1016/j.foodhyd.2022.107483
11. Müller-Maatsch, J., Bencivenni, M., Caligiani, A., Tedeschi, T., Bruggeman, G., Bosch, M., Petrusán, J.-I., Droogenbroeck, B.V., Elst, K., Sforza, S.: Pectin content and composition from different food waste streams. Food Chem. **201**, 37 (2016). https://doi.org/10.1016/j.foodchem.2016.01.012
12. Munarin, F., Tanzi, M.C., Petrini, P.: Advances in biomedical applications of pectin gels [Review of Advances in biomedical applications of pectin gels]. Int. J. Biol. Macromol. **51**(4), 681 (2012). https://doi.org/10.1016/j.ijbiomac.2012.07.002
13. Gawkowska, D., Cybulska, J., Zdunek, A.: Structure-related gelling of pectins and linking with other natural compounds: a review. Polymers. **10**(7), 762 (2018). https://doi.org/10.3390/polym10070762
14. Meade, R.C., Fish, V.B., Dustman, R.B.: The determination of pectic materials in apples. The use of 19 percent hydrochloric acid and pectinol A. Plant Physiol. **23**(1), 98–111 (1948). https://doi.org/10.1104/pp.23.1.98
15. Emaga, T.H., Ronkart, S., Robert, C.A.M., Wathelet, B., Paquot, M.: Characterisation of pectins extracted from banana peels (Musa AAA) under different conditions using an experimental design. Food Chem. **108**(2), 463–471 (2007). https://doi.org/10.1016/j.foodchem.2007.10.078
16. Magdum, M.V., Salokhe, B.T., Mali, A.S.: Tensile strength measurement for foundry sand brick. Int. J. Eng. Manag. Res. **8**(4), 40 (2018). https://doi.org/10.31033/ijemr.8.4.3
17. Nakamura, N., Mori, K., Okada, H., Abe, Y.: Punching test for estimating tensile strength and total elongation of steel sheets. Int. J. Adv. Manuf. Technol. **114**(5), 1847 (2021). https://doi.org/10.1007/s00170-021-06898-7

18. Brittain, H.G.: Mid-infrared spectroscopy of pharmaceutical solids. Profiles Drug. Subst. Excip. Relat. Methodol. **43**, 321 (2018). https://doi.org/10.1016/bs.podrm.2017.12.002
19. Fahelelbom, K.M., Saleh, A., Al-Tabakha, M.M., Ashames, A.: Recent applications of quantitative analytical FTIR spectroscopy in pharmaceutical, biomedical, and clinical fields: a brief review. Rev. Anal. Chem. **41**(1), 21 (2022). https://doi.org/10.1515/revac-2022-0030
20. Ibrahim, S.S., Ionescu, D., Grossart, H.: Tapping into fungal potential: biodegradation of plastic and rubber by potent fungi. Sci. Total Environ. **934**, 173188 (2024). https://doi.org/10.1016/j.scitotenv.2024.173188
21. Ahmed, E.M.: Hydrogel: preparation, characterization, and applications: a review. J. Adv. Res. **6**(2), 105–121 (2013). https://doi.org/10.1016/j.jare.2013.07.006

A Study on Impact of Dopants on Piezoelectric Parameters of Aluminium Nitride

Elsa Sneha Thomas[1,2](✉) and R. Ranjith[1]

[1] Christ University, Bangalore, Karnataka, India
elsa.thomas@res.christuniversity.in
[2] Providence College of Engineering, Chengannur, Kerala, India

Abstract. Doping is a critical in piezoelectric base materials used as sensors or energy harvesters. It has a significant effect in determining the output parameters of an energy harvester. This paper studies the effect and impact on piezoelectric properties when materials are co doped and singly doped by considering some common dopants found in the literature. From the study, the elastic constant and c/a ratio are found to be decreasing and all other lattice parameters such as piezoelectric stress coefficient, electromechanical coupling constant, Born effective charge etc. are increasing, although some discrepancies are there. Scandium, magnesium, Lithium and its co-dopants are mainly considered for the study. Another factor that should be considered is doping concentration. Excessive doping with scandium (Sc) has been demonstrated to reduce the coercive field (Ec), enabling ferroelectric switching in AlScN thin films. The critical factor for achieving ferroelectric switching lies in sufficiently lowering the energy barrier between the two polarization states made of the wurtzite structure. This can be accomplished either by increasing the proportion of the non-Group-III metal or through strain engineering.

Keywords: piezoelectricity · doping · piezoelectric lattice constant · co doping

1 Introduction

Piezo electricity and doping are often related in a way that doping can significantly affect the piezo electric nature of the same. The piezoelectric effect is an effect that simply describes the fact that a pressure applied to a piezoelectric material will generate a voltage. By the addition of dopants, this effect can be increased. In this study, piezoelectric Aluminium Nitride (AlN) is the topic of interest. Since AlN is piezoelectric in nature, before a leap into it, its piezoelectric nature and parameters need to be considered. Some of the piezoelectric material properties that is concerned with doping includes piezoelectric coefficient ($\mathbf{d_{33}}$), elastic constant ($\mathbf{C_{33}}$), piezoelectric stress constant ($\mathbf{e_{33}}$), electromechanical coupling constant ($\boldsymbol{K^2_{33}}$), Born effective charge ($\mathbf{Z_{33}}$) etc.

A dopant's concentration affects the electronic properties of a material, but the consequences vary depending on the dopant and its concentration. An intrinsic region with very low electrical resistance is created by doping, which creates a heterogeneous region

R. Rajasekharan et al. (Eds.): MatZone 2024, SPPHY 316, pp. 228–235, 2025.
https://doi.org/10.1007/978-981-96-6541-9_21

in the material. Devices with semiconductors perform better due to doping. As a result of doping, hole mobility increases and threshold voltage decreases, improving the electrical characteristics of the devices. A semiconductor is doped when atoms of different atoms are added to change its electrical properties. Compared to other materials, AlN has got a moderate piezoelectric response but it outperforms in curie temperature which is required for high temperature performance. Based on this fact, AlN can be proposed as an apt candidate for doping inorder to enhance the piezoelectric response. When reviewing the literature, S. A. Anggraini et al. [1] Investigated Mg and Ti co doping effect on the piezoelectric response of aluminum nitride thin films. In their work, It has been established that the combination of Magnesium and titanium enhances the d_{33}, whereas each element separately as a single dopant does not. In their work, Qi Wang et al, presented the design, fabrication and characterization of a Scandium doped AlN thin film piezoelectric micromachined ultrasound transducers. Compared to pure AlN a crystal structure expansion has been indicated for doped structure [2]. By introducing Yb into AlN, a huge piezoelectric response (d_{33}) is observed in the work done by Kenji Hirata et al., but the K_{33}^2 shows only a small increase because of the relatively lower values of ε_{33} [3]. Another piezoelectric parameter is power generation figure of merit, which increased 3 times larger than that of AlN when Mg-Hf codoped AlN ((MgHf)xAl1-xN) piezoelectric films are employed in vibration based micro energy harvester by Hung H. Nguyen et al. [4].This paper presented a study on the impact of piezoelectric parameters while doping piezoelectric AlN and factors that should be carefully considered for the same.

2 Parameters

2.1 The Piezoelectric Stress Coefficient (e_{33})

It relates the induced electric field (in volts per meter, V/m) to the applied mechanical stress (in Pascals, Pa). It measures the electric field generated in the material per unit of mechanical stress applied to it.

2.2 Elastic Constant (c33)

It is the piezoelectric strain coefficient. It quantifies the strain or deformation in a material along the 3-direction when an electric field is applied to it.

2.3 Born Effective Charge (Z_{33})

It is a measure of how the atoms within a crystal lattice respond to an applied electric field. It characterizes the displacement of the atoms in response to the electric field.

2.4 Piezoelectric Coupling Constant

It is denoted as k, quantifies the efficiency with which a material converts applied mechanical stress into electric charge or vice versa. The ability of a material or sensor to detect and respond to mechanical deformation or strain quantifies how its electrical or mechanical properties change in response to an applied strain.

2.5 Piezoelectric Coefficient (d_{33})

It measures the volume change in a piezoelectric material when an electric field is applied or the polarization generated when mechanical stress is applied.

3 Dopants

Here we consider the most common dopants that are employed in the literature for the doping of AlN. Doping can be done by single element doping and co doping. The most common single element dopants include Mg, Ti, Sc, B, Ge, Zr etc. and co doping can also be done along with single element doping. Some of the co dopants are Mg-Zr, Mg-Hf, MgTi, LiN, LiV, LiTa, LiZr etc. The Table 1 below summarizes the changes observed in some of the piezoelectric parameters for both single and co doping (Fig. 1).

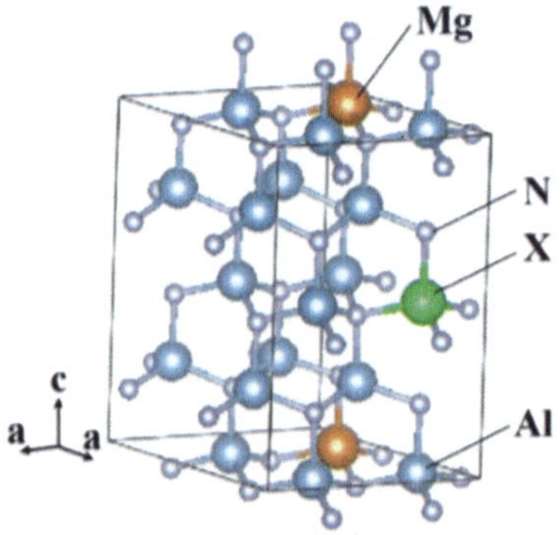

Fig. 1. AlN doped with Mg-X

The most common dopants of AlN are Sc and Mg. Calculated lattice parameters of after doping with them are shown in Table 1.

From the Table 2, it is clear that doping is such a sensitive process that dopant properties will affect the piezoelectric which led to changes in output parameters also (Figs. 2 and 3).

Stability criterion $(C_{11} + C_{12}) C_{33} - 2C_{13}{}^{2} > 0$ is satisfied when doped with Sc. A sharp drop in the c/a ratio for all of the co-doping above a concentration of 37.5%, indicating a phase transition from wurtzite to hexagonal (h-BN-like structure) phase. The

Table 1. Calculated Lattice Parameter of Wurtzite structure

Parameters	$Al_{16}N_{16}$	$Sc_2Al_{14}N_{16}$	$MgTiAl_{14}N_{16}$
c/a	1.603	1.587	1.590
e33	1.46	1.68	1.8
c33	1.8	1.44	1.5

Table 2. Change in piezoelectric parameters

AlN	Piezoelectric coefficient (d_{33}	elastic constant (C_{33})	piezoelectric stress constant (e_{33})	electromechanical coupling (K_{33}^2) constant	Born effective charge (Z_{33})	Young's modulus (Y	c/a ratio	Poissons ratio
Sc	Enhanced by500% [5] 149% increase in d_{33} [6]	Decreased [5] 87% decrease in C_{33} [6]	ScxAl1 – xN exhibits a 130% increase [5]	Enhanced [5]	Increased [5]	Reduced [7]	Decreased [5] sharp drop for all the co-doping above a concentration of 37.5% [7]	Sc doping above 25% transforms brittle w-AlN to ductile [5]
Mg-Zr	146% increased [8]	87% decreased	130%increased	Increased	–	–	Decreased	–
Mg-Hf	163%increased	89% decreased	145%increased	–	–	–	Decreased	–
Mg-Ti	152%increased [8] 135% increases in e_{33} [6]	90% decreased [8] 90% decreases in C_{33} [6]	135%increased [8]			Reduced [7]	Decreased [6] Decreased [8]	
Mg-X	Increased	Decreased	Increased	Increased	Increased	Increased	Decreased	Increased
Li-V	Increased	Decreased	Increased				Decreased	
Li-Nb	Increased	Decreased	Increased				Decreased	
Li-ta	Increased	Decreased	Increased				Increased	
Li-Zr	Increased	Decreased	Increased				Decreased	

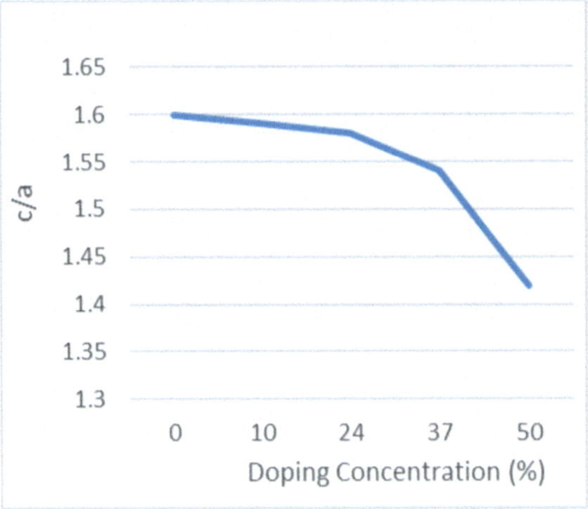

Fig. 2. C/A ratio vs doping concentration doped with Sc

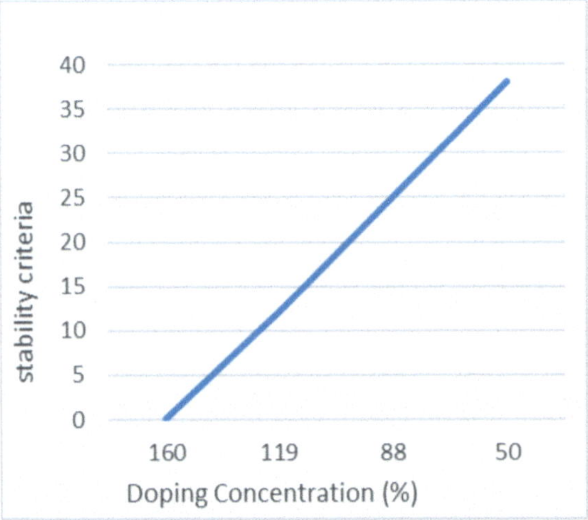

Fig. 3. Stability criterion vs doping concentration

elastic constants of the codoped or Sc-doped w-AlN should satisfy the stability criteria: C11 > 0, (C11 − C12) > 0, C44 > 0, (C11 + C12) C33 > 2C13.A large enhancement in piezo-response (d33) and electromechanical coupling ($\boldsymbol{K}_{33}^{2}$) has been discovered in Sc doped w-AlN (Figs. 4 and 5).

3.1 Scandium

Upon Sc doping, the d_{33} of pure AlN thin films has been enhanced by 500%,which was initially 5.5 pC N^{-1}, [1] and an increase of more than 5 times that of AlN for the Sc added AlN (ScAlN) films in [2]. Along with d_{33}, $K_{33}{}^{2}$ also enhanced in [7]. There is an

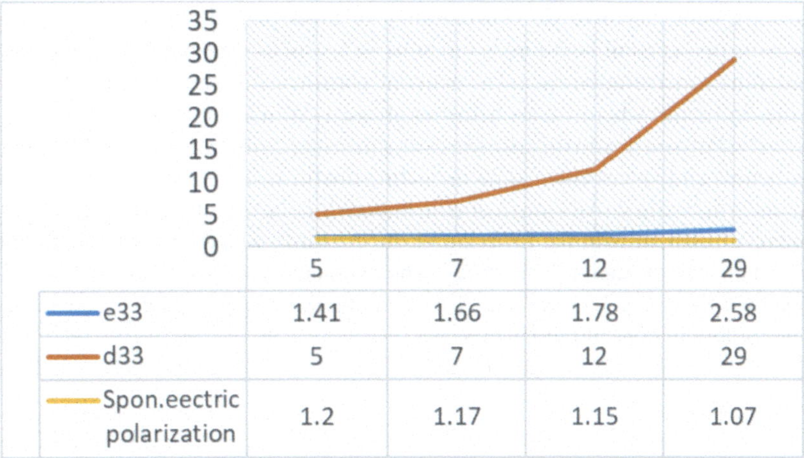

Fig. 4. Variation of lattice parameters with doping concentration doped with Sc

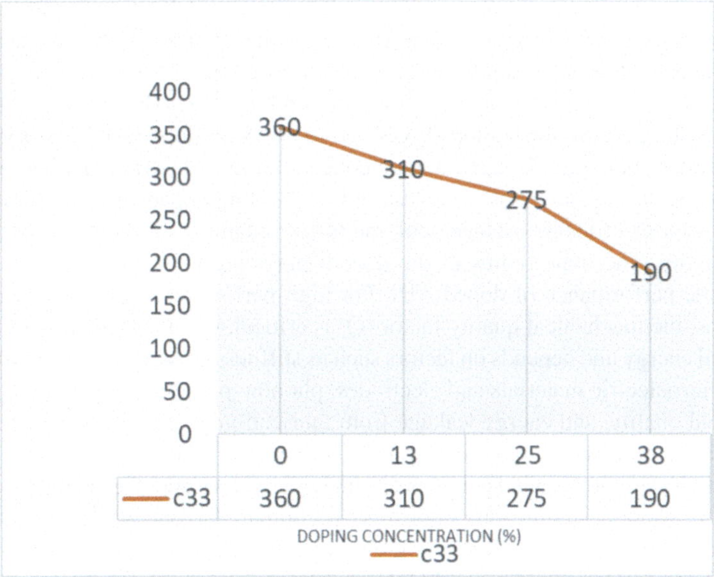

Fig. 5. Variation of C_{33} with doping concentration doped with Sc

increase of 130% in e_{33} for ScxAl1 − xN in [5] and an 87% decrease in C_{33} compared to pure AlN. Sc doping results in a 149% increase in d_{33}, which can be approximated as d33 ≈ e33 = C_{33} [9]. This behavior aligns with observations in ScxAl1 − xN alloys, where elastic softening and local structural instability occur simultaneously [10]. Theoretical studies on ScAlN films indicate that Sc atoms preferentially form hexagonal rather than tetrahedral bonds with nitrogen, reducing the c/a ratio. The energy difference between the wurtzite and hexagonal phases ($\mathbf{\Delta E = E_{hexagonal} - E_{wurtzite}}$) diminishes significantly with increased Sc concentration or co-doping. Both Sc and co-dopants decrease the polycrystalline Young's modulus (Y), which corresponds to a reduction in C_{33} as doping concentration rises. The material's brittleness or ductility can be evaluated using the B/G ratio, where values below or above 1.75 indicate brittleness or ductility, respectively. Poisson's ratio (v) also serves as a measure, with values below 0.26 indicating brittleness and values above 0.26 indicating ductility. Sc or co-doping levels above 25% transform brittle w-AlN into a ductile material. Polarization decreases with higher Sc doping, consistent with the lowering of the c/a ratio, which brings polar structures closer to nonpolar hexagonal structures. The estimated coercive field (Ec) for 37.5% Sc-doped www-AlN is 2.35 MV/cm. This Ec reduction aligns with the decrease in ΔE as Sc or co-doping concentrations increase [7].

3.2 Magnesium

The d_{33} of Mg-doped films decreases with increasing Mg concentration, reaching zero at Mg levels above 0.15 in MgxAl1 − xN. However, co-doping with Mg and Nb achieves the maximum d_{33} value when the Mg/Nb cathode power ratio is 1:2. All components of e_{33}, including clamped e_{33}, the Born effective charge Z_{33}, and the strain sensitivity du/dε of the internal parameter, improve when Al is substituted with Mg + X. The decrease in C_{33} and the enhancement in du/dε suggest elastic softening in Mg + X co-doped AlN. This softening is attributed to weaker covalent bonding in Mg–N and X–N pairs. The reduction in the c/a ratio for Sc-doped AlN and Mg + X co-doped AlN also arises from weaker covalent bonding. At equal dopant concentrations, Mg–Ti co-doping leads to a 135% increase in e_{33} and a 90% decrease in C_{33}. Adding dopants to AlN films generally alters their electrical resistance, with the films remaining insulating and exhibiting high resistivity. The ionic radius of the dopant plays a crucial role in enhancing the piezoelectric performance of doped AlN. For high-performance resonators with wide bandwidths, the mechanical quality factor (Q) is critical. Q reflects the loss of inserted mechanical energy and depends on factors such as stiffness constants, interface damping between piezoelectric materials and electrodes, phonon–phonon dissipation, piezoelectric material quality, and energy leakage from fabrication processes. As C_{33} decreases with doping, a corresponding reduction in Q is expected. The transverse piezoelectric coefficient (d_{31}) is another vital parameter, with larger values preferred for enhanced performance. Co-doping significantly improves d_{31} compared to pure w-AlN. For instance, d_{31} reaches −10.67 pC/N at 37.5% Sc doping. Higher d_{31} values are typically associated with improved figures of merit (FOM), with co-doping yielding substantially better FOM than pure w-AlN.

3.3 Li and Its Co Dopants

The crystal structure and properties of (LiyM1 − y) xAl1 − xN (where M = V, Nb, Ta, Zr were confirmed through first-principles calculations and experimental studies [11]. Theoretical analyses revealed that LiM doping increased d_{33} and lower C_{33}. These findings align with prior research for M = V, Nb, Ta. Additionally, LiNb doping reduces resistivity and induces polarity inversions.

4 Conclusion

The impact of single and co doping in AlN with various dopants are studied in this paper. Upon doping, variation in the lattice parameters along with piezo electric parameters can be observed from literature. The common dopants of AlN are Sc and Mg as separately and in co doped form along with various elements like Li, Nb, Va, Ta etc. A detailed crystal variation study is performed and the inferences are summarized in the table. From the observations it is clear that even tiny deviations can impact the sensitivity and accuracy of measurements, leading to erroneous outputs or decreased reliability. Doping enhances the material properties of AlN and extended its application in wide fields including sensing, and energy harvesting.

References

1. Anggraini, S.A., Uehara, M., Yamada, H., Akiyama, M.: Mg and Ti codoping effect on the piezoelectric response of aluminum nitride thin films. Scr. Mater. **159**, 9–12 (2018)
2. Wang, Qi & Lu, Yipeng & Fung, Stephanie & Jiang, Xiaoyue & Mishin, Sergey & Oshmyansky, Yury & Horsley, David. Scandium Doped Aluminum Nitride Based Piezoelectric Micromachined Ultrasound Transducers(2016)
3. Hirata, K., Mori, Y., Yamada, H., Uehara, M., Anggraini, S.A., Akiyama, M.: Significant enhancement of piezoelectric response in AlN by Yb addition. Materials. **14**, 309 (2021). https://doi.org/10.3390/ma14020309
4. Nguyen, H.H., Van Minh, L., Oguchi, H., Kuwano, H.: Development of highly efficient micro energy harvesters with MgHf-codoped AlN piezoelectric films. In: IEEE Micro Electro Mechanical Systems (MEMS), Belfast, pp. 222–225 (2018)
5. Hirata, K., Yamada, H., Uehara, M., Anggraini, S.A., Akiyama, M.: First-principles study of piezoelectric properties and bonding analysis in (Mg, X, Al)N solid solutions (X = Nb, Ti, Zr, Hf). ACS Omega. **4**(12), 15081–15086 (2019)
6. Iwazaki, Y., Yokoyama, T., Nishihara, T., Ueda, M.: Highly enhanced piezoelectric property of co-doped AlN. Jpn Soc. Appl. Phys. Express. **8**(6), 061501 (2015)
7. Olsson, R.H., Tang, Z., D'Agati, M.: Doping of aluminum nitride and the impact on thin film piezoelectric and ferroelectric device performance. In: 2020 IEEE Custom Integrated Circuits Conference (CICC), pp. 1–6 (2020)
8. Yokoyama, T., Iwazaki, Y., Onda, Y., Nishihara, T., Ueda, M.: Highly piezoelectric co-doped AlN thin films for bulk acoustic wave resonators. In: 2013 IEEE International Ultrasonics Symposium (IUS), Prague, Czech Republic, pp. 1382–1385 (2013)
9. Wang, C.Y., Vaicaitis, R.: Active control of vibrations and noise of double wall cylindrical shells. J. Sound Vib. **216**(5), 865–888 (1998)
10. Iwazaki, Y., Yokoyama, T., Nishihara, T., Ueda, M.: Highly enhanced piezoelectric property of co-doped AlN. Appl. Phys. Express. **8**, 061501 (2015). https://doi.org/10.7567/APEX.8.061501
11. Terada, T., Kimura, J., Sasaki, T., Umeda, Y., Inoue, Y.: Piezoelectric properties of LiM-doped (M = V, Nb, ta, or Zr) AlN thin films. Jpn. J. Appl. Phys. **60**, SFFB08 (2020)
12. Martínez-Cisneros, E., Velosa-Moncada, L.A., Del Angel-Arroyo, J.A., Aguilera-Cortés, L.A., Cerón-Álvarez, C.A., Herrera-May, A.L.: Electromechanical modeling of MEMS-based piezoelectric energy harvesting devices for applications in domestic washing machines. Energies. **13**, 617 (2020)

Cellulose Based Biosensors for Food Quality Evaluation and Health Monitoring

Gaziza Zhussipnazarova[1], Devika Padmakumar[2], Deepa Thomas[3], Akmaral Darmenbayeva[1], and R. Reshmy[4](✉)

[1] M.Kh. Dulaty Taraz Uuniversity, Taraz 080000, Kazakhstan
[2] Department of Food Science and Nutrition, Amrita Vishwa Vidyapeetham Amritanagar, Ettimadai 641112, Tamil Nadu, India
[3] Post Graduate and Research Department of Chemistry, Bishop Moore College, Mavelikara 690110, Kerala, India
[4] Department of Science and Humanities, Providence College of Engineering, Chengannur 689122, Kerala, India
reshmypkumar@gmail.com

Abstract. Biosensors are emerging nowadays because of their significance in various biotechnological sectors. These play a crucial role in early diagnosis of diseases and to provide a regular monitoring of the progress of the status of curing. Advantages of cellulose biosensors include biocompatibility, ease of immobilization, quick response, and precision. Many cellulose derivatives, including carboxymethyl cellulose, cellulose acetate, cellulose nitrate, hydroxyethyl cellulose, methylcellulose, and hydroxy methylcellulose have been applied in sensing applications.

Keywords: Biosensor · Cellulose acetate · Hydroxyethyl cellulose · Biocompatible

1 Introduction

A biosensor is an analytical tool for monitoring chemical or biological reactions because it generates signals that correspond to the quantity of particular analytes in the reaction. It is composed of a biological detecting element such as DNA, proteins, enzymes, antibodies, or carbohydrates, and a physical transducer such as an optical or electrochemical component arranged in close proximity to each other. Biosensors are essential to healthcare because they can quickly and accurately detect biological molecules, which help with early diagnosis, illness monitoring, and individualised treatment.

The utilization of cellulose as a potential substitute for biosensors has been made possible by the latest discoveries about cellulose-based smart materials. Hybrid materials based on cellulose have been used to fabricate sensors for a variety of uses, including humidity sensing, physical sensing such as proximity, strain and pressure sensors and biochemical sensing [1]. The cellulose-based biosensors exhibit several favorable attributes, including high sensitivity, quick response, precision, and affordability. As a

R. Rajasekharan et al. (Eds.): MatZone 2024, SPPHY 316, pp. 236–251, 2025.
https://doi.org/10.1007/978-981-96-6541-9_22

result, they are important for applications related to food processing safety, medical diagnostic devices, forensic science, and environmental monitoring. Because cellulose is non-toxic, biocompatible, and biodegradable, it is a desirable material for use in biosensors [2]. Natural cellulose possesses a variety of characteristics, including chirality, hydrophilicity, and the ability to undergo extensive chemical modifications in order to produce a wide range of cellulose derivatives. It may also produce semi-crystalline fibre morphologies. These qualities have led to a significant growth in interest in cellulose and the promotion of multidisciplinary research on materials and products based on cellulose [3]. This article deals with the emerging standpoints of cellulose derivatives as biosensors for optical and electrochemical applications.

2 Cellulose Derivatives for Biosensing Applications

A variety of cellulose derivatives have been used in sensing applications, such as cellulose acetate (CA), methylcellulose (MC), cellulose nitrate (CN), carboxy methyl cellulose (CMC), hydroxyethyl cellulose (HEC), and hydroxy methylcellulose (HMC). This can be achieved by adding different functional substitutions to the cellulose polymer chains or by changing the inherent hydrogen bonding network. Many new functionalized cellulose-based polymers have been introduced for use in biosensors [4].

CA possesses hydrolytic stability and high mechanical qualities. In addition, it is insoluble in water. Its natural origins are related to its biodegradability, biocompatibility, non-toxicity, and low cost. Because it is easily obtainable and simple to manufacture, CA and its blends with other polymers are widely used in medical diagnostic kits. For use in biosensors and bioanalysis, composite materials with a CA matrix and different kinds of functional fillers have been developed. Yu and coworkers, developed an innovative optical fibre biosensor for glucose detection by coating the end face of a 600 μm optical fibre with a sensitive film composed of carbon quantum dots and glucose oxidase that was distributed inCA [5]. A sensing probe for the electrochemical recognition of several significant catecholamines such as epinephrine and dopamine, was created by adding multiwalled carbon nanotubes to CA and then drop-casting the resulting aqueous dispersion onto a glassy carbon electrode [6].

For biomedical applications, CN with less than 12.6% nitrogen is thought to be suitable because to its strong physico-chemical stability and biocompatibility. The nitro groups present in polymer and the amino groups in the protein structure interact non-covalently, which is why CN membranes have been shown to be quite successful at retaining proteins. Paper-based biosensors have extensively utilized CN membrane as a popular and appropriate substrate material because of its porous structure, excellent absorption of biomolecules, remarkable hydrophilicity, and superior flexibility [7]. High fluorescence carbon dots were discovered to be best synthesized using CMC. Metal ion detection was carried out using a multifunctional structure based on carbon dots adorned CMC and hydroxyapatite [8].

The open long chains and loose network structure of MC are advantageous for macromolecular compounds' penetration and adsorption. There are some hydrophilic zones with low substituting degrees and some hydrophobic zones with high substituting degrees in the molecular structure of MC. These characteristics allow MC to build soft, translucent, and compact thin-films over a broad pH range. Organophosphorus pesticide detection was made possible with the development of a potentiometric enzymatic membrane biosensor based on MC. Urine sample uric acid was measured using a modified glassy carbon electrode coated with MC/graphene oxide/iron oxide nano hydrogel [9].

Superior biocompatibility, low toxicity and non-immunogenicity nature of HEC made it suitable for variety of biotechnological, biophysical, and industrial applications. HEC possesses polymeric network with the capacity to take in and hold a large amount of water inside its structure without dissolving. It is thought to be a hydrogel-like substance with characteristics of both a liquid and a solid. As a result, solvent or low molar mass solute can permeate the gel's bulk, and the network's elastic strength preserves the gel's structure while permitting it to grow in size. These properties suggest that proteins and enzymes could be included into a biosensor matrix using HEC [10, 11].

2.1 Optical Biosensors

Optical biosensors can be very advantageous in the areas of biotechnology, environmental research, disease detection, and medical applications because of their speed, accuracy, and high sensitivity in measurements. Figure 1 represents the components of an optical biosensor. Table 1 depicts the working principle, advantages and applications of cellulose based optical biosensors.

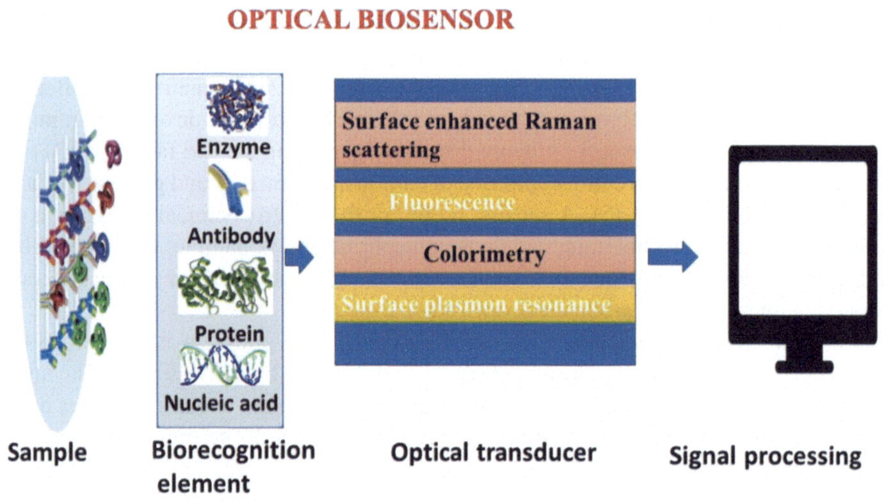

Fig. 1. Components of optical biosensor

2.2 Surface Enhanced Raman Scattering (SERS)

As a result of its exceptional sensitivity and selectivity, SERS is currently regarded as one of the most potent instruments for bio detection. Specifically, cellulose can function as a host of nanoparticles for SERS applications and as a reducing agent for the

Table 1. General optical techniques: Principle and applications as biosensor

Optical technique	Working principle	Signal processing	Advantages	Application/detection	Limit of detection	Reference
Surface enhanced Raman scattering	Emission wherein incident laser energy is inelastically scattered, producing spectral peaks that are frequency-shifted from the incident energy	Raman scattering signal	Versatility, sensitivity, selectivity, and biocompatibility	Tetracycline	0.1 ppm	[12]
				2-naphthalenethiol	1 ppb	[13]
				Rhodamine 6G	10 pM	[14]
				Thiram	10^{-9} M	[15]
				Melamine	1 ppm	[16]
Fluorimetry	Fluorophores, or molecules with fluorescent labels, absorb electromagnetic radiation and cause the fluorescence phenomena	Fluorescence intensity	Simple operation, quick reaction time, ease of use, numerous analysis capabilities, good sensitivity, and excellent selectivity	Listeria monocytogene	10^2 CFU/mL	[17]
				Cd^{2+}	10.70 μg/L	[18]
				Troponin I	1.28–1.40 ng/mL	[19]
				Glucose	25.79 nM	[20]
				DNA	0.4 fM	[21]
Colorimetry	Finds a colour shift related to a particular chemical reaction that occurs between sensing materials and an analyte	Absorbance intensity	Low cost, ease of operation, and easy readout, ability to read distinct color changes with the naked eye	Glucose	25 μM	[22]
				Cd^{2+}	0.01 mg/L	[23]
				Fe^{2+}	50 ppb	[24]
				Kanamycin	2.5 nM	[25]
				Phenol	0.082 μM	[26]
Surface plasmon resonance	The light absorbs its energy and converts it into a reflected light	Changes in refractive index	Excellent reliability, selectivity and reproducibility	Hg^{2+}	2 ppb	[27]
				Glucose	0.005 μM	[28]
				Ochratoxin A	3.8 ng/mL	[28]
				Thiram	0.5 ppm	[29]
				Zn^{2+}	0.01 ppm	[30]

preparation of plasmonic nanoparticles. In fact, there have been reports of strong surface plasmon resonances produced by nanostructured metals in the visible and near-infrared spectrums, including Au, Ag, and Cu. These nanoparticles cause local surface plasmon resonance, which in turn causes the creation of SERS hotspots—highly concentrated, locally intense areas where the adsorbed molecule signal is amplified by factors greater than 10^6. Crucially, the noble metal nanoparticle's composition, size, shape, and interparticle spacing all have a significant impact on signal enhancement. For these reasons, research has been done on increasing the SERS hotspots by inserting metal nanoparticles into a porous and flexible substrate [31, 32].

Fibrillated celluloses, are good candidates to inhibit the agglomeration of metal nanoparticles as they may form a flexible 3D network with customized porous nature and shape [32]. For instance, it has been investigated that Ag/Au nanoparticles may be deposited on paper or cellophane to provide convenient SERS substrates that can be used to detect chemical molecules such acetamiprid, benzoic acid, 4-aminothiophenol and acyclovir [33–36]. The capacity of cellulose derivatives, such as CA [37], MC [38], CN [39], HEC and HMC [4, 15] to dissolve in water or organic solvents or and then form hydrogel or membranes, has led to their use in the creation of SERS substrates. The first report on the application of bacterial cellulose as a SERS substrate was for the creation of a sensor to measure thiosalicylic acid and 2,2-dithiodipyridine [31] demonstrating the bacterial cellulose nanofibrillar structure to be an efficient matrix for the nanoparticles' nucleation and growth. Flexible and high sensitivity nanocellulose SERS substrates were fabricated and applied for the detection of pesticides directly on apple pie. This showed that the bacterial cellulose nanofibrillar framework was an effective matrix for the nucleation and growth of the nanoparticles. For the purpose of directly detecting pesticides on apple peels, flexible and highly sensitive nanocellulose SERS substrates were created and applied [40].

2.3 Fluorimetry

Photodiodes are used to detect light coming from a fluorescence process in fluorescence optical biosensors. Fluorescent biosensor design has recently been drawn to 2D materials because of their remarkable electrical, optical, mechanical, and other features [41]. The high hydroxyl content of cellulose makes it a perfect matrix for fluorescence-based sensing since it can integrate fluorescent groups with high yields. Biomolecules can be detected with a unique label-free, inexpensive, and color-indicating cholesteric liquid crystal. The interaction between the fluorescent molecule and the analyte greatly influences the sensitivity of fluorescence sensors. Analytes and fluorophores interact better with cellulose because of its excellent processability and huge surface area as compared to other polymers. Davis et al. provide compelling evidence of the strategic application of cellulose properties, such as its fibrillar structure, capacity to create a 3D network, functionalization/defunctionalization ability, and processability, to enhance sensor performances [42]. Bioactive, fluorescent nanocrystalline cellulose attached with peptide and enzyme were utilized for biosensing applications. Terpyridine-modified pyrylene cellulose is a self-assembling supramolecular molecule that exhibits a great affinity for transition metals and may find application in biosensing [43]. Numerous fluorescent dyes combined with cellulose have been studied. Duong et al. fabricated a ratiometric fluorescent urea biosensors based on ethyl cellulose and oxazine 170 perchlorate [44]. In order to construct ratiometric pH-sensing nanoparticles, cellulose nanocrystals were dual fluorescently labelled using a simple one-pot process. The resulting pH sensor could sense pH in the range 3.5 to 8.0 by monitoring the fluorescence emission [45]. The majority of the structural and morphological characteristics of "real" cellulose materials are present in well-characterized dispersions of cellulose microfibrils from bacteria or algae, making them more dependable substrates for the characterization of cellulases. When it came to the identification and characterization of minute amounts of cellulases, fluorescent cellulose microfibrils were employed for cellulase activity detection. Cellulase enzymes were detected with high precision when fibrillated bacterial cellulose was coupled with 5-(4,6-dichlorotriazinyl) amino fluorescein [46]. Optical biosensors for the recognition of methyl viologen and cytochrome c have been developed using electrospun CA in combination with hydrolyzed poly[2-(3-thienyl) ethanol butoxy carbonyl methyl urethane] [47].

Human neutrophil elastase (HNE), a biomarker for the halted inflammatory state of chronic wounds, can be detected using peptide-cellulose fluorescence sensors, which are cellulose and nanocellulose matrices' bioactivity and transducer surface characteristics [48]. It should be emphasized that cellulose can actively participate in the process of recognition in addition to being able to generate functionalized films suitable as transducer surfaces. For instance, a chiral fluorescence sensor for aromatic nitro compounds with central and axial chirality has been designed using cellulose's inherent chirality [49]. It is a well-known fact that cellulose derivatives including benzoate and phenylcarbamate exhibit superior chiral selectivity for racemates with varying chirality, including amino acids and basic or acidic aromatic chemicals. In order to create a chiral fluorescence sensor that can distinguish between different forms of chirality in aromatic nitro compounds, a cellulose derivative containing conjugated terthienyl pendants as fluorescent signaling units was devised [50].

2.4 Colorimetry

Since optical visual sensors make it possible to quickly identify analytes in screening applications, they are crucial sensing systems. Crucially, in these systems, color changes that are visible to the naked eye are used to enable signal transduction. Cellulose is a great substrate for chromophore anchoring as well as a platform for the embedding of nanoparticles and biomolecules. For instance, red cabbage pigment was anchored into CNF [51] and CMC [52], alone or in composite systems, to enable pH detection. When combined with the red cabbage pigment, electrospun non-woven cellulose fibers displayed distinctly different colors throughout a broad pH range. Moreover, optical sensing has made use of the aggregation or etching of implanted metal nanoparticles into cellulose scaffolds. Utilizing cellulose as a scaffold in these systems can improve sensor performance by adjusting the nanoparticles' size, composition, shape and interparticle distance. A sizable surface area that promotes metal nanoparticles–analyte interaction is presented by cellulose, together with a 3D network that has adaptable porosity, a high concentration of hydroxyl groups or other functional groups that can be employed as reducing and/or anchoring components for metal nanoparticles synthesis and stabilization [1].

Ag nanoparticle- bacterial cellulose and Au nanoparticle- bacterial cellulose composites were utilized for the sensing of methimazole, thiourea or cyanide [53]. High sensitivity and selectivity were demonstrated by a colorimetric sensor based on an Ag nanoparticle-carbon nanocellulose (CNC) composite for the detection of H_2O_2 in matrices of milk and water [54].

It has recently been revealed that glucose can be detected colorimetrically by using CMC as a scaffold to carry out a nanozyme–enzyme cascade reaction. The catalytic oxidation of glucose by glucose oxidase into H_2O_2 and the subsequent catalytic reduction of H_2O_2 by Prussian blue nanoparticles in the presence of 2,2′-azinobis(3-ethylbenzothiazolinesulfonic acid) diammonium salt were the sensing strategies employed in the developed biosensor [55].

Paper made of cellulose has unique geometry and a combination of hydrophilic and hydrophobic properties, making it a potential material for supporting catalysts and enzymes. The amount of phenylalanine in a urine sample is measured using a paper-based colorimetric biosensor that uses phenylalanine ammonia lyase. When CNC films are exposed to high relative humidity settings, they absorb water and change color, which is why they have been utilized for humidity sensing [56]. By assisting the close packing of CNCs with 1-butyl-3-methylimidazolium, an ionic liquid, CNC films with various color patterns were produced. Amine groups were added to the films so that formaldehyde and propanal gases could be detected clearly without interruption from other molecules. These colorimetric sensors provide on-site monitoring capabilities together with quick, low-cost volatile organic compound detection [57].

2.5 Surface Plasmon Resonance (SPR)

Label-free recognition of the analyte-receptor binding interaction is made possible by SPR, a surface spectroscopic method. In addition to being economical, they are sensitive and precise. The free electrons at the metal-dielectric contact oscillate in their surface charge density, which is what causes surface plasmon resonance. Evanescent waves can be produced by efficiently excitation it with an incident light beam. By matching the target molecule's absorption curve to the capture molecule, we may utilize SPR sensors to measure the concentrations of the target molecule [58].

It has been shown that cellulose is a remarkable material for the creation of SPR immunosensor interfaces. The ability of SPR sensors to measure direct binding without labelling is an exceptional feature. Cellulose may generate films or membranes with customized porosity due to its wide processability and simplicity of functionalization. This is particularly essential for SPR analysis since it allows for the highly sensitive monitoring of the binding of the analyte on the multilayers immobilised antibody.

Biosensors that use cellulose derivatives have been created to detect human immunoglobulin G (IgG) by employing immobilized anti-human IgG that is derived from non-homogenized milk of goats. Using a ready-to-use measurement chip put on an SPR apparatus, a raw milk sample can be subjected to an on-site examination utilizing a patterned cellulose membrane. By employing a simple technique that involved inserting a raw sample solution into the entrance hole with a mechanical pipette, sample flow and antigen identification were accomplished. Using human IgG as a model antigen that was ensured to be absent from a model raw sample of non-homogeneous milk, the detection performance was examined in this work [59].

2.6 Electrochemical Biosensors

Electrochemical biosensors have been used for an extended period in a variety of fields, not just healthcare. The dielectric characteristics, magnitude, shape, and charge distribution changes that occur during the formation of the antibody-antigen combination on the electrode surface are monitored by these biosensors. Many biological targets, such as proteins, cancer biomarkers, nucleic acids, etc., can be found using electrochemical biosensors. In electrochemical sensing, the signal is generated as a result of direct or indirect electrochemical reduction or oxidation of the substance to be detected. Electrochemical biosensors are stable, high sensitivity, and also low cost [60]. Despite these advantages, they presented difficulties in planting biochemicals. In order to eliminate such shortcomings, great attention was paid to the search and preparation of alternative materials. Accordingly, it was found that cellulose and its derivatives have the ability to immobilize biological molecules and show high biocompatibility [61, 62].

Currently, several important research works have been published on the synthesis of composite materials based on cellulose and its derivatives, their properties and applications. For example, Shi Z. et al. [63] developed a method of making electrochemical biosensors based on cellulose with high conductivity and achievements in the use of conductive materials. One of the first electrochemical biosensors was designed to determine glucose. The sensor device was formulated in 1962 by Clark and Lyon [64]. The first -generation biosensors were based on the determination of glucose by reaction with oxygen, catalyzed by a natural substrate, and to determine the excreted hydrogen peroxide. Such a simple principle was mainly used in the miniaturization of sensors. In general, glucose oxidase was sandwiched between an internally permeable cellulose acetate membrane and an outer biocompatible polycarbonate membrane that limited diffusion. In recent years, many attempts have been made to create a sensor modified with nanostructures of metals and their derivatives [65]. It is assumed that metal nanostructures and carbon nanotubes [66] can serve as reaction catalysts, both through direct electron transfer and using a mediator, which leads to improved analytical characteristics of the biosensor. Any conductive materials on which nanostructures can be modified due to specific interactions are used as substrates.

In the research [67], a nanocosposite was obtained by grafting graphene oxide onto cellulose. Morphological structure, composition, elasticity, solubility and sensitivity of graphene/cellulose nanocomposite were studied using modern physicochemical methods. In the obtained nanocomposite, it was found that the particles do not form agglomeration, and the morphological surface is uniformly dispersed, which means that the connection between graphene oxide and cellulose is good. The improved Young's modulus to 3.18 GPa suggested that the resulting biosensor could be used in many electronic applications.

Tun W.S.T. et al. [68] prepared an electrochemical biosensor based on CNF, graphene oxide and acetylcholinesterase. The preparation scheme of the cellulose-based biosensor is presented in Fig. 2. Cellulose nanofibers are synthesized from sugarcane bagasse. Cellulose nanofibers were studied by TEM and the average size was between 83–85 nm.

The obtained biosensor is characterized by a sizable surface area, high electrical conductivity and electron transfer rate due to graphite oxide. Biosensors based on cellulose nanofibers showed high selectivity and stability for 21 days in the detection of pesticides in water and fruit juice. Recent studies show that there is a growing demand for the use of nanocellulose as an electrochemical biosensor [69, 70]. In this regard, it is worth mentioning bacterial nanocellulose, whose frame is stable and strong. This type of nanocellulose exhibits the flexibility and miniaturization properties typical of nanoscale composites and is increasingly being used in the field of bioelectronics.

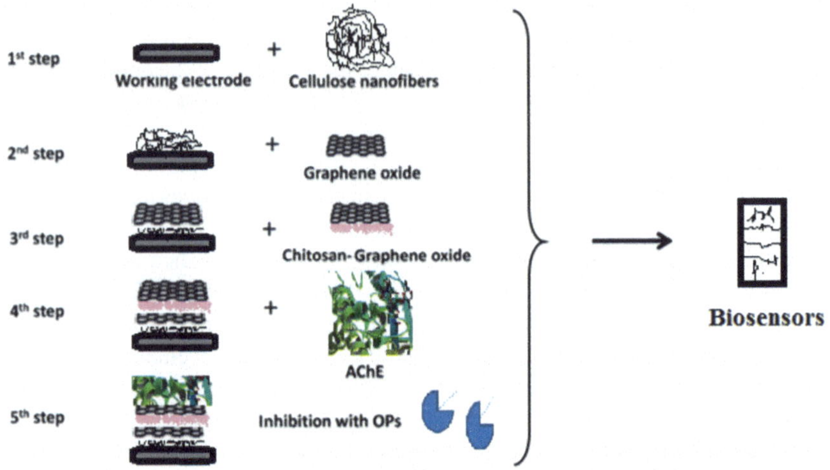

Fig. 2. Schematic diagram of the cellulose-based biosensor preparation

Electrochemical immunosensors have recently been formulated by scientists and are used to identify harmful bacteria [71–73]. The generation of the antigen-antibody complex in electrochemical immunosensors results in the conversion of the reaction product into an electrical signal, such as an electric current, a potential difference, or a change in resistivity [74].

To diagnose infections, potentiometric immunosensors have been developed based on the LAPS method (light-addressable potentiometric sensors) [75]. This method is a light-controlled immunosensor, a sandwich-type immunoassay and the test material is pre-filtered through a nitrocellulose membrane. Bacteria are detected by changes in pH on the membrane. The samples studied are *E. coli*, staphylococcal enterotoxin, and the causative agents of *brucellosis melitensis* [76]. Recently, to create the most efficient electrodes, nano-sized materials have begun to be used, which have many features important for the operation of the sensor; in particular, such materials can significantly increase the analytical signal due to electrocatalysis. The stability and intensity of the analytical signal of electrochemical sensors is enhanced by such promising nanomaterials as Au, Pt and Pd nanoparticles.

Liu C., et al. [77] obtained nanocellulose electrochemical immunosensors with gold nanoparticles. The immunosensors were investigated using TEM and XRD. Cellulose fibers have a mean diameter of 15 nm and a length of ~300 nm. Gold nanoparticles range in size from 5 to 10 nm. In addition, the nanocellulose-based electrochemical immunosensor with Au composite showed high sensitivity in the detection of *avian leukosis virus*. In addition, the selectivity and performance stability of the electrochemical immunosensor was high. Wang W. et al. [78] obtained a nanosensor fixed to the surface of a carbon electrode altered with nanocomposites of gold and bacterial cellulose. The supported heme proteins showed high electrocatalytic activity in the process of hydrogen peroxide reduction.

Lucas F. de Lima [79] developed an environmentally friendly, highly selective, accurate, low-cost, and simple diagnostic test for SARS-CoV-2 detection based on bacterial cellulose and a carbon electrode using green technology. Cyclic voltammetry and electrochemical impedance spectroscopy were used to evaluate the electrochemical behavior of the developed tests. First, the conductivity of the bare carbon screen-printed electrode was examined, which showed a high peak current (ip) and low resistance (RCT). After modification with bacterial cellulose, the ip value decreased to 7.9 mA, and RCT, on the contrary, increased to 1.466 U, which is consistent with high electrical conductivity. The obtained diagnostic test was studied for SARS-CoV-2 and showed high selectivity.

In research [80], to improve the analytical parameters of a biosensor based on bacterial cellulose, it was modified with a highly conductive polymer—poly(3,4-ethylenedioxythiophene): polystyrene sulfonate (PEDOT:PSS). To study the effect of a highly conductive polymer on the surface of graphite electrodes, 2 prototypes of a biosensor for determining glucose were formed. Bacterial cells were immobilized on the surface of working electrodes in a matrix of bacterial cellulose, modified and not modified with a highly conductive polymer. It was found that the highly conductive polymer has a positive effect on the analytical performance of the biosensor based on the printed electrode, increasing sensitivity to glucose by 4.7 times.

Gomes N.O. et al. developed an electrochemical biosensor from bacterial cellulose and proved that cellulose is a biocompatible material with high mechanical properties [81]. Rather than immobilising the biomolecules on the electrode surface, the investigators employed an enzymatic immobilisation approach that involves the direct covalent binding of the biomolecules to the bacterial cellulose substrate.

3 Conclusions

In conclusion, cellulose and its derivatives due to its exceptional characteristics such as availability, sensitivity, stability, biocompatibility, high permeability and accuracy are an indispensable material in the preparation of biosensors by immobilizing molecules and have high prospects for future use. Since cellulose exhibits minimal interference with SERS signals, good stabilization capabilities, and the capacity to regulate the growth of metal nanoparticles of different sizes and shapes when functionalized, it has been shown to be an ideal platform for optical biosensors. The fabricated cellulose based biosensors demonstrated excellent amperometric response, leading to excellent mechanical stability and biocompatibility. Cellulose based biosensors have been found to be suitable for describing human physiological changes and sensing food quality.

References

1. Chang, M., Song, T., Liu, X., Lin, Q., He, B., Ren, J.: Cellulose-based biosensor for biomolecules detection in medical diagnosis: a mini-review. Curr. Med. Chem. **27**, 4593–4612 (2020). https://doi.org/10.2174/0929867327666200221145543
2. Chandel, N., Jain, K., Jain, A., Raj, T., Patel, A.K., Yang, Y.H., Bhatia, S.K.: The versatile world of cellulose-based materials in healthcare: from production to applications. Ind. Crop Prod. **201**, 116929 (2023). https://doi.org/10.1016/J.INDCROP.2023.116929
3. Akki, A.J., Jain, P., Kulkarni, R., Rao Badkillaya, R., Kulkarni, R.V., Zameer, F., Anjanapura, R., Aminabhavi, T.M.: Microbial biotechnology alchemy: transforming bacterial cellulose into sensing disease—a review. Sens. Int. **5**, 100277 (2024). https://doi.org/10.1016/j.sintl.2023.100277
4. Gabrielli, V., Frasconi, M.: Cellulose-based functional materials for sensing. Chemosensors. **10**, 352 (2022). https://doi.org/10.3390/CHEMOSENSORS10090352
5. Yu, S., Ding, L., Lin, H., Wu, W., Huang, J.: A novel optical fiber glucose biosensor based on carbon quantum dots-glucose oxidase/cellulose acetate complex sensitive film. Biosens. Bioelectron. **146**, 111760 (2019). https://doi.org/10.1016/J.BIOS.2019.111760
6. Casella, I.G., Gioia, D., Rutilo, M.: A multi-walled carbon nanotubes/cellulose acetate composite electrode (MWCNT/CA) as sensing probe for the amperometric determination of some catecholamines. Sens. Actuators B. **255**, 3533–3540 (2018). https://doi.org/10.1016/J.SNB.2017.09.188
7. Tang, R., Xie, M.Y., Li, M., Cao, L., Feng, S., Li, Z., Xu, F.: Nitrocellulose membrane for paper-based biosensor. Appl. Mater. Today. **26**, 101305 (2022). https://doi.org/10.1016/J.APMT.2021.101305
8. Sarkar, C., Chowdhuri, A.R., Kumar, A., Laha, D., Garai, S., Chakraborty, J., Sahu, S.K.: One pot synthesis of carbon dots decorated carboxymethyl cellulose- hydroxyapatite nanocomposite for drug delivery, tissue engineering and Fe3+ ion sensing. Carbohydr. Polym. **181**, 710–718 (2018). https://doi.org/10.1016/J.CARBPOL.2017.11.091
9. Sohouli, E., Khosrowshahi, E.M., Radi, P., Naghian, E., Rahimi-Nasrabadi, M., Ahmadi, F.: Electrochemical sensor based on modified methylcellulose by graphene oxide and Fe3O4 nanoparticles: application in the analysis of uric acid content in urine. J. Electroanal. Chem. **877**, 114503 (2020). https://doi.org/10.1016/J.JELECHEM.2020.114503
10. Liu, X., Chen, T., Liu, L., Li, G.: Electrochemical characteristics of heme proteins in hydroxyethylcellulose film. Sens. Actuators B. **113**, 106–111 (2006). https://doi.org/10.1016/J.SNB.2005.02.029
11. Li, Y., Chen, C., Cui, G., Liu, L., Zhou, C., Wu, G.: Hydroxyethyl cellulose-based stretchable, antifreeze, ion-conductive hydrogel sensor. Eur. Polym. J. **202**, 112603 (2024). https://doi.org/10.1016/J.EURPOLYMJ.2023.112603
12. Marques, A., Veigas, B., Araújo, A., Pagará, B., Baptista, P.V., Águas, H., Martins, R., Fortunato, E.: Paper-based SERS platform for one-step screening of tetracycline in milk. Sci. Rep. **9**(1), 1–8 (2019). https://doi.org/10.1038/s41598-019-54380-y
13. Kim, D., Kim, J., Henzie, J., Ko, Y., Lim, H., Kwon, G., Na, J., Kim, H.J., Yamauchi, Y., You, J.: Mesoporous au films assembled on flexible cellulose nanopaper as high-performance SERS substrates. Chem. Eng. J. **419**, 129445 (2021). https://doi.org/10.1016/J.CEJ.2021.129445
14. Zhang, S., Xiong, R., Mahmoud, M.A., Quigley, E.N., Chang, H., El-Sayed, M., Tsukruk, V.V.: Dual-excitation nanocellulose plasmonic membranes for molecular and cellular SERS detection. ACS Appl. Mater. Interfaces. **10**, 18380–18389 (2018). https://doi.org/10.1021/ACSAMI.8B04817/SUPPL_FILE/AM8B04817_SI_001.PDF

15. Hu, B., Pu, H., Sun, D.W.: Multifunctional cellulose based substrates for SERS smart sensing: principles, applications and emerging trends for food safety detection. Trends Food Sci. Technol. **110**, 304–320 (2021). https://doi.org/10.1016/J.TIFS.2021.02.005
16. Xiong, Z., Chen, X., Liou, P., Lin, M.: Development of nanofibrillated cellulose coated with gold nanoparticles for measurement of melamine by SERS. Cellul. **24**, 2801–2811 (2017). https://doi.org/10.1007/S10570-017-1297-7/METRICS
17. Capo, A., D'Auria, S., Lacroix, M.: A fluorescence immunoassay for a rapid detection of Listeria monocytogenes on working surfaces. Sci. Rep. **10**, 1–12 (2020). https://doi.org/10.1038/s41598-020-77747-y
18. Tümay, S.O., Şanko, V., Demirbas, E., Şenocak, A.: Fluorescence determination of trace level of cadmium with pyrene modified nanocrystalline cellulose in food and soil samples. Food Chem. Toxicol. **146**, 111847 (2020). https://doi.org/10.1016/J.FCT.2020.111847
19. Natarajan, S., Jayaraj, J., Prazeres, D.M.F.: A cellulose paper-based fluorescent lateral flow immunoassay for the quantitative detection of cardiac Troponin I. Biosensors. **11**, 49 (2021). https://doi.org/10.3390/BIOS11020049
20. Kamel, S., Khattab, T.A.: Recent advances in cellulose-based biosensors for medical diagnosis. Biosensors. **10**, 67 (2020). https://doi.org/10.3390/BIOS10060067
21. Takalkar, S., Baryeh, K., Liu, G.: Bioelectronics, undefined 2017, Fluorescent Carbon Nanoparticle-Based Lateral Flow Biosensor for Ultrasensitive Detection of DNA. Elsevier, Amsterdam (n.d.)
22. Siripongpreda, T., Somchob, B., Rodthongkum, N., Hoven, V.P.: Bacterial cellulose-based re-swellable hydrogel: facile preparation and its potential application as colorimetric sensor of sweat pH and glucose. Carbohydr. Polym. **256**, 117506 (2021). https://doi.org/10.1016/J.CARBPOL.2020.117506
23. Jiang, X., Xia, J., Luo, X.: Simple, rapid, and highly sensitive colorimetric sensor strips from a porous cellulose membrane stained with Victoria blue B for efficient detection of trace Cd(II) in water. ACS Sustain. Chem. Eng. **8**, 5184–5191 (2020)
24. Krishnan, S., Syed, Z.Q.: Colorimetric visual sensors for point-of-needs testing. Sens. Actuators Rep. **4**, 100078 (2022)
25. Abedalwafa, M.A., Tang, Z., Qiao, Y., Mei, Q., Yang, G., Li, Y., Wang, L.: Anaptasensor strip-based colorimetric determination method for kanamycin using cellulose acetate nanofibers decorated DNA–gold nanoparticle bioconjugates. Microchim. Acta. **187**, 1–9 (2020)
26. Manan, F.A.A., Hong, W.W., Abdullah, J., Yusof, N.A., Ahmad, I.: Nanocrystalline cellulose decorated quantum dots based tyrosinase biosensor for phenol determination. Mater. Sci. Eng. C. **99**, 37–46 (2019)
27. Ramdzan, N.S.M., Fen, Y.W., Omar, N.A.S., Anas, N.A.A., Liew, J.Y.C., Daniyal, W.M.E.M.M., Hashim, H.S.: Detection of mercury ion using surface plasmon resonance spectroscopy based on nanocrystalline cellulose/poly(3,4-ethylenedioxythiophene) thin film. Measurement. **182**, 109728 (2021)
28. Rosddi, N.N.M., Fen, Y.W., Omar, N.A.S., Anas, N.A.A., Hashim, H.S., Ramdzan, N.S.M., Fauzi, N.I.M., Anuar, M.F., Daniyal, W.M.E.M.M.: Glucose detection by gold modified carboxyl-functionalized graphene quantum dots-based surface plasmon resonance. Optik (Stuttg). **239**, 166779 (2021)
29. Xiao, L., Feng, S., Hua, M.Z., Lu, X.: Rapid determination of thiram on apple using a flexible bacterial cellulose-based SERS substrate. Talanta. **254**, 124128 (2023). https://doi.org/10.1016/J.TALANTA.2022.124128
30. Daniyal, W.M.E.M.M., Fen, Y.W., Anas, N.A.A., Omar, N.A.S., Ramdzan, N.S.M., Nakajima, H., Mahdi, M.A.: Enhancing the sensitivity of a surface plasmon resonance-based optical sensor for zinc ion detection by the modification of a gold thin film. RSC Adv. **9**, 41729–41736 (2019)

31. Marques, P.A.A.P., Nogueira, H.I.S., Pinto, R.J.B., Neto, C.P., Trindade, T., Marques, P.A.A.P., Nogueira, H.I.S., Pinto, R.J.B., Neto, C.P., Trindade, T.: Silver-bacterial cellulosic sponges as active SERS substrates. J. Refract. Surg. **39**, 439–443 (2008). https://doi.org/10.1002/JRS.1853
32. Ogundare, S.A., van Zyl, W.E.: A review of cellulose-based substrates for SERS: fundamentals, design principles, applications. Cellul. **26**, 6489–6528 (2019). https://doi.org/10.1007/S10570-019-02580-0
33. Eskandari, V., Sahbafar, H., Karooby, E., Heris, M.H., Mehmandoust, S., Razmjoue, D., Hadi, A.: Surface-enhanced Raman scattering (SERS) filter paper substrates decorated with silver nanoparticles for the detection of molecular vibrations of acyclovir drug. Spectrochim. Acta A Mol. Biomol. Spectrosc. **298**, 122762 (2023). https://doi.org/10.1016/J.SAA.2023.122762
34. Wei, W., Huang, Q.: Preparation of cellophane-based substrate and its SERS performance on the detection of CV and acetamiprid. Spectrochim. Acta A Mol. Biomol. Spectrosc. **193**, 8–13 (2018). https://doi.org/10.1016/J.SAA.2017.11.062
35. Wei, H., Vikesland, P.J.: pH-triggered molecular alignment for reproducible SERS detection via an AuNP/Nanocellulose platform. Sci. Rep. **5**, 1–10 (2015). https://doi.org/10.1038/srep18131
36. Kang, Y., Kim, H.J., Lee, S.H., Noh, H.: Paper-based substrate for a surface-enhanced Raman spectroscopy biosensing platform—a silver/chitosan nanocomposite approach. Biosensors (Basel). **12**, 266 (2022). https://doi.org/10.3390/BIOS12050266/S1
37. Ke, Y., Meng, G., Huang, Z., Zhou, N.: Electrosprayed large-area membranes of ag-nanocubes embedded in cellulose acetate microspheres as homogeneous SERS substrates. J. Mater. Chem. C Mater. **5**, 1402–1408 (2017). https://doi.org/10.1039/C6TC04579K
38. Zhang, Q., Xu, G., Guo, N., Wang, T., Song, P., Xia, L.: In-situ synthesis of methyl cellulose film decorated with silver nanoparticles as a flexible surface-enhanced Raman substrate for the rapid detection of pesticide residues in fruits and vegetables. Materials. **14**, 5750 (2021). https://doi.org/10.3390/MA14195750/S1
39. Bishnoi, S.W., Lin, Y.J., Tibudan, M., Huang, Y., Nakaema, M., Swarup, V., Keiderling, T.A.: SERS biodetection using gold-silica nanoshells and nitrocellulose membranes. Anal. Chem. **83**, 4053–4060 (2011). https://doi.org/10.1021/AC103195E
40. Wu, J., Xi, J., Chen, H., Li, S., Zhang, L., Li, P., Wu, W.: Flexible 2D nanocellulose-based SERS substrate for pesticide residue detection. Carbohydr. Polym. **277**, 118890 (2022). https://doi.org/10.1016/J.CARBPOL.2021.118890
41. Devi, M.: Application of 2D nanomaterials as fluorescent biosensors. ACS Symp. Ser. **1353**, 117–141 (2020). https://doi.org/10.1021/BK-2020-1353.CH006
42. Davis, B.W., Niamnont, N., Hare, C.D., Sukwattanasinitt, M., Cheng, Q.: Nanofibers doped with dendritic fluorophores for protein detection. ACS Appl. Mater. Interfaces. **2**, 1798–1803 (2010)
43. Hassan, M.L., Moorefield, C.M., Elbatal, H.S., Newkome, G.R., Modarelli, D.A., Romano, N.C.: Fluorescent cellulose nanocrystals via supramolecular assembly of terpyridine-modified cellulose nanocrystals and terpyridine-modified perylene. Mater. Sci. Eng. B. **177**, 350–358 (2012). https://doi.org/10.1016/J.MSEB.2011.12.043
44. Duong, H., Talanta, J.R.: Development of a Ratiometric Fluorescent Urea Biosensor Based on the Urease Immobilized onto the Oxazine Perchlorate-Ethyl Cellulose Membrane. Elsevier, Amsterdam (2015)
45. Nielsen, L.J., Eyley, S., Thielemans, W., Aylott, J.W.: Dual fluorescent labelling of cellulose nanocrystals for pH sensing. Chem. Commun. **46**, 8929–8931 (2010). https://doi.org/10.1039/C0CC03470C

46. Helbert, W., Chanzy, H., Husum, T.L., Schülein, M., Ernst, S.: Fluorescent cellulose microfibrils as substrate for the detection of cellulase activity. Biomacromolecules. **4**, 481–487 (2003). https://doi.org/10.1021/BM020076I
47. Wang, X., Kim, Y.G., Drew, C., Ku, B.C., Kumar, J., Samuelson, L.A.: Electrostatic assembly of conjugated polymer thin layers on electrospun nanofibrous membranes for biosensors. Nano Lett. **4**, 331–334 (2004). https://doi.org/10.1021/NL034885Z/ASSET/IMAGES/MEDIUM/NL034885ZN00001.GIF
48. Fontenot, K.R., Edwards, J.V., Haldane, D., Graves, E., Citron, M.S., Prevost, N.T., French, A.D., Condon, B.D.: Human neutrophil elastase detection with fluorescent peptide sensors conjugated to cellulosic and nanocellulosic materials: part II, structure/function analysis. Cellul. **23**, 1297–1309 (2016) 10.1007
49. Ikai, T., Suzuki, D., Kojima, Y., Yun, C., Maeda, K., Kanoh, S.: Chiral fluorescent sensors based on cellulose derivatives bearing terthienyl pendants. Polym. Chem. **7**, 4793–4801 (2016). https://doi.org/10.1039/C6PY00967K
50. Ikai, T., Okamoto, Y.: Structure control of polysaccharide derivatives for efficient separation of enantiomers by chromatography. Chem. Rev. **109**, 6077–6101 (2009)
51. Devarayan, K., Kim, B.S.: Reversible and universal pH sensing cellulose nanofibers for health monitor. Sens. Actuators B. **209**, 281–286 (2015). https://doi.org/10.1016/J.SNB.2014.11.120
52. Liang, T., Sun, G., Cao, L., Li, J., Wang, L.: A pH and NH3 sensing intelligent film based on Artemisia sphaerocephalaKrasch. Gum and red cabbage anthocyanins anchored by carboxymethyl cellulose sodium added as a host complex. Food Hydrocoll. **87**, 858–868 (2019). https://doi.org/10.1016/J.FOODHYD.2018.08.028
53. Morales-Narváez, E., Golmohammadi, H., Naghdi, T., Yousefi, H., Kostiv, U., Horák, D., Pourreza, N., Merkoçi, A.: Nanopaper as an optical sensing platform. ACS Nano. **9**, 7296–7305 (2015). https://doi.org/10.1021/ACSNANO.5B03097/SUPPL_FILE/NN5B03097_SI_001.PDF
54. Teodoro, K.B.R., Migliorini, F.L., Christinelli, W.A., Correa, D.S.: Detection of hydrogen peroxide (H2O2) using a colorimetric sensor based on cellulose nanowhiskers and silver nanoparticles. Carbohydr. Polym. **212**, 235–241 (2019). https://doi.org/10.1016/J.CARBPOL.2019.02.053
55. Baretta, R., Gabrielli, V., Frasconi, M.: Nanozyme-cellulose hydrogel composites enabling Cascade catalysis for the colorimetric detection of glucose. ACS Appl. Nano Mater. **5**, 13845–13853 (2022). https://doi.org/10.1021/ACSANM.2C01609/ASSET/IMAGES/LARGE/AN2C01609_0004.JPEG
56. Giese, M., Blusch, L.K., Khan, M.K., Hamad, W.Y., Maclachlan, M.J.: Responsive mesoporous photonic cellulose films by supramolecular Cotemplating. Angew. Chem. Int. Ed. **53**, 8880–8884 (2014). https://doi.org/10.1002/ANIE.201402214
57. Song, W., Lee, J.K., Gong, M.S., Heo, K., Chung, W.J., Lee, B.Y.: Cellulose nanocrystal-based colored thin films for colorimetric detection of aldehyde gases. ACS Appl. Mater. Interfaces. **10**, 10353–10361 (2018). https://doi.org/10.1021/ACSAMI.7B19738/SUPPL_FILE/AM7B19738_SI_001.PDF
58. Daniyal, W.M.E.M.M., Fen, Y.W., Eddin, F.B.K., Abdullah, J., Mahdi, M.A.: Surface plasmon resonance assisted optical characterization of nickel ion solution and nanocrystalline cellulose-graphene oxide thin film for sensitivity enhancement analysis. Phys. B Condens. Matter. **646**, 414292 (2022)
59. Miura, T., Horiuchi, T., Iwasaki, Y., Seyama, M., Camou, S., Takahashi, J.I., Haga, T.: Patterned cellulose membrane for surface plasmon resonance measurement. Sens. Actuators B. **173**, 354–360 (2012). https://doi.org/10.1016/J.SNB.2012.07.019
60. Teodoro, K.B., Sanfelice, R.C., Migliorini, F.L., Pavinatto, A., Facure, M.H., Correa, D.S.: A review on the role and performance of cellulose nanomaterials in sensors. ACS Sens. **6**, 2473–2496 (2021)

61. Frey, W.M.: Electrospinning cellulose and cellulose derivatives. Polym. Rev. **48**, 378–391 (2008)
62. Klemm, D., Heublein, B., Fink, H.P., Bohn, A.: Cellulose: fascinating biopolymer and sustainable raw material. Angew. Chem. Int. Ed. **44**, 3358–3393 (2005)
63. Shi, Z., Phillips, G.O., Yang, G.: Nanocellulose electroconductive composites. Nanoscale. **5**, 3194–3201 (2013)
64. Clark, L.C., Wolf, R., Granger, D., Taylor, Z.: Continuous recording of blood oxygen tensions by polarography. J. Appl. Phys. **6**, 189–193 (1953)
65. Das, P., Das, M., Chinnadayyala, S.R., Singha, I.M., Goswami, P.: Recent advances on developing 3rd generation enzyme electrode for biosensor applications. Biosens. Bioelectron. **79**, 386–397 (2016)
66. Liu, Q., Lu, X., Li, J., Yao, X., Li, J.: Direct electrochemistry of glucose oxidase and electrochemical biosensing of glucose on quantum dots/carbon nanotubes electrodes. Biosens. Bioelectron. **22**, 3203–3209 (2007)
67. Kafy, A., et al.: Cellulose/graphene nanocomposite as multifunctional electronic and solvent sensor material. Mater. Lett. **159**, 20–23 (2015)
68. Tun, W.S.T., Saenchoopa, A., Daduang, S., Daduang, J., Kulchat, S., Patramanon, R.: Electrochemical biosensor based on cellulose nanofibers/graphene oxide and acetylcholinesterase for the detection of chlorpyrifos pesticide in water and fruit juice. RSC Adv. **13**, 9603–9614 (2023)
69. de Assis, S.C., et al.: Review of bacterial nanocellulose-based electrochemical biosensors: functionalization, challenges, and future perspectives. Biosensors. **13**, 142 (2023)
70. Deisingh, A.K., Thompson, M.: Biosensors for the detection of bacteria. Can. J. Microbiol. **50**, 69–77 (2004)
71. Salam, F., Tothill, I.E.: Detection of Salmonella typhimurium using an electrochemical immunosensor. Biosens. Bioelectron. **24**, 2630–2636 (2009)
72. Sharma, M.K., Rao, V.K., Agarwal, G.S.: Highly sensitive amperometric immunosensor for detection of *Plasmodium falciparum* histidine-rich protein 2 in serum of humans with Malaria: comparison with a commercial kit. J. Clin. Microbiol. **46**, 3759–3765 (2008)
73. Moina, C., Ybarr, G.: Fundamentals and applications of immunosensors. In: Chiu, H.L., Christopoulos, T.K. (eds.) Advances in Immunoassay Technology, pp. 65–80. InTech, Rijeka, Croatia (2012)
74. Yoshinobu, T., Miyamoto, K.I., Werner, C.F., Poghossian, A., Wagner, T., Schöning, M.J.: Light-addressable potentiometric sensors for quantitative spatial imaging of chemical species. Annu. Rev. Anal. Chem. **10**, 225–246 (2017)
75. Nikolelis, D.P., Nikoleli, G.-P.: Biosensors for Security and Bioterrorism Applications. Springer, Cham (2016)
76. Liu, C., Dong, J., Waterhouse, G.I., Cheng, Z., Ai, S.: Electrochemical immunosensor with nanocellulose-au composite assisted multiple signal amplification for detection of avian leukosis virus subgroup J. Biosens. Bioelectron. **101**, 110–115 (2017)
77. Wang, W., et al.: Amperometric hydrogen peroxide biosensor based on the immobilization of heme proteins on gold nanoparticles–bacteria cellulose nanofibers nanocomposite. Talanta. **84**, 71–77 (2011)
78. de Lima, L.F., et al.: A bacterial cellulose-based and low-cost electrochemical biosensor for ultrasensitive detection of SARS-CoV-2. Cell Rep. Phys. Sci. **4**, 101476 (2023)
79. Reshetilov, A.N., Kitova, A.E., Tarasov, S.E., Plekhanova, Y.V., Bykov, A.G., Sundramoorthy, A.K.: Highly conductive polymer PEDOT: PSS-application in biomedical and bioelectrochemical systems. RENSIT. **12**, 471–482 (2020)
80. Gomes, N.O., et al.: Bacterial cellulose-based electrochemical sensing platform: a smart material for miniaturized biosensors. Electrochim. Acta. **349**, 136341 (2020)

81. Chen, S., Wang, Y., Fei, B., Long, H., Wang, T., Zhang, T., Chen, L.: Development of a flexible and highly sensitive pressure sensor based on an aramid nanofiber-reinforced bacterial cellulose nanocomposite membrane. Chem. Eng. J. **430**, 131980 (2022)
82. Solin, K., Beaumont, M., Rosenfeldt, S., Orelma, H., Borghei, M., Bacher, M., Opietnik, M., Rojas, O.J., Solin, K., Beaumont, M., Borghei, M., Rojas, O.J., Bacher, M., Rosenfeldt, S., Orelma, H., Opietnik, M.: Self-assembly of soft cellulose nanospheres into colloidal gel layers with enhanced protein adsorption capability for next-generation immunoassays. Small. **16**, 2004702 (2020). https://doi.org/10.1002/SMLL.202004702

Potential Pharmacological and Therapeutic Components of *Ixora coccinea L.* for Value Added Biomaterials

Rekha Unni[1](✉), Tiju Joseph Mathew[2], G. Zhussipnazarova[3], and A. Darmenbayeva[3]

[1] Department of Chemistry, Research Centre of University of Kerala, Christian College, Chengannur 689122, Kerala, India
rekhaunni08@gmail.com
[2] Department of Physics, Research Centre of University of Kerala, Christian College, Chengannur 689122, Kerala, India
[3] Department of Chemistry, M.Kh. Dulaty Taraz University, Taraz, Kazakhstan

Abstract. *Ixora coccinea* Linn. (Rubiaceae) is a traditional evergreen shrub that withholds a priceless reservoir of bioactive compounds. It grows throughout most of India and Sri Lanka. The flowers, leaves, roots, and stem of the plant are utilized in Indian ethnomedicine, naturopathy, and various folk remedies to treat myriad ailments. Phytochemical analysis has revealed that this plant is composed of beneficial secondary metabolites namely Quercetin, Kaempferitin, lupeol, β-sitosterol, Stigmasterol, oleanolic acid, Epicatechin, sitosterol, ursolic acid, rutin, leucocyanadin, Ixorene, proanthocyanidins, anthocyanins, myricetin etc. which proves the medicinal activities of the plant. Nanocellulose extracted from *I. coccinea* L. has been shown to be extremely effective in wound healing due to the presence of these bioactive components. These findings support the bioconversion of the whole plant to value added biomaterials which aimed to explore its medicinal applications.

Keywords: *Ixora coccinea* L. · Phytochemicals · Biomedical · Nanocellulose · wound healing

1 Introduction

Ixora Coccinea L. is a shrub in the Rubiaceae family that is commonly identified as Jungle flame, flame of wood, or Jungle shrub. It is native to Sri Lanka and Asia. The name "Ixora" is believed to have originated from the Indo-Aryan term "ikvana," which means a Malaysian divine soul, or probably from the word "iswara" which refers to Lord Shiva. The species name "coccinea" refers to the scarlet colour of the flowers. It has historically been used for antimicrobial, hepatoprotective, anti-nociceptive, anti-inflammatory, antimitotic, antioxidant, and chemoprotective properties. A decoction of the plant roots is utilized to treat anorexia, nausea and hiccups. Powered plant roots are applicable to treat wounds and ulcers. A root decoction is exploited for urine clarification

R. Rajasekharan et al. (Eds.): MatZone 2024, SPPHY 316, pp. 252–265, 2025.
https://doi.org/10.1007/978-981-96-6541-9_23

in Indo-China, and fresh leaves and stems are poulticed for eczema, contusions, boils and sprains. Modern researchers examined this plant according to ethnobotanical study and discovered numerous phytochemicals that provide the foundation for its therapeutic properties.

Different plants have different amounts of active chemicals extracted from flowers, leaves, roots, and stems promoting their uses for a range of applications. Phytochemical analysis has revealed that *I,coccinea* L. plant is composed of beneficial secondary metabolites namely Quercetin, Kaempferitin, lupeol, β-sitosterol, Stigmasterol, oleanolic acid, Epicatechin, sitosterol, ursolic acid, rutin, leucocyanadin, Ixorene, proanthocyanidins, anthocyanins, myricetin, Ixoroid etc. Fig. 1 depicts the major phytochemical components present in the plant parts and its therapeutic activities. The purpose of this review was to gather and compile data regarding the ornamental plant *I. coccinea* L. and to categorize the plant according to studies of its phytochemical composition. This would enable the scientific community to carry out additional research on the recommended plant [1].

2 Bioactive Value-Added Materials in *I. coccinea* Plant

I. coccinea plant is a reservoir of numerous bioactive components. Lupeol was isolated from the petroleum ether fraction of ethanolic extract of *I. coccinea* plant leaves. Pharmacological studies proved its anti-inflammatory activities [2]. *I. coccinea* leaves were used to isolate quercitrin, a dominant flavonoid by fractionating the hydro-alcoholic extract using various polarity solvents, then purifying the mixture using column chromatography. The in vitro antioxidant activity analysis of quercitrin demonstrated its difficult-to-scavenge DPPH and nitric oxide free radicals with a very low IC value.

Ixorene, a dammarane skeleton triterpene, was extracted from the leaves of *I. coccinea* Linn by Ikram et al. in 2013. They also evaluate the impact of floral extract and its fraction on the cardiovascular system and reported a new tirucallane triterpene from

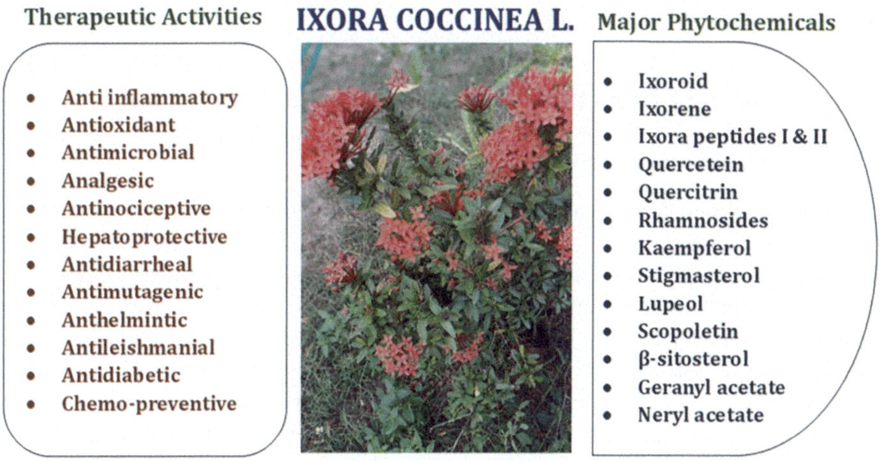

Fig. 1. *Ixora coccinea* L: phytochemicals and therapeutic activities

the plant's bloom—the first time this effect has been reported on *Ixora coccinea*. Additionally, they extracted four compounds: D-mannitol, lupeol, ixorene, and β-sitosterol. It was assumed that these separated chemicals were likely responsible for a large portion of the therapeutic activities of this plant because they all exhibited anti-inflammatory, antimitotic, antioxidant, antitumor, anticancer, chemoprotective, cardio-protective and also diuretic properties [3]. The flower of *I. coccinea* yielded three recognized compounds: stigmast-5-en-3-O-β-D-glucoside, 5-O-caffeoylquinic acid, and D-mannitol, as well as ixoroid, a novel triterpenoid. The wide range of biological functions of the plant was further demonstrated by the isolation of these chemicals [4]. Yadav in 1989 isolated stearic palmitic oleic and linoleic acids from the roots of *I. coccinea* proves the presence of fatty acids in the plant [5]. Ixorapeptide I and II along with 28 known chemicals were separated from the methanol extract of the aerial portions of the plant. The substantial anti-platelet aggregation and anti-inflammatory properties of *I. coccinea* L. were demonstrated by the isolation of these bioactive substances.

In 2010, Oyebode et al. isolated ixoratannin A-2, an A-type trimeric proanthocyanidin epicatechin, along with seven other known compounds: epicatechin, procyanidin A2, cinnamontannin B-1, and four flavon-3-ol rhamnosides: kaempferol-7-O-a-L-rhamnnoside, kaempferol-3-O-a-L-rhamnoside, quercetin-3-O-a-Lrhamnopyranoside, and kaempferol-3,7-O-a-L-dirhamnoside. The separation of these chemical components demonstrates the antibacterial and antioxidant properties of the leaves. The flowers of the plant were used to treat catarrhal bronchitis. By using silica gel chromatography, the air-dried flowers of *I. coccinea*Linn. produced two new cycloartenol esters as well as the well-known substances sitosterol, ursolic acid, oleanolic acid, lupeol, and lupeol fatty ester. These substances encourage the growth of its blossoms for therapeutic purposes [6]. The numerous chemical components found in plant parts demonstrate the use of plant in a variety of therapeutic fields. The aforementioned chemical ingredients were supported by the following research. The chemical compounds isolated from different parts are summarised in Tables 1, 2 and 3.

The numerous chemical components found in plant parts demonstrate the use of plant in a variety of therapeutic fields. The aforementioned chemical ingredients were supported by the following research.

3 Application of Bioactive Materials in Biomedical Sector

3.1 Anti-Inflammatory Activities

Lupeol, a compound, was obtained from the petroleum ether portion of the ethanolic extract of *I. coccinea* leaves [2]. Further research on *I. coccinea* was conducted by Ratnasooriya et al. in 2005, who investigated the anti-inflammatory properties of *I. coccinea* leaves [17]. In their study, the aqueous extract of *I. coccinea* leaf was utilized to examine the anti-inflammatory activity in rats through the carrageenan-edema procedure. The methanolic leaf extract of *I. coccinea* showed anti-inflammatory activity [18]. In addition, Aher et al. in 2013 reported the activities such as anti-inflammatory and analgesic of the methanolic extract of *I. coccinea* flowers. Rats with carrageenan-induced paw edema served as a sensitive anti-inflammatory model to assess the in vivo effectiveness. After oral administration of the *I. coccinea* flower extract in carrageenan addition, it

Table 1. Phytochemical constituents isolated from the roots and leaves along with the pharmacological activities of *I. coccinea* L.

Plant parts	Phytochemicals	Compounds isolated	Pharmacological activity	References
Root	Fatty acids	Methyl esters of oleic, stearic, palmitic, linoleic acids and mannitol	Anti-bacterial Anti-oxidant Heapatotoxicity Anthelmintic Anti-inflammatory	[5, 7–9]
	Essential oil	β- sesqui phellandrene Di-n-octyl phthalate, Octadecadienoic acid, Kaempferitin, β-Amyrin Kaempferol-7-glucoside Quercetin		
	Triterpenoids	β-Amyrin		
Leaves	Alkaloid	Camptothecin	Antibacterial Antileishmanial Anti-inflammatory Antioxidant Antinociceptive Anti-tumor	[3, 4]
	Flavonoids	Epicatechin, Kaempferol, Quercetein, Rhamnosides		
	Triterpenoid	Lupeol. Ixoroid		
	Triterpene	Ixorene		
	Proanthocyanidins	Ixora tannin A-2, Procyanidin A-2, Cinnamtannin B-1		

was discovered to have a significant analgesic effect and to decrease paw volume [19]. Adnan et al., 2014 conducted a study on rodents using the acetic acid-induced writhing, brewer's yeast-induced pyrexia, carrageenan-induced paw edema tests to evaluate the analgesic, anti-inflammatory, and antipyretic characteristics of the ethanolic extract of *I. coccinea* leaves. The anti-inflammatory activity of root extract of *I. coccinea* was also studied in rats [20].

3.2 Antioxidant Activities

Saha et al., 2008 assessed the antioxidant property of *I. coccinea* L. methanol extract using the phosphor-molybdenum process. The extract demonstrated considerable antioxidant activity in every antioxidant assay with a typical dose-dependent mode, and the significant ability to scavenge oxygen group may be credited to the increased concentration of hydrophilic polyphenols [21]. Torey et al., 2010 tested the antioxidant activity of methanol extracts of *I. coccinea* leaf, flower, and stem using in vitro. The activity was investigated using the 2,2-diphenyl-1-picrylhydrazyl (DPPH) radical scavenging activity. The ability of Ixora coccinea flower extract to offer protection against arsenic-induced genotoxicity and cytotoxicity in the CHO cell lines was investigated. The MTT assay confirmed its efficacy in reducing arsenic-induced cellular toxicity. These findings suggest that *I. coccinea* flowers could be used as a healthier option with antioxidant

Table 2. Phytochemical constituents isolated from stem along with the pharmacological activities of *I. coccinea* L

Plant parts	Phytochemicals	Compounds isolated	Pharmacological activity	References
Stem	Coumarin	Erythro-1,2-albiflorin Scopoletin	Anti-bacterial Anti-inflammatory Antioxidant	[14, 15]
	Quinones	Tocopheryl quinine 1,4-Dihydroxy-3-methylanthraquinone		
	Peptides	Ixora peptides I, Ixora peptides II		
	Diterpenoid	16α-hydro-kauran-17-oic acids		
	Proanthocyanidins	2β-*O*-7-ent-epicatechin 4β-8-epicatechin		
	Flavonoids	Kaempferol, Kaempferitin, luteolin, (−)epicatechin, (+)catechin, Kaempferol-7-*O*-α-rhamnoside		
	Sterol	6β-hydroxy stigmastone, Stigmasterol, Sitosteryl-β-d-glucoside, β-sitosterol		
	Triterpenoid	Lupeol, Oleanonic acid, 3-acetyl betulic acid Butenolic acid, α-and β-Amyrin, Ursolic acid, Acetyl ursolic acid		

activities and protective qualities against arsenic-induced toxicology effects [22]. Nair et al., 2020 investigated the antioxidant activity of an ethanolic extract of stem, leaf, and flower of *I. coccinea* L. plant using a hydrogen peroxide scavenging test. Muhammad et al., 2020 assessed the content and constituents of *I. coccinea* L. root methanolic extract and were tested for antioxidant capacity using radical scavenging and reducing power procedures. Furthermore, antimicrobial properties of root against pathogenic bacterial strains were determined [23]. The in vitro antioxidant potential of methanol, ethanol and aqueous extracts of *I. coccinea* L. was evaluated, wherein the methanol extract was showing the highest activity. The fruit extract had anti-cancer activity. In vitro antioxidant and antityrosinase actions of *I. coccinea* roots were assessed using a 1,1-diphenyl-2-picrylhydrazyl (DPPH) test and a ferric ion-reducing power analysis. Methanolic and aqueous extracts of the roots were tested and showed that methanolic extract has high antioxidant and antityrosinase activities [24, 25].

3.3 Hepatotoxicity Activity

In rats with hepatotoxicity triggered by carbon tetrachloride, an ethanolic extract of the *I. coccinea* flower was evaluated as a hepatoprotective and anti-oxidative agent [26]. According to the study, an ethanolic *I. coccinea* flowers extract confirmed hepatoprotective against carbon tetrachloride-induced liver injure in rats when tested under specific experimental conditions. In order to study the hepatotoxic effects of aflatoxin

Table 3. Phytochemical constituents and the pharmacological activities of *I. coccinea* L

Plant parts	Phytochemicals	Compounds isolated	Pharmacological activity	References
Flowers	Triterpenoids	Cycloartenol esters, Ursolic acid, Oleanolic acid, Lupeol esters	Antibacterial Antileishmanial Anti-inflammatory Antioxidant Antinociceptive Anti-tumor Anthelmintic Anti-diarrheal Antiulcer Antioxidant Hepatotoxicity	[6, 11, 16]
	Monoterpenes	Geranyl acetate Neryl acetate Linalyl acetate α-Terpeniol acetate Borneol acetate		
	Ester	Ethyl cinnamate		
	Sesquiterpenes	Cyperene, α-copaene		
	Flavonoid	Rutin, biochanin A, myrecetin, quercitrin, quercetin, daidzein formononetin, kaempferol-3-rutinoside, leucocyanidin glycoside, monoglycosides of cyanidin and delphinidin		

B1-intoxicated livers of albino male Wistar rats, the roots of *I. occinensis*, *Rhinacanthusnasuta*, and *Spilanthesciliata* were utilized [27]. It was discovered that the extracts exhibited therapeutic action against aflatoxin B1-toxicity.

3.4 Anti-Ulcer Activity

I. coccinea L. methanolic floral extracts, *Psidium guajava* roots, and *Neolamarckiacadamba* leaves were studied for antiulcer and in-vitro antioxidant properties by Patibandla et al. (2014), and they found substantial ulcer curative potential activity [28]. This study demonstrated the strong antiulcer and antioxidant properties of methanolic extract as well as the polyherbal extract's ability to treat rat stomach ulcers caused by aspirin.

3.5 Antinociceptive and Anthelmintic Activities

In 2005, Ratnasooriya et al. investigated the antinociceptive property of *I. coccinea* L. leaves using three nociception models (formalin, tail flick, hot plate tests) on male rats. The rats were providing 1 mL of distilled water, and the antinociceptive property of the extract was analyzed. The study found that the extract exhibited significant antinociceptive activity [29]. In 2011, Another study conducted by Maniyar et al. in 2013 investigated the antidiarrheal property of the water extract of *I. coccinea* L. flowers using albino Wistar rats with castor oil-induced diarrhea. The findings confirmed the aqueous extract antidiarrheal properties and its historical application in the management of diarrhea [30].

Surana et al. investigated the anthelmintic property of *I. coccinea* L. roots using the earthworm *Pheritumaposthuma*, with Albendazole as the standard drug. The chloroform extract was initiate to have good anthelmintic property [31]. In a 2014 study, Sowjanya et al. investigated the in vitro anthelmintic potency of acetone, n-Hexane, ethanol, methanol, and aqueous extracts of *I. coccinea* L. flowers using the Indian earthworm (*Pheretimaposthuma*) and found that the results were comparable to those of Albendazole [32].

3.6 Antileishmanial Activity

Using the MTT technique, the antileishmanial activity of *I. coccinea* L. leaf extracts in ethyl acetate and methanol was examined in vitro. In this investigation, both extracts significantly and concentration-dependently reduced the proliferation of microorganisms. Methanol and ethyl acetate were the two most effective extracts [33].

3.7 Antitumour Activity

I. coccinea L. flowers were tested antitumor activity against lymphoma and the findings proved the inhibition of the growth of leukemic lymphocytes while sparing normal cells [34]. Latha and Panikkar (2011) investigated the antitumor activity of *I. coccinea* L. flower against chemically-induced cancel cells in mice for the first time [35]. In mice, n-hexane extract of *I. coccinea* L. demonstrated chemoprotective property against cisplatin-induced toxicity. Additionally, it maintained blood urea nitrogen levels close to normal and dramatically extended the life of mice given cisplatin, suggesting chemoprotective effects.

The ethyl acetate fraction of *I. coccinea* leaf was chosen to assess the wound healing assay because to its high flavonoid content and antioxidant activity. The active fraction can cleave apoptotic melanoma cancer cells, according to the results of a DNA damage experiment. The reported chemicals camptothecin, quercetin, and rutin have the ability to bind to the skin cancer protein 2vcj, according to in silico docking experiments. As a result, this study strongly suggests that more research be done using the topical formulations of *I. coccinea* for the treatment of skin cancer [36].

I. coccinea L. has been found to possess chemoprotective properties due to its antioxidant activity. Its extracts have been reported to absorb free radicals, shield against DNA damage and lipid peroxidation, and enhance the activities of antioxidant enzymes. These activities are believed to contribute to its potential for preventing or mitigating the toxic effects of chemotherapy drugs, radiation therapy, and other carcinogenic agents. However, more studies are required to fully understand the chemoprotective mechanisms of *I. coccinea* L. and its potential clinical applications.

3.8 Antibacterial Activity

Latha et al., (1995) studied the antibacterial action of *I. coccinea* L in 50% ethanolic extract. The significant inhibitory content against both the bacterial and fungus test organisms studied was 125 μg mL^{-1}, after which the inhibitory activity declined and the microbes began to revive due to the antimicrobial principle's impact [37]. Annapurna et al., (2003) tested the antimicrobial property of methanol and ether and extracts of *I. coccinea* L. dry leaves and was discovered that the ether fraction was more active on comparing with the methanol fraction. Marimuthu and coworkers (2014) examined the antibacterial inhibitions of methanolic extracts of *I. coccinea* L. stems, leaves, and flowers [38]. The methanol extract of *I. coccinea* L flowers demonstrated significant antibacterial activity against bacterial fish pathogenic microbes [39]. The flowers and leaves extract of *I. coccinea* L. in methanol was studied for antimicrobial activity and results showed that both parts were active against bacteria and fungus. The antimicrobial activities of *I. chinensis*, *I. lutea*, *I. coccinea*, and *I. parviflora* leaf extracts were studied, and only *I. parviflora*was found to be effective against some organisms [40]. The phytochemical, minerals and proximate analysis of ethanolic extract of *I. coccinea* leaf extract showed the presence of various phytochemicals, metals such as sodium, potassium, manganese, iron, calcium, magnesium, copper and phosphorus and also moisture, protein, fiber and carbohydrates and also the antimicrobial activity was also analyzed [41]. The phytochemical constituents of essential oil isolated from *I. coccinea* L. leaf were examined using GC-MS and the antimicrobial activity was tested against various microbes using broth microdilution procedure and showed significant antimicrobial effects against the test microbes. Using the cup-plate way, extracts of *I. coccinea* L. flowers in chloroform, methanol, acetone, ethanol, and water were tested for antimicrobial property against various microbes [42]. *I. coccinea* L. and Datura Metel powdered leaf extracts were macerated and extracted with isopropanol, acetone, and methanol using agar well diffusion procedure and tested against bacterial strains of *Escherichia coli* and *Staphylococcus aureus*. While Datura Metel had demonstrated strong antibacterial activity against *Staphylococcus aureus*, the most effective area of inhibition of the growth was seen in the instance of *I.coccinea* L. against the bacterial strains of *Escherichia coli* and *Staphylococcus aureus* [43]. The methanolic extract of *Ixora coccinea* flowers has anti-diabetic activity in vitro, which could be related to the high concentration of bioactive molecules such as polyphenolics [44]. Unni et al. (2022) found that nanocellulose extracted from *I. coccinea* L. root blended with glycerol, gelatin, and PEG 600, with the 1% PEG 600 plasticized nanocellulose scaffold showing good tensile strength, cytocompatibility, and active cell proliferation, making it a potential tissue engineering scaffold. Glutaraldehyde treatment enhanced the tensile property of cellulose extracted from root and its 1% PEG 600 nanocellulose scaffold, indicating potential uses in tissue engineering, adsorbent paper, and other industries [45].

In a rat model using the dead space wound technique, the alcoholic extract of *I. coccinea* L. flowers were investigated for its effect on wound healing. The results showed rapid wound healing and increased fibroblast cell proliferation [46]. Extracts of *I. coccinea* were tested for their in vitro growth inhibition properties against various bacterial strains, and the ethanolic extract was found to have considerable antibacterial including *Bacillus pumilius*, *Enterococcus faecalis*, *Pseudomonas aeruginosa*, *Escherichia coli*, *Salmonella typhi*, and *Staphylococcus aureus*. It is also renowned for their antifungal, and wound healing properties [47]. In another study, the antioxidant, antimicrobial, and fibroblast proliferation activities of different extracts of *I. coccinea* L. leaves were investigated in vitro and in vivo in Wistar rats, with superior activity observed. Recently, Rescober et al. (2022) investigated the antimicrobial and wound contraction properties of a preformulated hydrogel containing the methanolic leaf extract of *I. coccinea* L. in New Zealand rabbits and found it to be effective against staphylococcal infections commonly found in wounds, with no acute dermal irritation observed. First of all, an open wound gives germs a way into the body. With the reduction of bacterial colonization and growth at the wound site, rapid wound healing can reduce the risk of infection [48]. During the process of healing a wound, the immune system is also active to fight off any invasive microorganisms. As part of this immune reaction, chemicals that can help prevent bacterial infections and antimicrobial peptides are released. Moreover, certain plant compounds, such as those found in *I. coccinea* L., have direct antibacterial actions that can aid in preventing bacterial infections in wounds.

The plant extracts made up of active ingredients like flavonoids, triterpenes, tannins, alkaloids and anthraquinones encourage the healing process of wounds.Tannins function as scavengers of free radicals, while flavonoids, triterpenoids, have antimicrobial properties which promote wound healing. On the other hand, saponins hold antimicrobial and antioxidant activity that seem to be in charge of wound closure and an increased epithelialization. The antioxidant enzymes' level in granuloma tissue is increased by flavonoids' potent free radical-scavenging and antioxidant properties. Because of their ability to scavenge free radicals and their antioxidant character, these are known for their lower lipid peroxidation and improve vascularity, steroids and polyphenols are also constitutes for wound healing [49]. Figure 2 displays the active phytochemicals and biological activities *I. coccinea* L. plant.

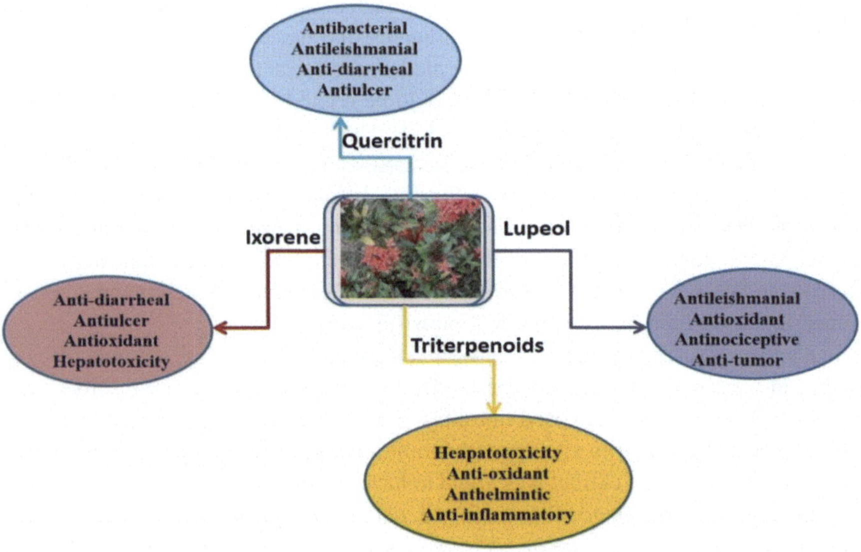

Fig. 2. Major phytochemicals and biological activities of *I. coccinea* L. plant

3.9 Biosynthesis of Nanoparticles and Its Therapeutic Activities

Coccinea leaf extract was utilized to synthesize silver nanoparticles that displayed superior photocatalytic and antimicrobial properties [50]. The synthesis of zinc oxide nanoparticles using *I. coccinea* L. leaf extract has potential applications in biomedical, cosmetic, and biosensor fields [51]. In a study by Baskar et al. (2016), zinc oxide nanoparticles were synthesized and shown to possess anti-cancer activity against MCF-7 tumor cell line. Another study utilized the MTT procedure to evaluate the anti-cancer properties of *I. coccinea* flower extract against human breast cancer cells, demonstrating the IC50 value to be 31.3 g/mL [52]. Lad et al. (2019) utilized *Momordica charantia* and *I. coccinea* leaf extracts to synthesize copper nanoparticles, which demonstrated effective antimicrobial properties against *Escherichia coli* and *Salmonella paratyphi*A [53]. In addition, iron oxide nanoparticles were biosynthesized via *I. coccinea* L. leaf extract and were employed for the removal of toxic metals such as Cd, Ni, and Pb from water [54]. The aqueous extract of *I. coccinea* L. flowers and silver nanoparticles synthesized using the extract was studied by Mizhusima procedure and found that both the samples were having comparable anti-microbial property. Metal nanoparticles can also modulate the inflammatory response, improve angiogenesis, and provide structural support to the wound site. Additionally, some nanoparticles possess antimicrobial properties that can help prevent bacterial infections, which can delay wound healing. The specific mechanisms by which nanoparticles promote wound healing depend on the type of nanoparticle and the wound micro-environment.

4 Prospects, Challenges and Limitations of *I. coccinea* as Herbal Medicine

Ixora grown in full sun produces the most beautiful flower displays. A pH of moderate to slightly acidic and soil with good drainage are essential for the growth of Ixora bushes. When planted in alkaline soils, the plant gets chlorosis. Like all other herbal medicines *I. coccinea* phytochemicals and plant part extracts have some difficulties and restrictions.

- A key concern is the extraction, purification, and stability of these bioactive substances. Maintaining their efficacy throughout manufacturing, transportation, and storage presents the most significant challenge.
- The processing conditions that affect these bioactive materials include temperature, pH, light, and excipients.

We hope to bring some light on the developments in the applications of *I. coccinea* in the fields of medicine to the scientific community through this review.

5 Conclusions

In conclusion, *I. coccinea* L., a shrub native to India and Sri Lanka, has been used for centuries in medicinal field in conventional way. Its antimicrobial, hepatoprotective, antinociceptive, chemoprotective, antimitotic, antioxidant, and anti-inflammatory properties have prompted extensive phytochemical research, resulting in the identification and analysis of numerous compounds derived from various plant parts. Nanocellulose derived from *I. coccinea* L. has been shown to be extremely effective in wound healing due to the abundance of bioactive materials. This review paves the way for the shrub *I. coccinea* L. to be used in a variety of applications, as well as providing a novel source of nanocellulose that is both biocompatible and compatible with cells, as well as a medicinal agent with no side effects.

References

1. Roja, P., Nandini, M., Deepthi, B.: A review on Ixora coccinea plant. Int. J. Pharm. Res. Appl. **6**, 729–734 (2021). https://doi.org/10.35629/7781-0603729734
2. Zachariah, R., Sudhakaran Nair, C.R., Velayudha Panicker, P.: Anti-inflammatory and anti-mitotic activities of lupeol isolated from the Ixora coccinea Linn. Indian J. Pharm. Sci. **15**, 129–132 (1993)
3. Ikram, A., Versiani, M.A., Shamshad, S., et al.: Ixorene , a New Dammarane Triterpene from the Leaves of Ixora. Rec. Nat. Prod. **4**, 302–306 (2013)
4. Versiani, M.A., Ikram, A., Khalid, S., et al.: Ixoroid: a new triterpenoid from the flowers of Ixora coccinea. Nat. Prod. Commun. **7**, 831–834 (2012). https://doi.org/10.1177/1934578x1200700706
5. Yadava, R.N.: Analysis of the fixed oil from the roots of Ixora Coccinea Linn. Asian J. Chem. **1**, 307–308 (1989)
6. Ragasa, C.Y., Tiu, F., Rideout, J.A.: New Cycloartenol esters from Ixora Coccinea. Nat. Prod. Res. **14**, 37–41 (2004). https://doi.org/10.1080/14786410310001630519
7. Joshi, A.B., Surlikar, P.M., Bhobe, M.: Ixora Coccinea Linn: phytochemical investigation. Int. J. Res. Pharm. Chem. **3**, 691–696 (2013)
8. Baliga, M.S., Kurian, P.J.: Ixora coccinea Linn.: traditional uses, phytochemistry and pharmacology. Chin. J. Integr. Med. **18**, 72–79 (2012). https://doi.org/10.1007/s11655-011-0881-3
9. Owolabi, M.S., Akintayo, L., Satyal, P., et al.: Essential oil constituents and study of Antioxdiant, antidiabetic activities of Ixora coccinea L. extract from south west, Nigeria. Am. J. Essent. Oils Nat. Prod. **10**, 20–26 (2022)
10. Oyebode, T., Oguntuga, A., Oladimeji, A., et al.: Phytochemistry doubly linked, A-type proanthocyanidin trimer and other constituents of Ixora coccinea leaves and their antioxidant and antibacterial properties. Phytochemistry. **71**, 2092–2098 (2010). https://doi.org/10.1016/j.phytochem.2010.08.018
11. Okhale, U.O., Ib-rahim, E., Kunle, E.: Chemical constituents and antimicrobial activity of the leaf essential oil of Ixora coccinea L (Rubiaceae) collected from north Central Nigeria. Int. J. Bioassays. **5**, 5630–5637 (2018)
12. Maniyar, Y., Bhixavatimath, P.: Evaluation of the hypoglycaemic and hypolipidaemic activities of the aqueous extract of the leaves of Ixora Coccinea Linn in diabetic rats. J. Clin. Diagn. Res. **5**, 1381–1384 (2011)
13. Vadivu, R., Jayshree, N., Kasthuri, C., et al.: Pharmacognostical standardization of leaves of Ixora coccinea Linn. J. Pharm. Sci. Res. **2**, 164–170 (2010)
14. Kharat, A.R., Nambiar, V.V., Tarkasband, Y.S., Pujari, R.R.: A review on phytochemical and pharmacological activity of genus Ixora. Int. J. Res. Pharm. Chem. **3**, 628–635 (2013)
15. Jaiswal, R., Elsadig Karar, M.G., Gadir, H.A., Kuhnert, N.: Identification and characterisation of phenolics from Ixora coccinea L. (Rubiaceae) by liquid chromatography multi-stage mass spectrometry. Phytochem. Anal. **25**, 567–576 (2014). https://doi.org/10.1002/pca.2530

16. Sumathy, H., Sangeetha, J., Vijayalakshmi, K.: Chromatographic fingerprint analysis of Ixora coccinea methanolic flower extract. Int. J. Pharm. Sci. Drug. Res. **3**, 327–330 (2011)
17. Ratnasooriya, W.D., Deraniyagala, S.A., Galhena, G., et al.: Anti-inflammatory activity of the aqueous leaf extract of Ixora coccinea. Pharm. Biol. **43**, 147–152 (2005). https://doi.org/10.1080/13880200590919483
18. Handunnetti, S.M., Kumara, R.R., Deraniyagala, S.A., Ratnasooriya, W.D.: Anti-inflammatory activity of Ixora coccinea methanolic leaf extract. Pharm. Res. **1**, 80–90 (2009)
19. Aher, A.N., Pal, S.C.: Analgesic and anti-inflammatory activities of Ixora coccinea flower. Indian Drugs. **50**, 64–67 (2013)
20. Rajendra, K., Nitin, G., Mahavir, G., et al.: Evaluation of anti-inflammatory and antioxidant potential of Ixora coccinea, Linn ethanolic root extract. Int. J. Drug. Dev. Res. **5**, 1–7 (2013)
21. Saha, M.R., Alam, A., Akter, R., Jahangir, R.: In vitro free radical scavenging activity of Ixora coccinea L. Bangladesh Pharmacol. Soc. **88**, 90–96 (2008). https://doi.org/10.3329/bjp.v3i2.838
22. Salmataj, S.A., Kamath, S.U., Murty, V.R., Pai, S.R.: Amelioration of arsenic-induced oxidative stress in CHO cells by Ixora coccinea flower extract. 3 Biotech. **8**, 1–7 (2018). https://doi.org/10.1007/s13205-018-1446-1
23. Muhammad, H., Qasim, M., Ikram, A., et al.: Antioxidant and antimicrobial activities of Ixora coccinea root and quanti fi cation of phenolic compounds using HPLC. S. Afr. J. Bot. **135**, 71–79 (2020). https://doi.org/10.1016/j.sajb.2020.08.012
24. Sindhu Rani, J.A.: Evaluation of in vitro antioxidant and Antityrosinase activities of Ixora coccinea Linn. roots. J. Pharmacogn. Phytochem. **11**, 144–150 (2022)
25. Kumar, V.R., Shankar, K.R., Priyabhandhavi, P.: In vitro evaluation of antioxidant activity of ethanolic activity of ethanolic leaf extract of Ixora Coccinea. Int. J. Pharamaceut. Chem. Biol. Sci. **3**, 380–387 (2013)
26. Soma, B., Reshma, S., Anjan, A., et al.: Hepatoprotective activity of Ixora Coccinea L. in animal model. Int. J. Res. Ayurveda Pharm. **5**, 339–342 (2014). https://doi.org/10.7897/2277-4343.05370
27. Shyamal, S., Latha, P.G., Suja, S.R., et al.: Hepatoprotective effect of three herbal extracts on aflatoxin B1-intoxicated rat liver. Singapore Med. **51**, 326–331 (2010)
28. Patibandla, N.B., Limited, E.A., Bandaru, N.: Evaluation of antiulcer and in-vitro antioxidant activities of Ixora coccinea flowers and Polyherbal extract in Wistar albino rats. Int. J. Pharm. Pharm. Sci. **6**, 339–344 (2014)
29. Ratnasooriya, Deraniyagala, S.A., Bathige, Jayakody, C.L.G.: Antinociceptive action of aqueous extract of the leaves of Ixora Coccinea. Acta Biol. Hung. **56**, 21–34 (2005)
30. Maniyar, Y., Bhixavatimath, P., Agashikar, N.V.: Antidiarrheal activity of flowers of Ixora Coccinea Linn. in rats. J. Ayurveda Integr. Med. **1**, 287–291 (2010). https://doi.org/10.4103/0975-9476.74422
31. Surana, A.N.A., Deore, S.P.: Evaluation of anthelmintic activity of Ixora coccinea. Int. J. Pharm. Life Sci. **2**, 813–814 (2011)
32. Sowjanya Pulipati, S.T., Srinivasa Babu, P.: Invitro evaluation of anthelmintic activity of flower extracts of invitro evaluation of anthelmintic activity of flower extracts of Ixora Coccinea Linn. Indiana Pharm. **25**, 50–54 (2014)
33. Naskar, M.: Antileishmanial effect of Ixora coccinea leaf extracts on the in vitro growth of Leishmania donovani promastigotes. J. Adv. Pharm. Educ. Res. **3**, 471–474 (2013)
34. Latha, P.G., Panikkar, K.R.: Cytotoxic and antitumour principles from Ixora coccinea flowers. Cancer Lett. **130**, 197 (1998)
35. Latha, P.G., Panikkar, K.R.: Inhibition of chemical carcinogenesis in mice by Ixora Coccinea flowers. Pharm. Biol. **38**, 152–156 (2011). https://doi.org/10.1076/1388-0209(200004)3821-1FT152

36. R, J.S., Chandrasekar, V., D, C., et al.: In vitro wound healing and anticancer effects of Ixora coccinea in malignant melanoma cell lines. Cureus. **16**, 1–14 (2024). https://doi.org/10.7759/cureus.58958
37. Latha, P.G., Abraham, T.K., Panikkar, K.R.: Antimicrobial properties of Ixora Coccinea L. Anc. Sci. Life. **14**, 286–291 (1995)
38. Marimuthu, M.M., Aruldass, C.A., Sandrasagaran, U.M., et al.: Antimicrobial activity and phytochemical screening of various parts of Ixora coccinea. J. Med. Plant Res. **8**, 423–429 (2014). https://doi.org/10.5897/JMPR11.1281
39. Nithiyasoundari, M., Parimala, K.S., Balaji, S.: Efficacy of Ixora Coccinea against common fish pathogens. Int. J. Res. Ayurveda Pharm. **6**, 489–492 (2015). https://doi.org/10.7897/2277-4343.06493
40. Akter, S., Haque, T., Irine, E.J., et al.: Comparative antimicrobial activities of different species of Ixora. J. Pharmacogn. Phytochem. **3**, 103–105 (2015)
41. Christy, O., Temitope, O.O., Bamidele, A., et al.: Antimicrobial activity , chemical compositions and proximate analysis of Ixora coccinea L. leaves on some clinical pathogens. Int. J. Curr. Res. **10**, 72555–72561 (2018)
42. Karishma, S.K., Brahma Reddy, D.R., Manasa, P., Mariyamma, P., Jyothi, S., Asma, S.K.: Studies on antimicrobial activity from flowers of Ixora Coccinea. World J. Pharm. Res. **8**, 1292–1298 (2019). https://doi.org/10.20959/wjpr20193-14323
43. Krishnan, S.: Comparative anti-microbial activity analysis of Ixora coccinea and Datura metel. J. Bioprocess. Biotech. **11**, 6–8 (2021)
44. Anila, L., Hashim, M.S.: Evaluation of in vitro anti-diabetic activity of methanolic extract of Ixora coccinea flowers. J. Pharm. Sci. Res. **11**, 2367–2372 (2019)
45. Unni, R., Reshmy, R., Latha, M.S., et al.: Enhancement of mechanical and thermal properties of Ixora coccinea L. plant root derived nanocellulose using polyethylene glycol-glutaraldehyde system. Chemosphere. **298**, 134324 (2022). https://doi.org/10.1016/J.CHEMOSPHERE.2022.134324
46. Nayak, B.S., Udupa, A.L., Udupa, S.L.: Effect of Ixora coccinea flowers on dead space wound healing in rats. Fitoterapia. **70**, 233–236 (1999)
47. Selvaraj, N., Lakshmanan, B., Mazumder, P.M., et al.: Evaluation of wound healing and antimicrobial potentials of Ixora coccinea root extract. Asian Pac. J. Trop. Med. **4**, 959–963 (2011). https://doi.org/10.1016/S1995-7645(11)60226-5
48. Rescober, A.A.S., Blanco, I.F.M., Janio, J.A.T., et al.: Antimicrobial and wound healing activity of Ixora coccinea leaf extract in hydrogel formulation. Acta Med. Philipp. **56**, 12–17 (2022)
49. Thakur, R., Jain, N., Pathak, R., Sandhu, S.S.: Practices in wound healing studies of plants. Evid. Based Complement. Alternat. Med. **2011**, 1–10 (2011). https://doi.org/10.1155/2011/438056
50. Vinay, C.N.: Ixora coccinea extract-mediated green synthesis of silver nanoparticles: photodegradative and antimicrobial studies. Int. J. Biosens. Bioelectron. **5**, 100–105 (2019). https://doi.org/10.15406/ijbsbe.2019.05.00161
51. Yedurkar, S., Maurya, C., Mahanwar, P.: Biosynthesis of zinc oxide nanoparticles using Ixora coccinea leaf extract—a green approach. Open J. Synth. Theor. Appl. **5**, 1–14 (2016). https://doi.org/10.4236/ojsta.2016.51001
52. Baskar, G., Devi, A.G., Nishanthini, T., Aiswarya, R.: Anticancer activity of Ixora coccinea Linn flower extracts against human breast adenocarcinoma cells. Int. J. Mod. Sci. Technol. **2**, 288–291 (2017)

53. Lad, R., Singh, S.: Green synthesis of copper nanoparticles from Ixora coccinea leaves and Momordica charantia fruit extracts. J. Emerg. Technol. Innov. Res. **6**, 307–316 (2019)
54. Ali, A., Jameel, B., Baig, A., et al.: Biosynthesis and analytical characterization of iron oxide nanobiocomposite for in-depth adsorption strategy for the removal of toxic metals from drinking water. Arab. J. Sci. Eng. **4**, 1–14 (2022). https://doi.org/10.1007/s13369-022-07477-y

Author Index

R. Rajasekharan et al. (Eds.): MatZone 2024, SPPHY 316, pp. 267–268, 2025.
https://doi.org/10.1007/978-981-96-6541-9

The manufacturer's authorised representative in the EU is Springer Nature Customer Service Centre GmbH, Europaplatz 3, 69115 Heidelberg, Germany. If you have any concerns regarding our products, please contact ProductSafety@springernature.com

Printed and bound by CPI Group (UK) Ltd, Croydon, CR0 4YY
01/10/2026
02225922-0006